THE NATURE OF SCIENCE

THE NATURE OF SCIENCE

An A–Z Guide to the Laws and Principles Governing Our Universe

JAMES TREFIL

Houghton Mifflin Company

Boston New York 2003

Contents

TO KIM

Acknowledgements

No book of this magnitude is the work of one person, and it is with a profound sense of gratitude that I acknowledge the contributions of the many individuals who have been part of the project. Richard Milbank, Publisher at Cassell, conceived the idea of a book devoted to an explication of the laws of nature and saw that it became a reality. Rosie Anderson at Cassell, in her inimitable and invaluable way, oversaw the daily operations that brought the book to print. John Woodruff far exceeded his duties as editor, and became a valuable colleague whose help was critical in turning the draft manuscript into its final form.

The laws of nature cover the entire gamut of scientific subjects, and no one can be an expert on everything. I would, then, like to thank a number of my colleagues who read the parts of the manuscript in their areas of expertise and shared their wisdom and experience with me. Of course, no two scientists agree on everything, and in these cases the opinions that appear are mine. With the understanding that any mistakes that remain in the manuscript are my responsibility, and my responsibility alone, I would like to thank Peter Atkins, Chris Kitchin, David Nelson, Victoria Pocock and Nick Rogers.

The Nature of Science

The laws of nature are the skeleton of the universe. They support it, give it shape, tie it together. Taken as a whole, they embody a vision of our world that is both breathtaking and awe-inspiring. But perhaps most important of all, they tell us that the universe is a place we can know, understand, approach with the power of human reason. In an age that seems to be losing confidence in its ability to manage things, they remind us that even the most complex systems around us operate according to simple laws, laws easily accessible to the average person. But before we embark on a tour of these laws, we should take a few moments to think about how they arise and what role they play in the enterprise we call science.

The story of science

Think for a moment about the 20th century, the one in which almost all of us have spent most of our lives, and ask yourself a simple question: What made the century we've just completed so palpably different from anything that came before? It was, to be sure, a century in which old political patterns crumbled and new ones took their place, but the same could be said of almost any century since humans started to record their activities. It was a century that saw great writers and artists, but again that's nothing new. It even produced art forms (jazz and cinema spring to mind) that might come to be regarded as the equals of grand opera and the symphony. I doubt this will happen, but in any case it wasn't the first time that new art forms have been created, nor will it be the last.

In my view, what stamped itself on the 20th century and made it unique was the advance of science and technology. If you were making a list of important achievements of the century, here are some things you might include:

> antibiotics, the Apollo Moon landings, computers, the Internet, open-heart surgery, jet airplanes, frozen food, skyscrapers.

The enormous increase in human numbers and in the global economy over the past hundred years is a direct consequence of an enormous increase in our knowledge of the universe we inhabit.

At one level, there is nothing particularly novel about this view. Throughout history, the really deep changes in the human condition have been driven by new knowledge. For example, about 10,000 years ago someone—probably a woman in the Middle East—figured out that it was possible to raise and tend plants to obtain food, rather than gathering it in the wild. This was the beginning of agriculture, a development without which modern civilization wouldn't have got off the ground (and a development that was reproduced independently in many parts of the world). A few centuries ago, a Scottish engineer by the name of James

Watt developed a practical version of the steam engine, an essential part of the Industrial Revolution. Someday, scholars may speak of the invention of the transistor in 1947 and the recently completed Human Genome Project in the same way, as fundamental milestones in human history.

Of course, this way of thinking about science tends to concentrate our thoughts on the practical benefits that flow from it, on improvements in human health and comfort. There is, however, another dimension to science. At the same time that it is improving the quality of our lives, it is providing us with a magnificent intellectual window on the universe. It allows us to see that everything around us operates according to the dictates of general rules and principles—rules and principles that can be discovered by the methods of science. Rules that have been tested and verified most stringently are elevated to the status of "laws of nature," although, as we shall see, scientists and philosophers are far from being in agreement about how this term should be used. Those laws form an intellectual framework into which every phenomenon in the universe can be fitted.

Humans have always been curious about the world around them, not least because an individual's survival often depended on the ability to make judgments about how a particular situation was going to develop. Farmers long ago developed lore about weather and climate to guide them to a successful harvest, hunters studied the habits of their prey, and sailors learned to read the sea and sky for warnings of impending storms. However, it wasn't until a few hundred years ago that the specialized technique and methodology we call science was developed. Why it developed when it did, and why it developed in Europe instead of elsewhere, are questions I would prefer to leave to historians. For our purposes, the important thing is to understand what science is and how it leads us to things we call laws of nature.

A word of warning is in order before we start this discussion. You will often find, particularly in textbooks, a stepwise procedure that is said to constitute something called "the scientific method." Typically, this is something like, "A scientist first does X, then proceeds to Y, then to Z," and so on. It makes doing science sound like making a batch of cookies from a recipe. The problem with this approach isn't so much that it's completely wrong—scientists often *do* carry out steps X, Y, and Z. Rather, it's that it leaves no room for the human creativity, ingenuity, and just plain cussedness that are, and always have been, essential components of the scientific enterprise. Describing the process of science by a "method" is like describing a painting by Rembrandt or Van Gogh solely in terms of where different colors have been applied to the canvas. Science is not the equivalent of painting by numbers.

So, when it comes to discussing how science is done and how scientists get to laws of nature, I prefer to borrow a concept from the legal profession. That is, the elements I'm about to describe shouldn't be thought of as rigid parts of a formulaic procedure, but rather as steps in a process that scientists go through. By this I mean that in thinking about science you do have to consider all of these elements, but you have to decide how important each is in a given context (or even whether they are all there). There is not, in other words, a fixed, rigid procedure by which you can decide if something is or isn't science.

Most of the time most scientists proceed through these steps in pretty much the order listed below, which is the order usually presented in textbooks. Occasionally, though, comes the kind of wild leap of intuition or blazing new insight you might not associate with the conventional sober, white-coated image of scientists. That's good, because what I'd like most of all for you to take away from this book is the notion that science, like art, is one of the great outlets for the human creative urge, and that scientists share human traits and foibles with the rest of us. With that caveat in mind, here are the elements of the scientific process:

Observation or experiment?

If you want to know about the world, you need to look at it to see what it is like. That seems such an obvious statement, and you may well wonder why I even bothered to write it down, but it is the basic cornerstone of science. It is also something that is not universally accepted even today, and certainly hasn't been universally accepted throughout history.

Confronted with a conflict between an observation of the real world and an interpretation of a religious doctrine, for most of recorded history human beings have consistently chosen the latter over the former. An incorrect interpretation of the Bible, for example, is what led the 17th-century Catholic hierarchy to force Galileo to recant the notion that the Earth moves around the Sun. A similar viewpoint can be seen in the United States today, where school boards often ignore the massive evidence supporting evolution and the big bang in order to honor an interpretation of the Book of Genesis that is not accepted by most Christian or Jewish scholars.

And it's not only religious people who refuse to look at data or accept the notion that complexity and ambiguity are often part of our world. Many modern environmentalists, for example, confronted with massive evidence that natural carcinogens produced by plants far outweigh those in artificial pesticides, choose to ignore the data and stick with the "Natural good, artificial bad" mantra of their youth. In both of these examples, the comfort of a self-contained belief system outweighs any benefit that might be gained from looking at the world as it actually is.

Nevertheless, observing the world is the first step toward science, and one that was first taken long ago. Ever since the introduction of agriculture, farmers have saved the seeds from their biggest, most productive plants, having realized that this will give them a better harvest next year. Craftsmen noted and recorded (perhaps in an oral tradition) how different blends of metals behaved when handled and heated in a specific way. The forerunners of today's medical workers noticed that extracts of certain plants helped with certain kinds of sicknesses, and thereby laid the foundations for our modern pharmaceutical industry. In all of these examples, the memory of observations and experiments lingered because they allowed human beings to meet their needs. In short, they worked. This is a notion to which we shall return when we come to discuss other ways of knowing.

When I talk about the open-mindedness of the scientific community, then, I mean this ability to ignore preconceptions and follow the data wherever they lead, regardless of where you think they ought to point.

According to a common bit of folklore, a scientist is supposed to approach the world with an open mind—with no preconceived ideas about how an observation or experiment will turn out. This idea runs all the way back to the English friar, philosopher, and scientist Roger Bacon (*c.* 1220–92), but like certain customs in medieval Denmark, it is more honored in the breach than in the observance. In my entire career I have encountered only one person who operated this way. He was a field geologist who liked to go out and "let the rocks talk to him." Everybody else I've known went into experiments having a pretty good idea about how things would turn out. The point is, though, that if the results weren't what they expected, they were able to ignore their preconceptions and follow the data. When I talk about the open-mindedness of the scientific community, then, I mean this ability to ignore preconceptions and follow the data wherever they lead, regardless of where you think they ought to point.

There are many examples of individual scientists, even entire scientific communities, who have followed this path. For example, in 1964, Arno Penzias and Robert Wilson (*see* BIG BANG), two scientists at Bell Labs in New Jersey, were making measurements of microwave radiation from space. At the very beginning of the development of satellite communication, such measurements were routine—after all, if you're going to build receivers to detect signals from satellites in orbit, you'd better know what's coming into your detectors that's not from the satellite. As they swept the skies with their receiver, Penzias and Wilson detected interference from many known sources, but they also found something totally unexpected. No matter which way they turned their equipment, they always picked up a faint signal from incoming microwaves (it showed as a faint hiss in their earphones). Try as they might, they couldn't make it go away. They even went so far as to evict a pair of pigeons which had nested in the apparatus and, as they delicately put it, had coated parts of the receiver with a "white dielectric substance." Eventually, though, Penzias and Wilson simply had to accept the fact that the universe was, quite unexpectedly, pervaded by microwave radiation. Today, we recognize this so-called cosmic microwave background as an important piece of evidence for the BIG BANG, our best theory of the origin of the universe. For believing their completely unexpected data, Penzias and Wilson shared the Nobel Prize for Physics in 1978.

Having made my point about the central place of observation and experiment in science, I should point out that the two terms, though similar, imply slightly different ways of working. An astronomer cannot make a star and let it age to see how it behaves. An evolutionary biologist cannot create a new vertebrate organism and then wait for a few million years to see how it evolves. A geologist cannot speed up the motion of tectonic plates at the Earth's surface to see how a particular rock formation will change. In all of these cases, scientists have to content themselves with observing nature, as they are unable to manipulate the objects of their study.

An experimenter, on the other hand, tries to control the system being studied, often changing only one thing at a time to see what happens. For example, in a classic application of the experimental method, ecologist David Tilman of the University of Minnesota divided a large tract of Midwestern prairie into a grid of squares of a few meters. In one experiment he kept everything the same in all the squares except for the amount of nitrogen fertilizer he added. In this way, he was able to separate the effects of one element, nitrogen, from all of the other things that can affect plant growth. Other experimenters do the same sort of thing. A nuclear physicist colliding subatomic particles at enormous speeds will keep everything in each collision the same except for the energy of the incoming particle; a chemist will keep the proportions of all the ingredients in a reaction the same except for one; a cancer researcher will keep all elements of the treatment of tumors in experimental animals the same except for one; and so on. In these and many other experiments, scientists subdue the complexity of a system in order to make a detailed examination of one piece in isolation.

The distinction between observation and experiment, though important, does not divide sciences into two completely distinct enterprises. Astronomers, for example, may observe a star, but may also use the results of experiments involving nuclear reactions to understand how that star generates its energy. Evolutionary biologists may look at experimental work on mutation in short-lived fruit flies to guide their thinking about the long-term process of evolution, and geologists will almost certainly look at laboratory experiments on the formation of mineral compounds to help them think about rocks in the landscape. Whether to use experiment or observation is not an either/or question. Every science uses a judicious mix of the two.

Many if not most new ideas in the sciences come from unexpected results in an experiment or observation, so this can be thought of as a starting point of the scientific method. There are, however, exceptions to this general rule. The development of the theory of RELATIVITY in the first decades of the 20th century, for example, arose from Albert Einstein's musing about the fundamental theories of science in existence at the time. As I pointed out above, science doesn't always follow a neat, predetermined path.

Regularities

The next element of the process of science comes into play after a series of experiments or observations have been done and scientists begin to get a feel for how some aspect of nature is behaving. This new understanding is generally expressed as a kind of regularity in nature. For example, in the ecological experiments on the effects of nitrogen mentioned above, Tilman found that as more nitrogen is added to a plot, the amount of plant material (or *biomass*) increases, while at the same time the number of species (*biodiversity*) drops. In essence, a few species are able to take advantage of the increased availability of nitrogen and crowd out the species that can't.

These newly discovered regularities can be stated in simple sentences, as above, but more often they are cast in mathematical terms—"if you increase the nitrogen by x%, the biomass will go up by y%"—or in a formula. As a science educator, I have come to dread the moment when I have to abandon the comforts of the English language and write an equation on the blackboard. In a very real sense, when scientists lapse into mathematics they are speaking in a foreign language. Perhaps it will help you in dealing with the use of mathematics to remember that an equation is just a compact way of saying what could be said more clumsily or at greater length in ordinary English.

There is an important historical example that illustrates the role of regularities. In the 17th century, one of the central questions faced by scientists concerned the place of the Earth in the cosmos. Was it at the center, as the ancient Greek scholars had taught, or did it orbit the Sun (*see*

COPERNICAN PRINCIPLE), as Nicholas Copernicus had suggested in 1543? This is a question that had profound religious and philosophical implications, as Galileo was to learn to his cost (see ACCELERATED MOTION). From the scientific point of view, however, there is only one way to go about answering it. The scientist has to look at the skies and see whether the movements of celestial objects are consistent with a stationary or an orbiting home planet.

The man who spent a lifetime developing instruments and making the necessary observations was the Danish astronomer Tycho Brahe (1546–1601). By the end of his life he had compiled a huge list of accurate measurements of the positions of the planets in the sky—a list, incidentally, that had great commercial value at the time because of its role in casting horoscopes. After Tycho's death, his assistant, the German mathematician Johannes Kepler, took these measurements and, in a brilliant *tour de force* of mathematical deduction, showed that all the data could be explained by three simple rules. These rules, now called KEPLER'S LAWS of planetary motion, say that

— all the planets move around the Sun in elliptical orbits,
— planets move faster when they're near the Sun than when they're farther away, and
— the farther away from the Sun a planet orbits, the slower it moves and the longer is its "year."

The laws are also stated in mathematical form so that, for example, if you know how far a particular planet is from the Sun, the third law will allow you to calculate the length of its year. Kepler's laws are actually a good example of the point made above: Instead of having to wade through volumes of data, a few simple rules, which could be written on the back of an envelope, summarize the observational information.

There is another point to be made here: Scientists are remarkably sloppy about their use of the word "law." In a book devoted to the elucidation of laws of nature, this is an issue that needs to be addressed. It would be very convenient if there were a simple rule about how words like "theory," "principle," "effect," and "law" were used in the sciences. It would be nice, for example, if something that had been verified a thousand times was called an "effect," something verified a million times a "principle," and something verified 10 million times a "law," but things just don't work that way. The use of these terms is based entirely on historical precedent and has nothing to do with the confidence scientists place in a particular finding.

For example, NEWTON'S LAW OF GRAVITATION is one of the best-verified statements in the sciences. It is, however, contained within

Einstein's theory of RELATIVITY. Every verification of Newton's *law*, in other words, is also a verification of Einstein's *theory*. Yet there are verifications of general relativity that lie outside of universal gravitation. Here we have a "theory" that is better confirmed than a "law." This is far from an isolated example. Certain behaviors of ideal gases are governed by CHARLES'S LAW and BOYLE'S LAW, but those *laws* can be derived from (and are therefore less general than) a *theory*—the KINETIC THE-ORY of gases. One of the best-verified ideas in science, that forms the foundational bedrock of all the life sciences, sets out how life developed on our planet. Despite all its verification, though, scientists still talk about the *theory* of EVOLUTION.

The word "theory," then, can indicate a new idea that has yet to be thoroughly tested. It can also be referring to an idea for which that was once the case, but which has since been so thoroughly verified that it can now be ranked with the most secure known truths about the universe. Scientists simply aren't too interested in how ideas are named or what they are called. What matters is what the ideas are, how well they work. As a result of this inattention to nomenclature, you will find all manner of titles in this book. There are "effects," "theories," "laws," and "principles" mixed together with little regard for status. Regard this, as well as the tendency of many laws to be named after people who didn't actually discover them, as a reflection of the way scientists go about their business.

> Once the regularities are stated in equation form, they can be manipulated by the normal rules of mathematics—the language's grammar, if you will.

Hypothesis

Once experiments have been done and new regularities in nature discovered, the time has come for scientists to pause and take stock. Do the newly uncovered regularities suggest something about the way that nature works? Do they fit with regularities already known to give a broader picture of part of science? With these sorts of questions, scientists embark on the creation of hypotheses—guesses, or conjectures, about how the universe works. It is here that the grand ideas about the forces that govern the universe are spun out.

It is also at this stage that the language of mathematics comes into its own. Once the regularities are stated in equation form, they can be manipulated by the normal rules of mathematics—the language's grammar, if you will. Often, these sorts of manipulations lead to stunning new insights. For example, Isaac Newton combined NEWTON'S LAW OF GRAVITATION with NEWTON'S LAWS OF MOTION and produced a totally new picture of the solar system in which the planets circle the Sun, kept from flying off into space by the force of the Sun's gravity. As we shall see later, this development had profound intellectual implications. For one thing, Kepler's laws, previously thought of as summaries of observational data, became consequences, statements that could be derived logically from Newton's deeper theory. For another, the motion of comets, previously thought to be unpredictable, was brought into the realm of rational discourse, a development that in 1705 led Edmund Halley (1656–1742) to the discovery of the orbit of the comet that now bears his name.

In this stage of the scientific process, we encounter another variation on our linguistic theme—yet another shading on the meaning of the word "theory." In the physical sciences, this word is often used to denote a mathematical formulation of ideas about how the universe works. It can be used to refer to anything from an explanation of a very minor phenomenon—a footnote to nature—to a grand, overarching structure that explains a vast assortment of known results. Once again, however, the word can be (and is) often applied to ideas that are as well established as an idea can be. The *theory* of QUANTUM CHROMODYNAMICS, for example, is arguably one of the most precisely verified theories in physics. Some of its predictions have been verified by experimentalists to 16 decimal places, and it is known to apply to structures that range from single electrons to clusters of galaxies. Despite this enormous range of verification, however, scientists use the same word to refer to it as they do to the newest untried hypotheses of a novice postgraduate student!

Prediction

No matter how complex or elegant a theory is, it is still only as good as the experimental and observational evidence that backs it up. A good theory, however, will do more than incorporate facts already known—it will make predictions about phenomena that haven't been seen before. A good theory, in other words, will "put its head on the chopping block" by making clean, testable predictions. In a process that closes the circle of the scientific method, we can then go back to experiment and observation to see whether these predictions are borne out. If they are, we look for new predictions to test the theory. If they aren't, we must go back to the drawing board, change the theory, and try again. Either way, the ultimate test of the theory is how well its predictions fare.

Predictions by theories can be very precise. For example, when Halley used Newton's theories to calculate the orbits of comets, he made a very unforgiving prediction: At some time in 1758 or 1759, the comet would reappear in the sky. No excuses, no maybes—the comet either had to be there or the entire theory would collapse. We now know, of course, that the recovery of Halley's comet on Christmas night, 1758, marks one of the great verifications of the Newtonian concept of the universe, but it's important to remember that it could have been otherwise. Today, of course, the return visits of periodic comets can be predicted with greater accuracy.

Predictions can also be very broad. For example, the story of EVOLU- TION was first worked out from the fossil record. This process gave scientists a sense of how different organisms were related to one another, how long it had been since they had a common ancestor, and so on. In recent years, a new way of working out the relationships between living organisms has been developed, known as the MOLECULAR CLOCK. The technique is centered on the analysis of DNA: the more differences there are between the DNA of two organisms, the farther back in time their evolutionary paths must have diverged. The theory of EVOLUTION would predict that there is only one family tree for living organisms, so that the stories told by the fossils and by DNA should be the same. This is a clear prediction of the theory, though not one that has received a lot of attention. The fact that the two stories of the past are the same is one of the observational facts that buttress the theory of evolution.

This reliance on testing is, I think, exactly what makes science different from other forms of intellectual endeavor. To state the difference in its most blunt and unfashionable form: In science there *are* right answers. It doesn't make any difference how elegant a theory is or how high a status its proposers have in the community, if the theory doesn't work, it has to be abandoned or modified. Period. This reliance on testing is also what creates the great divide between the sciences and the humanities. In fields such as philosophy or literary criticism, there is no outside objective arbiter that plays the role of nature. Interpretations of a work of art, for example, can't really be compared in this way. Consequently, people in the humanities and the sciences often have a great deal of difficulty understanding each other's intellectual framework, a point to which we'll return in a moment.

The reliance on experimental testing of ideas also gives rise to an interesting bit of sociology in the scientific community. It often happens that after a theory has failed, a small group—perhaps even a single individual—will spend a great deal of time trying to salvage it. It has been my experience that there is no scientist so lonely as those who labor to resurrect theories that haven't met the experimental test. Abandoned by their colleagues, they soldier on, often for a lifetime, vainly trying to reverse nature's verdict. Science can be a hard taskmaster, because it demands this relentless questioning of ideas, this relentless return to the tribunal of observation before anything can be accepted.

This fact has an important corollary. If an idea can't be tested against experiment, if it can't be made to confront nature, it simply isn't science. To use a term popularized by the philosopher Karl Popper (1902–94), scientific ideas must be *falsifiable*—they must yield statements that can be tested. It must be possible, in other words, to imagine an experimental or observational result that would show the theory (the law of universal gravitation, for example) to be wrong, even if such results never show up in practice. Halley's Comet might never have reappeared. The fact that it did verified Halley's theory, of course, but the fact that it might not have shows that the theory was falsifiable. In the same way, it could have turned out that the DNA of a fish is closer to human DNA than is the DNA of chimpanzees. That would have falsified the theory of EVOLUTION. The result didn't turn out this way, of course—human and chimpanzee DNA have about a 98% overlap—but it could have. This shows that the theory of evolution is falsifiable.

Compare this situation to a theory that is popular these days with creationists—the doctrine of created antiquity. According to this theory, the Earth was created a few thousand years ago, but it was created with evidence that it was much older. Rocks were created with fossils already in them, for example; trees were created with tree rings; light from stars more than a few thousand lights years away was created en route to the Earth; and so on. The first (and very elegant) elucidation of this point of view was in a book titled *Omphalos,* written shortly after Darwin published his work. *Omphalos* means "navel" in Greek, and the central idea of the book was that Adam was created with a navel, even though he never was in a womb and so never needed an umbilical cord.

The point of this notion is that it is impossible to imagine an experiment or observation that could prove it false. Any contrary evidence would simply be waved away with, "Well, that's the way the Earth was created." The theory is not falsifiable, so however engaging it may be, it simply isn't scientific. A good deal of what has come to be called alternative science suffers from this kind of defect in that it fails the falsifiability test. The TV series *The X Files* (of which, incidentally, I am an avid fan) deals with a vast conspiracy whose sole function is to erase evidence of an extraterrestrial presence on Earth. Lack of evidence is always explained by the mantra "They don't want you to see it." It's good television, but bad science.

Before moving on, I'd like to mention that accusations of non-falsifiability are sometimes thrown around in the debate over the efficacy of classical Freudian psychotherapy. Some critics argue that no matter what outcome the treatment has, Freudian theory can explain the result away. If this is true (and I'm not sure that it is) then this theory, too, would lie outside of science.

The grand cycle

Scientific inquiry, then, goes through a cycle, from doing experiments to discovering regularities to creating theories to making predictions, then the return to experiment to test the predictions. Most scientists spend most of their lives pushing their field around this cycle. This is what philosophers call "normal science." Occasionally, as we have seen, events break with this comfortable pattern, but that is only to be expected in a profoundly human enterprise. So at any stage in its development, each field of science is trying to move from one stage to another. One way to compare different kinds of science is to look for where on the cycle they are stuck at the present time. What, in other words, are the practitioners trying to do right now to move their field ahead?

I began my career in elementary particle physics, the branch of science devoted to understanding the fundamental constituents of matter. At the moment, this field is hung up on the step between prediction and testing. There are a number of plausible theories available, and many of them make predictions about what will happen when particles collide at extremely high energies. Unfortunately, we lack the means of testing these predictions because we do not have machines capable of accelerating particles to high enough energies. In 1993, the United States Congress, in its wisdom, ended the construction of a machine called the Superconducting Super Collider, thereby ensuring that the ordered progression of theory and experiment in this field could not continue. A smaller machine, the so-called Large Hadron Collider, is scheduled to come on line at the European Centre for Nuclear Research (CERN) in Geneva, Switzerland, by 2005, and it may be that particle physics will then start to move again at that time.

While some fields of science are hungry for data, others are suffering from a surfeit. In many areas of molecular biology, for example, new information is pouring in so fast that it cannot be digested. Living organisms are the most complex structures in the universe, and only now do we have the capabilities to investigate this level of complexity. Many areas in the life sciences are hung up at the step between experiment and the discovery of regularity, and researchers are devoting much effort to finding the counterpart of Kepler's laws for molecular processes.

A good example of this is the so-called *protein-folding problem*. Among their other functions, PROTEINS are the workhorses that control the chemical processes of life. They are large molecules that have complex, convoluted shapes, and these shapes allow the protein to act as a kind of molecular broker—facilitating chemical reactions without taking part in them (*see* CATALYSTS AND ENZYMES). A protein is constructed from smaller molecules called amino acids, much as a necklace can be built by sliding a succession of beads onto a string. Once the amino acids are strung together, complex interactions between atoms in neighboring amino acids, and between those atoms and the water in their environment, cause the protein to fold up and take on the complex shape that allows it to carry out its function.

The protein-folding problem, then, can be stated in this way: Given the sequence of amino acids along the protein "necklace," is it possible to predict the shape the molecule will have, and hence predict its chemical function? At the moment, the answer to this question is "no," simply because the problem is too complex to be solved by even the fastest computer. Presumably there are rules—molecular analogues of Kepler's laws—that will tell us how the folding process works, but we haven't been able to find them because of the complexity of the problem. It's a classic case of not being able to see the forest for the trees.

Complexity blocks progress in other areas of science as well. For example, a good deal of the current debate about the GREENHOUSE EFFECT and global warming arises because of the inability of climatologists to make firm predictions about the consequences of adding substances such as carbon dioxide to the atmosphere. The reason for this uncertainty isn't primarily a lack of understanding of the basic physical and chemical processes that determine the behavior of the atmosphere, but the fact that the real atmosphere is so complex that we can't put all of the required information into a computer program. At the moment, for example, two important influences on the climate, clouds and ocean currents, are poorly handled by these programs. In terms of our conceptualization of the scientific method, then, this field is currently hung up between theory and prediction.

To round off our examples, we can look at the field of evolutionary theory. Data in this field have been accumulating for centuries, and many regularities are known. Some evolutionary theorists are now turning their attention to the broader problem of establishing the general principles that drive the grand sweep of life's history. It's one thing to know, for example, how a particular species of flatworm or bird changed during a particular period of time, but quite another to understand how entire ecosystems respond to change, to be able to predict the fate of each species. In terms of our discussion, you could say that these scientists are trying to move their field from regularity to theory.

As we have seen, scientists are engaged in a constant effort to move their fields through the cycle from experiment to regularity to theory to prediction and back again to experiment. Their program of activities varies from one field to the next, depending on the subject matter and the maturity of the field. At each new turn of the cycle, the theories get more refined and more detailed, and our picture of nature becomes fuller. And although philosophers can (and do) disagree with me on this, I believe that each turn brings us closer to the truth about our universe.

A few points should be made about the orderly picture of scientific progress I have painted. The first, already made above, is that occasionally the whole system gets shaken up when new data or new theories come in. Although I believe that philosophers make rather more of this point than it warrants (you can count these "revolutions" on the fingers of one hand), they do happen and we need to be aware of them. The second point is that the process is open-ended. You never finish going around the cycle: You never finish testing your ideas against nature. This means that there is always room in science for new ideas and for extending the limits of knowledge in new directions. News reports fifty and a hundred years from now are likely to be as full of new scientific advances as they are now. Finally, the cycle has no fixed timeframe attached to it. Science is driven by its own internal logic, by the development of instruments and ideas, so it's not always possible to predict when specific problems will be solved or specific knowledge acquired. Sometimes progress happens rapidly, sometimes the process just stalls. Sometimes discoveries in one field will have profound effects on entirely different fields of science, by introducing new kinds of instruments—the laser, for example. In the end, progress is difficult to predict—a fact that gives research managers and policy makers many sleepless nights.

> Science is driven by its own internal logic, by the development of instruments and ideas, so it's not always possible to predict when specific problems will be solved or specific knowledge acquired.

Owing to these characteristics of the scientific process, the conduct of science and the activities of governments frequently fail to connect. It may be true, for example, that an important vote on a particular issue is coming up next Tuesday, and that the representatives of the people need certain scientific information to help them decide how to vote, but there is no guarantee that the internal logic of the sciences will produce that information on time. To a scientist, this is of no concern. As long as we get to the right answer eventually, there's no particular reason to worry about when the answer is achieved. To politicians, however, information that comes after next Tuesday is worse than useless. Not only does it not help them decide how to vote, it could embarrass them by showing that their vote was wrong.

In the same way, the function of the courts is to try questions of fact and provide closure on issues at conflict. If a company is being sued because a plaintiff claims that one of the company's products causes cancer, for example, then the decision on the question of fact has to be made immediately. Scientists can't ask the court to wait ten years while they get all the facts sorted out and all the relevant studies analyzed. The decision has to be made during the trial, with the information available at the time. And if new information turns up at a later date, it usually doesn't do the litigants much good, since it's very hard to reopen these sorts of cases.

The reality of the scientific method, then, is that it is a superb instrument for answering questions about the working of the physical universe. Ideas are not accepted until they have been submitted to rigorous testing, and hence they have a high level of reliability. It takes time to come to a scientific consensus, however, and this means that it is not always possible to use the best possible information to guide policy or settle disputes in court.

The role of natural laws

The range of phenomena and objects in the universe is huge, from stars thirty times as massive as the Sun to microorganisms invisible to the naked eye. These objects and their interactions make up what we call the physical world. In principle, each object could behave according to its own set of laws, totally unrelated to the laws that govern all other objects. Such a universe would be chaotic and difficult to understand, but it is logically possible. That we do not live in such a chaotic universe is, to a large extent, the result of the existence of natural laws.

It is the role of natural laws to order and arrange things, to connect the seemingly unconnected, to provide a simple framework that ties together the universe. The analogy I like to use is that of a spider's web. On the outside of the web are all the phenomena of the universe—blades of grass, mountains, comets, and so on. If you enter the web at any point on its periphery, choosing a single phenomenon to investigate, you can start asking questions about it. As you do so, you will find yourself being drawn deeper and deeper into the web, uncovering more profound explanations of the phenomenon you are investigating. Eventually, you will begin to uncover generalities which apply not only to your phenomenon, but which connect that phenomenon to others, even though those connections aren't obvious at first glance. These deeper explanations are what we call laws of nature.

The same sort of process goes on as you continue probing. You find that many laws of nature are themselves connected to others by still deeper laws, that those deeper laws have even deeper connections, and so on. Eventually, at the very center of the web, you find a relatively small number of laws that cement the whole framework together. Following the scientific practice of not being too careful about nomenclature, these are sometimes referred to as "laws of nature." But to avoid confusion I shall call them "overarching principles," to distinguish them from the other kinds of laws, principles, and effects that we shall be discussing.

So, to paraphrase *Animal Farm,* all laws of nature are equal, but some laws are more equal that others. As you might expect, of course, there is no universal agreement among scientists as to what the overarching principles of our craft are exactly, but you would be hard pressed to find a scientist who doesn't agree that they exist. There is also, I suspect, pretty much universal agreement that certain principles—the first law of THERMODYNAMICS, for example—are in this select group. At the fringes, however, there can be healthy disagreement. I am reminded of a time several years ago when *Science* magazine ran an article on this subject and asked readers to send in their Top Twenty lists. Over 800 replies later, the editors saw that they could easily name the top ten "great ideas," but that there were many candidates vying for the next ten slots. In what follows, I shall simply state the extent of each law's domain, among other things, and leave you to decide whether it belongs among the overarching principles.

As you might expect, of course, there is no universal agreement among scientists as to what the overarching principles of our craft are exactly ...

Let me illustrate the interlocking web of laws and principles by talking about a subject we have already referred to in passing—comets. There are many questions we can ask about them. One very old question is why they appear erratically in the sky—where do they come from and where do they go? This is really a question about the orbits of comets or, more generally, about the influence of the Sun and the planets on the motion of comets. To understand how a comet moves, we have to understand the forces that act on it and the laws that govern the action of forces on it. As it happens, our understanding of both of these is due to Isaac Newton. His law of universal gravitation tells us the force the Sun exerts on the comet, while his laws of motion tell us how that force affects the comet's motion. Together, these laws tell us how each comet will move as it moves around the Sun.

They also provide us with the first of the unifying insights we talked about above (it was this same procedure that had allowed Newton to explain the motion of the planets). In other words, it appears that the laws that govern the motion of comets are exactly the same as the laws that govern the motion of planets. The Sun's gravity pulls on both, and the differences between planetary and cometary orbits has to do with how the two different classes of objects formed (*see* NEBULAR HYPOTHESIS). Comets fall into the inner solar system from regions far outside the orbit of the farthest planet, so they come in toward the Sun at a grazing angle. Think of them as being analogous to kids playing the familiar game of running as fast as they can toward a post, then grabbing the post and swinging around. Planets, on the other hand, formed pretty much where they are found today, and consequently follow stately, almost circular, paths around the Sun.

This insight that the motions of planets and comets are governed by the same laws was both revolutionary and profoundly unexpected. After all, what could be more different from the stately, regular, predictable progression of a planet across the sky than the erratic, unpredictable appearance of comets? Yet these two seemingly different kinds of celestial phenomena are governed by exactly the same laws and respond to the Sun's gravity in exactly the same way.

It was this fact that allowed astronomers in the mid-20th century to figure out where comets come from. By looking at the paths of comets making their first entry into the solar system, they could use Newton's laws to track the orbits back to where the comets had started from. They found that the comets came from two reservoirs in the cold vastness of space: a flat disk just outside the orbit of Pluto called the Kuiper Belt, and a huge sphere reaching perhaps a light year and a half from the Sun, called the Oort Cloud. Thus, just as Newton's laws had allowed Edmond Halley to work out the orbit of a comet that had been captured by the Sun, they also allowed his successors to figure out where the comets came from, which is another of the sorts of unexpected connections we've been discussing.

If instead of asking *where* a comet is you ask *what* it is, you enter an entirely different part of the spider's web. A simple question in this category might be, "What chemical elements and compounds are there in comets?" Since most studies of comets have to be made from a great distance, astronomers who ask this question typically look at the light that is emitted and absorbed by a comet to try to deduce its chemical composition. Like all other material objects, comets are made from atoms, and atoms have a special way of interacting with light. Atoms of each chemical element and each chemical compound give off their own characteristic array of wavelengths, which you can think of as a sort of optical fingerprint (*see* SPECTROSCOPY). For visible light, we perceive these optical fingerprints as different colors. The bright blue you see when a bit of copper falls into your campfire or the vivid yellow of a sodium vapor street lamp are both examples of this phenomenon. The atoms in the comet give off light, the light traverses the great distances of space to a telescope, and astronomers work out the chemical composition of the comet, even though they can't get pieces of the comet to examine in the laboratory.

But of course there's nothing in this procedure requiring that the light travel great distances. It should work equally well if the light travels only a few meters or even a few millimeters. In fact, industrial chemists routinely use the same property of atoms—the fact that each type of atom emits a characteristic set of colors—for quality control in production processes. Products from pharmaceuticals to paints to drinks (and many more) are tested in this way. The fact that an industrial engineer checking the quality of a batch of gasoline and an astronomer studying a distant comet use the same laws of atomic behavior to do their jobs is yet another example of unexpected connections.

I could go on giving examples, but I think you get the picture. When you see the world as a coherent whole governed by natural laws rather than as a large number of disparate phenomena, your picture of the universe becomes more connected. You see relations between seemingly unrelated things, coherence in the vast diversity of natural phenomena. This, I think, is the ultimate gift of science to the human intellect, one of the greatest achievements of the human mind.

Science in the 20th century

The picture I've just painted for you, which I refer to as the "scientific world view," has a decidedly Newtonian cast to it. There is a common misconception that the 20th century was not kind to Isaac Newton, and that scientists no longer accept that the universe is an ordered place governed by natural laws. Philosophers who assert this view frequently argue that the theory of RELATIVITY, HEISENBERG'S UNCERTAINTY PRINCIPLE, and DETERMINISTIC CHAOS have rendered the old scientific world view obsolete. Nothing could be further from the truth!

Let's start by looking at relativity. As we shall see, Einstein came to his theory through an attempt to salvage the primacy of natural law in science. In particular, he concentrated on one particular aspect of Newton's laws of motion—the fact that different observers, even observers moving with respect to one another, will see the same laws operating in the universe. Upon this simple premise he built an elaborate theoretical framework that didn't so much displace Newton's laws as extend them into new regions. For example, one consequence of Einstein's work is that moving clocks run slower than stationary ones. For something moving at normal speeds (a car or an airplane, for example), this slowing down is so small that it would be impossible to measure, so for everyday purposes we can happily ignore relativity. For objects moving at appreciable fractions of the speed of light, however, the effects can be large and have to be taken into account.

Two points can be made about this so-called time-dilation effect: First, it has been amply verified by experiment, as required by the scientific method. Second, when the equations of relativity are applied to slowly moving objects, they reproduce Newton's laws of motion. Furthermore, in keeping with the notion that every law of nature is only as good as the experimental verifications that back it up, we need to appreciate that Newton's laws were originally verified only for normal objects moving at normal speeds—for which they make the same predictions as relativity. At speeds near the speed of light, however, where Newton's laws were never tested, the two theories diverge in their predictions, and the predictions of relativity have been verified experimentally. This tells us the domain in which Newton's laws are valid, but it also tells us that relativity doesn't contradict these laws, but provides a way of extending our theory into new regions.

The notion that relativity somehow upsets the Newtonian apple cart is based on a hidden assumption, the assumption that when Newton's laws were extended to encompass objects moving near the speed of light they would be unchanged. There is no logical reason why this should be so. Arguing in this way is a little like saying that because people in America speak English, people in Paris should do the same, and then claiming some great contradiction when it turns out that Parisians speak French.

The relation between Newton and Einstein does in fact exemplify how mature sciences progress. It is not a case of a new theory coming along and throwing an old one out, but of deeper theories extending and encompassing the old ones. We still use Newton's laws to guide spacecraft for the simple reason that they have been amply verified and tested in those sorts of situations. In this respect science grows like a tree, adding new branches all the time, but always retaining the heartwood.

The arguments from chaos theory and QUANTUM MECHANICS deal with another aspect of the Newtonian world view—the concept of DETERMINISM. In the language of physics, a system is deterministic if, given its starting conditions and the laws governing its behavior, you can predict its state at any time in the future. The classic example of a deterministic system is the collision between two billiard balls. Given their positions and velocities before the collision, it is easy to use the conservation of LINEAR MOMENTUM and the first law of THERMODYNAMICS to predict where the balls will be any time after the collision. Relativity is completely deterministic. Clocks on moving billiard balls may slow down, but you can use Einstein's equations in exactly the same way as you can use Newton's laws to predict their behavior. Give me the initial position and velocity of each ball, and I will tell you where that ball will be in the future.

The situation is somewhat different for chaotic systems. Discovered in the latter part of the 20th century, primarily through the use of computer simulation, these are systems in which the final state is extremely sensitive to the starting point. Whitewater rapids are a good example of a chaotic system. Start two chips of wood right next to each other on the upstream side of the rapids, and they will come out widely separated on the other side. This means that if you want to predict the future of a chaotic system you have to know its starting place very accurately. In a truly chaotic system, if you want to be able to predict future behavior into the infinite future, you have to know the initial state with infinite accuracy. For a wood chip in the rapids, this would translate into the requirement that we know the starting point of the chip with infinite accuracy. Since this is clearly impossible in any real system, the future of the system cannot, for all practical purposes, be predicted with Newtonian accuracy.

Does this mean that chaos has defeated the Newtonian world view? Not at all. Newtonian determinism is a classic If–then statement: *If* I know the initial state of a system, *then* I can predict its future. What chaos theory deals with is not this central tenet, but the relationship between the amount of error in the "if" part of the statement and the resulting error in the "then" part. If the error in the former is zero (i.e., if you know the initial state with infinite accuracy), then the error in the latter is also zero (i.e., you can predict its future course exactly). Chaotic systems, then, are deterministic in principle, but not in practice. Scientists have come to recognize this fact by referring to systems like whitewater rapids as exhibiting DETERMINISTIC CHAOS.

The confusion about chaos and Newton, I think, arises because some people assume that the future of classical Newtonian systems can always be predicted with infinite accuracy. This simply isn't true, even in very simple situations. When I described the idealized Newtonian billiard balls earlier, I glossed over the question of how accurately the positions and velocities of the balls could be known. In fact, in the real world there will always be some uncertainty in those numbers and hence some uncertainty in the predicted future positions of the balls. Indeed, I can remember being made to work through just such an example as a postgraduate student, just to drive home to me the fact that no system in the real world is infinitely predictable. In the end, then, the difference between chaotic and classical Newtonian systems is a matter of degree. Every system—even Newtonian billiard balls—will have uncertainties in its predictions if there are uncertainties in measuring its initial conditions. Chaotic systems just represent an extreme case of this statement.

The situation with quantum mechanics is a bit more complicated, mainly because of HEISENBERG'S UNCERTAINTY PRINCIPLE. This principle says, in essence, that it is not possible to know both the position and the velocity of a subatomic particle at the same time with complete accuracy. You can know one or the other exactly, or both with some degree of uncertainty, but not both together. This means that in the world of the atom we have to describe the state of a particle differently than we do in the Newtonian world. Instead of thinking of a particle as a specific thing (like a baseball, for instance) located at a certain place and moving with a certain speed, we have to think of it as a kind of wave.

Because of this, the only predictions you can get from quantum mechanics are probabilities (physicists call them wave functions). Some people assume that this makes quantum mechanics not deterministic, but that really isn't the case. What quantum mechanics does is tell you how to go from an initial state described in terms of probabilities to a final state which is also described in terms of probabilities. Virtually all of the difficulties that people have with "quantum weirdness" happen because they try to mix the Newtonian and quantum worlds. They might, for example, assume (usually without realizing it) that the initial state of an electron is described in Newtonian terms, and then use the fact that the final state is described in terms of probabilities to argue that we have somehow lost the ability to make deterministic predictions.

> Virtually all of the difficulties that people have with "quantum weirdness" happen because they try to mix the Newtonian and quantum worlds.

But the central point about quantum mechanics is this: *If you want to play the quantum game, you have to play by the quantum rules.* In other words, if you're going to describe the final state of a system in terms of probabilities, you have to describe its initial state in the same way. Once you realize this, then you can see that quantum mechanics is just another if–then deterministic system. If I know the initial state of the system (described in terms of probabilities), then I can predict with certainty the final state of the system (also described in terms of probabilities). The only thing different between Newtonian and quantum mechanics is what is meant by a "state." For Newton, a state was a collection of variables such as position and velocity; for the pioneers of quantum mechanics, a state was a wave function. The choice of definition is forced upon us by nature, but once the choice is made the statement of the ability to predict is exactly the same.

To summarize, then, the effects of the three great 20th-century discoveries that are supposed to have canceled out the Newtonian view of the universe are these:

— relativity extends and expands Newtonian theory to incorporate objects moving near the speed of light,

— chaos theory enables us to make statements about the effects of errors in determining initial states on the accuracy of predictions, and

— quantum mechanics redefines and broadens the notion of a physical state to include wave functions.

In other words, all of the new developments in science that were supposed to bring about the downfall of Newtonian physics are simply extensions and redefinitions of its basic ideas.

The central idea of science—that you can learn about the laws that govern natural events by experimentation and then produce predictive theories—is still in effect. That's good, because it makes a discussion of natural laws not only interesting, but essential to an understanding of our universe.

About This Book

One of the great truths that we have discovered is that we live in an ordered universe, a universe whose workings are accessible to the human mind. The enterprise we call science differs from other attempts to interpret the universe in that it does not seek absolute truth, but instead uses a method that produces successively better representations of physical reality. The scientific method begins with a question: Why do things happen this way and not some other way? The scientist explores, systematically observing and measuring, looking for correlations and anomalies. Once a pattern emerges, an explanation is framed. The more general is the explanation, the more predictions it will make about how other things should happen. The scientist continues to observe and measure, to test those predictions. If the explanation survives the tests, the result is a law of nature. This book is your guide to those laws.

The laws of nature are not always called "laws." Scientists haven't always concerned themselves overmuch with terminological precision, and a law is just as likely to be known as a "theory," a "rule," a "model," or a "principle," or, reflecting the fact that laws are often stated in the language of mathematics, as a "relation" or an "equation." I suspect that whether or not something is called a law has more to do with quirks of history than with the logic of the sciences. It is entirely conceivable, for example, that in some alternate universe the phenomenon of photosynthesis is called "Professor Branestawm's Law of Biological Energy Conversion." Consequently, this book contains explanations of natural phenomena that aren't usually called "laws," but which do have the same depth and explanatory power.

Each entry is meant to be freestanding and self-sufficient, so certain explanations and information are presented in more than one entry. The book is not designed to be read cover to cover, but to be dipped into for specific information, set down, and then dipped into again. If you find a particular entry interesting, you may want to take a look at related entries indicated by SMALL CAPITAL LETTERS, each of which will lead you further into the network of knowledge that constitutes our scientific understanding of the universe. You can trace all the entries in the same field of science from the colored tabs above each headword. One of these categories is "Rear View Mirror"—ideas which have now been superseded or discredited but which have their place in the story of science. Additional signposts are provided in the form of timelines,* which locate an entry temporally in relation to associated entries; at the end of the book is a chronology which puts all the entries into a single historical framework. Also at the end of the book is a glossary of terms which belong to the basic vocabulary of science.

Putting together a list of the entries for this book was, like any similar

* Sometimes the subject of an entry can be tied inarguably to a specific year (e.g., the DRAKE EQUATION). In other cases the year selected is that of a significant development, which may or may not be discussed in the entry. When a concept emerged over a period, perhaps with many contributors, we can be no more specific than, for example, "1880s" or even "20th C." Thus the timelines aren't necessarily strictly chronological threads (though many are).

exercise, a matter of judgment. While I was compiling it, I could easily imagine being confronted by critics asking in an accusing tone, "Why did you include X but not Y?" Even something apparently as objective as listing the laws of nature turns out to be prone to subjective criteria. To forestall at least some of the "Why did you?" questions, I should like to explain here how the entries were selected.

First, some entries selected themselves as being unquestionably appropriate for a book on the laws of nature. CHARLES'S LAW, which relates the temperature and pressure of a gas, has been used by physicists and chemists for over three centuries. It is clearly a "law of nature," no matter how that term is defined. These sorts of laws are included as a matter of course, with the caveat that any laws too obscure to be included in the index of standard university-level textbooks have been omitted. As mentioned above, there are firmly established "laws" of nature which for no good reason bear a different label. Many of these are automatic choices.

Another class of entries reflects the crucial historical relationship between scientific inquiry on the one hand, and experiment and observation on the other. To stress this point and to give some historical perspective, critical experiments and observations are described, usually as separate entries. Thus, for example, the RUTHERFORD EXPERIMENT, which established the existence of the atomic nucleus, and the VAN HEL-MONT EXPERIMENT, which showed that plants build up their biomass by extracting substances from the air and not the soil, are included.

Finally, we come to a small set of entries that are included simply because, in the judgment of the author and the editors, they will be of interest to the kind of person who would read a book on the laws of nature. This category includes some serious mathematical entries like the one on GÖDEL'S INCOMPLETENESS THEOREMS, some more offbeat entries like the one on the BEAUTY CRITERION, and some that are just plain fun, such as the entries on MURPHY'S LAW and the three laws of ROBOTICS.

Taken together, the entries map out our current understanding of the physical universe, from the BIG BANG to the HUMAN GENOME PROJECT.

James Trefil
Fairfax, Virginia

THE LAWS

Accelerated Motion, equations of

The velocity of a constantly accelerating object starting at zero velocity is proportional to the time

The distance traveled by a constantly accelerating body starting at zero velocity is proportional to the square of the time

Galileo Galilei is one of those people who are famous for the wrong reason. Many remember this Italian philosopher-scientist for his trial on suspicion of heresy near the end of his life, and being forced to recant the then-heretical notion that the Earth moves around the Sun. This trial had very little effect on the development of science, but work that Galileo had done earlier in his life had a crucial impact on everything that followed.

The study of the motion of physical objects was old long before Galileo was born. Simple problems on motion are routinely taught in schools today. For example, you know that a car traveling at 20 miles an hour will cover a distance of 20 miles in 1 hour, 40 miles in 2 hours, 60 miles in 3 hours, and so on. For as long as the car continues to move at a constant speed—for as long as the speedometer needle stays in one position—it's easy to work out the distance the car will have covered at any time in the future simply by multiplying the speed by the travel time. Indeed, this solution has been known for so long that the credit for solving it is lost in the mists of antiquity.

Things get more complicated when an object doesn't move at a constant velocity. For example, when you pull away from a traffic light, the speedometer needle moves all the time. It never really stays on one number, which means that the car's velocity is constantly changing. It will have one value at the start of each second, and another at the end, so there's no simple way of figuring out how far the car will travel during that second. This problem of accelerated motion was a major question for natural philosophers for centuries before Galileo came on the scene.

Galileo's approach was completely novel and pointed the way toward the modern scientific method. Instead of sitting down and thinking about how objects should move when they are accelerated, he designed ingenious experiments to see what happened. To us that might not seem much of an innovation, but until then the standard approach of "natural philosophy"—as the phrase suggests—was simply to think. Doing experiments was a truly radical idea. We can understand Galileo's method by thinking about an object falling under the influence of gravity. If you were to let go of this book it would initially have zero velocity, but it would also begin accelerating immediately and would continue to accelerate until it hit the ground. If we can work out how a dropped object falls, then we will know how to handle any other case of accelerated motion.

Today, measuring the fall of an object is simple—we time its arrival at various points on its downward path. We can do this because we can measure time extremely accurately, but in Galileo's day clocks were very primitive by modern standards. So he devised an experimental apparatus that got around the problem. First, he "diluted" the force of gravity, slowing the time of fall to something manageable, by having objects roll down

The Trial of Galileo

Like Newton's apple, the trial of Galileo by the Catholic Church is one of the enduring episodes of scientific mythology. And, like most myths, the story that is told is rather different from what actually happened. The myth of the trial is that Galileo had produced incontrovertible evidence that the Copernican view of the solar system, in which the Earth moves around the Sun, was correct, but that he was crushed by a Church intent on suppressing that theory. In fact, Copernicus, as a savvy church politician, had presented his theory in a way that was acceptable to the ecclesiastical authorities (essentially by calling it a hypothesis). The theory was widely discussed by scholars before Galileo's time, even in the Vatican itself.

In 1616, Galileo published a book called *The Starry Messenger* in which he summarized his telescopic observations and made a strong argument for the Copernican system. It was written in Italian, rather than Latin, which made it accessible to literate (but non-scholarly) readers. Responding to complaints that the book violated Church teachings, the College of Cardinals summoned Galileo to a meeting. Most of the controversy about subsequent events has to do with conflicting reports of what happened at this meeting. The official version is that Galileo was warned not to discuss Copernican ideas as anything other than hypotheses unless he could produce incontrovertible proof of their correctness. Galileo's stance was that he had never received such a warning.

In any case, in 1632 Galileo published his *Dialogue Concerning the Two Chief World Systems* in which he presented long arguments in favor of the Copernican system, putting the points that had been made publicly by the Pope into the mouth of a character named Simplicio. Galileo was then accused of an offense that went by the name of "suspicion of heresy," which bears roughly the same relation to actual heresy as a charge of involuntary manslaughter does to first-degree murder in the modern legal system. Galileo purged himself of suspicion by publicly denying that he believed what he had written, and spent the last few years of his life under what amounted to house arrest in Florence.

So what are we to make of this story? My own sense is that it describes a cumbersome bureaucracy being goaded into action by a man seeking a confrontation. (I imagine, for example, that the College of Cardinals had more important things to do than deal with an abstract point of cosmology.) It is also true that Galileo's arguments in support of the Copernican system were not very strong. From a modern perspective, we can see that Galileo came to the right conclusion on the basis of faulty arguments. This doesn't justify the trial, of course, but it does put the whole operation into a different—as well as less mythological—light.

A footnote: In 1992 the Catholic Church officially reversed the verdict of the trial on the grounds that the judges had not separated questions of faith from scientific fact.

a slope rather than falling straight down. Then he got around the inadequacy of the clocks available to him by stretching thin wires across the path of the ball, so that each wire would be plucked and emit a sound as the ball rolled by. By adjusting the positions of the wires and listening to the sounds, he could eventually arrange things so that it took the ball equal amounts of time to travel between successive pairs of wires.

In the end, Galileo had experimental information on how an accelerating body falls. For an object starting from rest, its subsequent motion is described as at the top of page 2. Translating the words into the symbols of mathematics, we get the equations of accelerated motion:

$$v = at \quad \text{and} \quad d = \tfrac{1}{2}at^2$$

where a is the acceleration, v is the velocity, and d is the distance traveled in time t. You can get a feeling for these equations by watching closely when something falls. Its velocity seems to be increasing all of the way through the fall. This comes from the first equation. It also seems to take the object longer to go through the first part of the fall than through the last part. This what the second equation describes, for it tells us that the longer an object has been accelerated, the greater the distance it will cover.

Galileo was also able to make an important statement about a body falling freely under gravity, even though he couldn't measure it directly. By extrapolating his results for an object rolling down a slope, he was able to determine how an object accelerates when it falls to the surface of the Earth. This quantity, which is represented by g, has the value

$$g = 32 \text{ feet per second per second}$$
$$= 9.8 \text{ meters per second per second}$$

An artist's portrayal of an experiment that probably never happened: Galileo drops two balls from the Leaning Tower of Pisa.

This means that if you drop something right now, its velocity will increase by 32 feet per second (or 9.8 meters per second) during each second of the fall. At the end of the first second it will be moving at a speed of 32 feet per second, at the end of the next second the speed will be 64 feet per second, and so on. g represents the rate at which a freely falling body will accelerate near the Earth's surface, so it is called the *acceleration of free fall*, or *acceleration due to gravity*.

There are two important points that can be made about this result. This first is that, for Galileo, this value of g is purely an experimental result, not something that could have been predicted. It was only later, with the work of Isaac Newton, that the value of g could be calculated from a combination of NEWTON'S LAWS OF MOTION and NEWTON'S LAW OF GRAVITATION. Galileo's pioneering work thus paved the way for the eventual triumph of Newtonian physics.

The second point is that the acceleration does not depend on the mass of the falling object. In effect, the greater gravitational pull on a more massive object is balanced by its greater inertia (its reluctance to move, if you like), so that, neglecting air resistance, every object falls at the same rate. This was in direct conflict with the predictions of ancient and medieval scholars, who argued that it was in the nature of all objects to seek to get to the center of the universe (which for them was the center of the Earth), and that more massive objects would seek to do so more vigorously than smaller ones.

Galileo backed up his prediction with data, of course, but he probably never did the experiment for which he is most famous. According to the folklore of science, he dropped objects of different masses from the Leaning Tower of Pisa to demonstrate that they hit the ground at the same time. Had he done so, he would have found that the more massive body actually hit the ground first. The reason has to do with the effects of air resistance. If the objects were of the same size, then the effect of air resistance would be the same on both of them. From Newton's laws of motion it follows that the less massive object would have been decelerated more and therefore hit the ground later than the more massive one. This, of course, would have been contrary to Galileo's prediction.

GALILEO GALILEI (1564–1642) Italian scientist. Born in Pisa, Galileo can rightly be called the father of modern experimental science. His father, Vincenzo, was a prominent musician who later moved the family to Florence. Galileo's education began at the University of Pisa where, though enrolled as a medical student, he spent a great deal of time studying mathematics. This preoccupation eventually resulted in his appointment to the chair of mathematics at that university.

After his father's death, he moved to Padua as professor of mathematics at the university there (mainly, it appears, because it paid better than the position at Pisa). At Padua, three important themes that were to run through his life became evident. First, he began doing experiments on falling bodies—work that was to revolutionize the science of mechanics. Next, he developed an interest in the new astronomical ideas of Nicholas Copernicus (see COPERNICAN PRINCIPLE). Finally, he invented an instrument—the proportional compass—whose sale greatly helped his financial position. (Like many of Galileo's inventions, it is still widely used today.)

In the winter of 1609/10, with a telescope he constructed for himself, using optical ideas then current in Holland, Galileo observed the heavens. He may not have been the first to do so, but he was the first to give wide publicity to what he had seen. He observed the moons of Jupiter, mountains on the Earth's Moon, the rings of Saturn (though he did not recognize them for what they were), and the phases of Venus. All of these new discoveries cast doubt on the old idea, enshrined in the teachings of Aristotle, that the Earth stood at the center of the universe, and supported the new Copernican view. His book, *Dialogue Concerning the Two Chief World Systems,* was an eloquent defense of the Copernican universe. His outspoken views on this subject would lead to his trial on suspicion of heresy.

After his trial, he wrote *Discourses and Mathematical Demonstrations Concerning Two New Sciences,* which summarizes his experimental discoveries in what we would now call materials science and kinematics. As in all his work, it emphasized the importance of experiment as a means of verifying theory.

Acid Rain

Sulfur and nitrogen compounds in the atmosphere increase the acidity of rain

Pure water, as we all learned in school, is made up of molecules consisting of two atoms of hydrogen and one of oxygen. At any moment, though, some of these molecules will be separating into positively charged hydrogen ions (the same thing as protons, H^+) and negatively charged hydroxide ions (OH^-), while elsewhere in the fluid these ions will be coming together to make water molecules. Thus, even in the purest water there will be a dynamic balance, an *equilibrium,* with a certain number of hydrogen ions (protons) present. Technically, these protons are associated with what chemists call a *hydronium ion*—three hydrogens and an oxygen. In pure water, the number of moles (*see* AVOGADRO'S LAW) of hydronium is 10^{-7} in every liter.

Chemists use a quantity called pH (for "power of hydrogen") to describe the amount of hydrogen ions in a fluid. By convention, the pH of pure water is taken to be 7. This is the definition of chemical neutrality (*see* ACIDS AND BASES). If the concentration of hydrogen ions is higher, the pH is lower and the liquid is defined as an acid. Because acids have all those extra hydrogen ions, they can react strongly with other substances.

The term "acid rain" has been in the scientific vocabulary since the middle of the 19th century, when British scientists noticed that air pollution in the industrial Midlands was causing the rain to become more acid than normal. It was not until the latter part of the 20th century, however, that acid rain became well recognized as an environmental threat.

One thing has to be made clear: Normal rain is itself acid, even in the absence of factories. This is because, as raindrops form and fall, they dissolve carbon dioxide in the air and react with it to produce carbonic acid (H_2CO_3). Pure rain, falling through unpolluted air, will be a fluid with a

Acid rain is particularly damaging to objects made from limestone, such as this statue of a lion outside Leeds Town Hall, England.

pH of 5.6 by the time it hits the ground. As we shall see, human activity is the cause of most of the acid rain that falls, but there are natural sources as well, from volcanoes and lightning strikes to the actions of bacteria. Generally, if we were to shut down all the factories and stop driving cars and trucks, we would expect the pH of rain to be about 5.0. Most scientists now define acid rain as having a pH lower than 5.0.

The extra acidity of rain in the modern industrialized world comes from two basic sources:

— *Sulfur oxides* These compounds can be injected into the atmosphere naturally in volcanic eruptions, but most atmospheric sulfur oxides originate from the burning of fossil fuels. Coal and petroleum contain small quantities of sulfur. When these fuels are burned, the sulfur enters the atmosphere combined with oxygen. These molecules enter water droplets, where they form sulfuric acid.

— *Nitrogen oxides* If the temperature is high enough, nitrogen in the air combines with oxygen to form nitrogen oxides. This can happen naturally, in lightning strikes, but it also happens when gasoline is burned in an internal-combustion engine (in a car, for example) or when coal is burned. These compounds, when dissolved in water droplets, form nitric acid.

So, acid rain falls when the nitrogen and sulfur compounds in the air are washed out. This has several detrimental effects. Many historic buildings in Europe, for example, are built from limestone, a material that reacts with acids. Acid rain, over time, literally dissolves the outer surfaces of these buildings. It also acidifies the soil, which can place forests under stress. For a time it was thought that forest diebacks in the eastern United States and Germany were caused by acid rain, but that view is no longer held. (The dieback is real, but is now attributed to other sources.) Finally, it can increase the acidity of the water in rivers and lakes, posing an obvious threat to aquatic life forms.

Measures to combat the formation of acid rain tend to concentrate on developing technology to remove sulfur compounds in industrial and power-plant smokestacks, usually by devices called scrubbers. The governments of some nations have introduced legislation to ensure that road vehicles emit less than a certain amount of pollutants.

Acids and Bases, theories of

An acid has been defined as a substance that produces a hydrogen ion in water, one that can transfer a proton, or one that can accept a pair of electrons

A base has been defined as a substance that produces a hydroxide ion in water, one that can accept a proton, or one that can donate a pair of electrons

We know from everyday experience that some substances are highly corrosive. If you get battery acid from your car on your clothes, for example, it will eat through them. We sometimes use substances such as ammonia for the more demanding of domestic cleaning tasks. These corrosive substances are known to chemists as acids and bases. On one level, it is quite easy to distinguish between the two. Acids taste sour and turn litmus paper red, whereas bases feel soapy and turn litmus paper blue. Chemists, however, are seldom satisfied with phenomenological definitions such as these. The question they ask is, "What is it about a certain molecule that makes it an acid or a base?" For more than a century, chemists wrestled with the definitions of acids and bases.

The first attempt to define an acid goes back to 1778. Antoine Lavoisier had hit upon what really happens during combustion, disproving the old theory of PHLOGISTON. The gas in the air that combines with substances when they burn he named *oxygine,* from the Greek for "producing acid," because he believed, wrongly as it turned out, that all acids contain oxygen.

Arrhenius's definition

The first modern approach to the problem was taken by the Swedish chemist Svante Arrhenius (1859–1927). His definition, put forward in 1877, was simple: If a particular substance produces a hydrogen ion (i.e., a proton, H^+) when it is dissolved in water, then it is an acid. If, on the other hand, it produces a hydroxide ion (OH^-) when it is dissolved in water, then it is a base. By this definition, battery acid, which is a solution of sulfuric acid (H_2SO_4) in water, is an acid because the hydrogen atoms in the sulfuric acid become hydrogen ions in solution. In the same way, sodium hydroxide (NaOH) is a base because it produces a hydroxide ion in water. This definition explains how acids and bases can neutralize each other. When a hydrogen ion meets a hydroxide ion, the two combine to form H_2O, ordinary water.

Incidentally, Arrhenius had a lasting impact on the debate on extraterrestrial intelligence (*see* FERMI PARADOX). He argued for a theory called *panspermia*—the notion that life could be transferred from one planet to another by microorganisms drifting through space, so that it only had to develop once, rather than on each planet on which life exists. More recent is a theory called *directed panspermia*, according to which somewhere in the Galaxy is a civilization that sends out seeds of life to colonize suitable planets. This does no more than pass the buck on the question of how life originated, because there remains the question of how life originated in the first place.

Brønsted–Lowry definition

The Arrhenius definition is fine as far as it goes, but it is rather restrictive—it applies only to aqueous solutions (substances dissolved in water). Here's a simple illustration of the limitations of this definition: If you place a beaker of hydrochloric acid (HCl) and ammonia (NH_3) next to each other, you will see a white haze form above the beakers. Fumes from the ammonia and hydrochloric acid are mingling in the air above the beakers, and the chemical reaction

$$NH_3 + HCl \rightarrow NH_4Cl$$

is taking place, in which the acid and base combine to form ammonium chloride. Since there is no water involved in this reaction, the Arrhenius definition simply doesn't apply.

In 1923, the Danish chemist Johannes Nicolaus Brønsted (1879–1947) and the British chemist Thomas Martin Lowry (1874–1936) proposed a new definition. An acid, they said, is a molecule or ion capable of donating a proton (i.e., a hydrogen atom, H^+), while a base is a molecule or ion capable of accepting one. When the reaction we are looking at takes place in water, this definition is essentially the same as that proposed by Arrhenius, but it extends to reactions that take place in the absence of water, such as the formation of ammonium chloride described above.

Lewis's definition

The final generalization freed the definition of acids and bases not only from the presence of water, but from the production of protons. It was put forward in 1923 by the American chemist Gilbert Newton Lewis (1875–1946). His definition concentrates on the way that CHEMICAL BONDS form in the chemical reactions between acids and bases, rather than on whether or not protons are donated or accepted. According to Lewis, an acid is a chemical species that can accept an electron pair in the formation of a covalent bond, whereas a base is any species that can donate a pair of electrons.

This definition incorporates both of the earlier ones, and also covers reactions in which no hydrogen takes part. For example, when sulfur dioxide reacts with an oxygen ion to form sulfur trioxide (a reaction important in ACID RAIN), the oxygen ion donates two electrons to form covalent bonds. It acts, in other words, as a base, while the sulfur dioxide accepts the electrons and is therefore acting as an acid. This reaction, which takes place without a proton and without water, is covered by the Lewis definition but not by either of the previous definitions.

pH: measuring acidity

For aqueous solutions there is a widely used system for characterizing the strength of an acid or base, a system that is best discussed in terms of the Brønsted–Lowry definition. In pure water, there will always be H_2O molecules that are separating into hydrogen (H^+) and hydroxide (OH^-) ions, while elsewhere H^+ and OH^- will be coming together to form water molecules. At any time, then, there will be hydrogen ions (protons) in the water. The molar concentration (*see* AVOGADRO'S LAW) of hydrogen in pure water is 10^{-7} moles per liter, which means that one H_2O molecule in every 10 million is in the form of ions.

We say that the pH, short for "power of hydrogen," of pure water is 7 (from "power" in the mathematical sense, i.e., the 7 in 10^{-7}). We can increase the concentration of hydrogen ions in water by adding acids. For example, if we add hydrochloric acid (HCl) to pure water, the concentration of hydrogen ions will increase. If we reach the point where the molar concentration is 10^{-1} moles per liter, we would have something the strength of stomach acid. The pH of this solution would be 1. Thus, a pH of less than 7 characterizes an acid, and the lower the pH, the stronger is the acid.

In a similar way, we can lower the concentration of hydrogen ions in pure water by adding a base (in essence, the OH^- ions from the base will react with the H^+ ions to form water molecules). In household ammonia, for example, the molar concentration of hydrogen ions is only about 10^{-11} moles per liter, and the pH of this solution is therefore 11. A pH greater than 7, then, characterizes a base.

Some sample pH values

	pH
Seawater	7.8–8.3
Human blood	7.3–7.5
Soft drink	2.5–3.5
Vinegar	2.4–3.4

Allen's Rule

*Endothermic animals
(those that generate heat
by their own metabolism)
from cold climates have
shorter extremities than
do similar animals from
warmer climates*

19th C. • HEAT
            TRANSFER

1850 • THERMODYNAMICS,
          SECOND
          LAW OF

1859 • EVOLUTION,
          THEORY OF

1877 • ALLEN'S
          RULE

Most of the regularities we see among plants and animals are fairly straightforward consequences of the theory of EVOLUTION, and Allen's rule is no exception. Endothermic animals are those that, like human beings, have an internal mechanism that maintains their body temperature at a more or less constant setting. In essence, these animals convert the energy in their food into heat to maintain their bodies at a constant temperature.

Obviously, heat will flow from the interior of a warm-blooded animal to its cooler outer surface from where it will be lost to the environment. This lost heat must be replaced by the animal's metabolism, so there is an obvious advantage to arranging things so that a minimum amount of heat is lost. This is why animals that live in polar regions have thick layers of fur or blubber to provide insulation and slow down the flow of heat to the surface.

It is obvious that the less surface area is exposed to the outside environment, the less heat will be radiated away at a given environmental temperature. The relative proportions of, say, a musk ox and a giraffe (the former adapted to cold climates, the latter to warm) illustrate this point. Thus, there is an evolutionary advantage to having short legs in a cold environment—there is simply less surface area through which heat can be lost.

The law illustrates some well-known laws of physics. The heat generated inside a warm-blooded animal will move toward the outside environment (*see* HEAT TRANSFER), where it is colder (*see* the second law of THERMODYNAMICS; STEFAN–BOLTZMANN LAW), and into which it will eventually be lost by either radiation or convection. The amount of heat an animal generates depends on its volume, whereas the amount it loses to its environment depends on its surface area. So, the more compact the animal—or, to use the technical term, the lower its *surface-to-volume ratio*—the less heat will be lost and the more heat retained. The adaptive value of low surface-to-volume ratios in northern climates, then, is obvious.

JOEL ASAPH ALLEN (1838–1921) American mammalogist and ornithologist. He was born in Springfield, Massachusetts, and studied under Jean Louis Agassiz (1807–73). After working as the first curator of birds at the Museum of Comparative Zoology, Harvard, he became in 1885 curator of mammals and birds at the American Museum of Natural History, New York. He recognized that populations of a single species could show variations, which led to the concept of the subspecies. Allen helped to found conservationism in America, and was one of the first to publicize the plight of the bison.

Ampère's Law

*Moving electrical charges
produce magnetic fields*

One of the great scientific developments of the early 19th century was the growing realization that there is a fundamental and deep connection between the seemingly unrelated phenomena of electricity and magnetism. Hans Christian Oersted (*see* OERSTED DISCOVERY) had established experimentally that a wire carrying an electrical current is capable of deflecting a compass needle. André-Marie Ampère was captivated by this discovery, and began a vigorous experimental and mathematical investigation of the connection between the two. The law that now bears his name is a result of that investigation.

The crucial experiment that Ampère did was a simple one. Two straight wires were laid down side by side, and electrical currents were run through them. He found that there was a force between the two wires. Actually, you don't have to be a rocket scientist to come to this conclusion, for in most versions of the experiment the wires actually start to bow toward or away from each other as soon as the current is turned on. But by careful measurement, Ampère was able to determine that the force between the wires was proportional to the strengths of the currents, and fell off as the distance between the wires was increased. From this information he was able to explain the observed force in terms of the creation of magnetic fields.

Ampère's explanation went like this: The current in one wire produces a magnetic field. The shape of this field is such that a cross-section through it, at right angles to the wire, would look like a set of concentric circles centered on the wire. This magnetic field extends past the second wire, and it exerts a force on the moving electrical charges in that wire. It is this force, transmitted to the atoms that make up the metal of the wire, that causes the bowing mentioned above. Ampère's experiment thus shows us two complementary facts about the nature of electricity and magnetism: First, electrical currents give rise to magnetic fields; and second, magnetic fields exert forces on moving electrical charges. It is the first of these that is known as Ampère's law (it is closely related to the BIOT–SAVART LAW). That electrical currents can create magnetic fields is a central tenet of MAXWELL'S EQUATIONS.

Taking things a little further, the law tells us that if we draw a closed loop in space, then when we add up the magnetic field around the loop we get a value that is proportional to the total electrical current flowing through the surface. For example, around a single straight wire there will be a magnetic field of strength B a distance r from the wire. Adding up the magnetic field (the technical term is *integrating*) around the circle gives $2\pi rB$, where $2\pi r$ is the circumference of the circle. According to Ampère's law, this quantity is proportional to the strength of the electrical current flowing in the wire.

Actually, you have frequently encountered André-Marie Ampère, probably without realizing it. Look at any electrical appliance you own and you will find somewhere a little plate that gives you technical information about it. If you live in the United States it will say something like "115 V, 3.2 A." This tells you that the appliance runs on 115 volts (the standard US household voltage), and the 3.2 refers to the amount of electrical current that will flow when the appliance is in use. The "A" stands for "ampere," the name of the unit of electrical current. You sometimes hear current expressed in "amps," which is, of course, short for "ampere."

The official definition of the ampere is related to Ampère's original experiment: It is the electrical current that flows in each of two parallel wires, placed 1 meter apart in a vaccum, that generates a force of 2×10^{-7} newtons per meter. (Definitions of the main units used in science have to be this rigorous. What's more, these wires are "ideal" wires: infinitely long and of infinitesimal cross-section.) For the record, one ampere corresponds to about 6×10^{23} electrons flowing past any point in the wire each second.

ANDRÉ-MARIE AMPÈRE (1775–1836)
French physicist. He was born in Lyons to a merchant family and received his education by being given the run of his family's library. (He taught himself Latin, for example, so that he could read works by prominent mathematicians.) He worked his way up through the French educational system, eventually being appointed by Napoleon to the post of inspector general of the new university system. His most famous work, published in 1827, summarized his work on Ampère's law and gave a precise mathematical formulation of it. Its title, in English, was *Notes on the Mathematical Theory of Electrodynamic Phenomena, Solely Deduced from Experiment*.

Angular Momentum, conservation of

The angular momentum of an isolated system is conserved

A rotating object tends to keep rotating. Physicists talk about this everyday phenomenon in terms of a quantity called *angular momentum*. The angular momentum of a rotating body depends on its speed of rotation, its mass, and its configuration. Increase either the speed of rotation or the mass, and the angular momentum goes up. Displace the object's mass away from the axis of rotation, and the same thing happens. In mathematical symbolism, the angular momentum *J* of an object rotating with angular speed ω is given by the equation $J = I\omega$, where *I* is a quantity known as the *moment of inertia*, which depends on the object's mass and configuration. In general, the moment of inertia is larger when more of an object's mass is located far from the axis of rotation.

A quantity that doesn't change throughout the course of an interaction is said to be *conserved*. The law of conservation of angular momentum says that unless a force is applied in such a way that it changes an object's rotation, its angular momentum will be conserved. The most familiar example of this law is an ice skater going into a spin. When she pulls her arms in, her mass is closer to her axis of rotation, and the moment of inertia becomes smaller. To compensate for the smaller value of *I*, the rate of spin has to increase.

But not just any force will have this effect. Imagine upending your bicycle and spinning one of the wheels. You know that if you then apply a force to the tire (e.g., a frictional force by putting the flat of your hand, against it), the wheel will slow down. Push on the axle, though, and it will just keep spinning. Thus, to change the wheel's angular momentum, the force has to act off center. Such a twisting force is called a *torque*. Another way of stating the conservation law is to say that only the presence of a torque can change an object's angular momentum.

In addition to conservation of angular momentum causing objects to speed up and slow down as their mass distribution changes, it also requires that the object continues to spin in the same plane. In other words, the axis of rotation can only be changed by the action of a torque. This is a major factor in the design of guidance systems for airplane and satellite systems. A gyroscope is basically a spinning disk mounted on very low friction bearings. To a very good approximation, it can be thought of as a disk rotating freely in space. The conservation law tells us that the angular momentum of the disk, and hence its axis of rotation, has to remain pointed in a fixed direction. The satellite's instruments typically read the directions of the axes of several gyroscopes and use them to determine the orientation of the satellite in space.

Animal Territoriality

Animals (or groups of animals) will often defend a territory against others of their own species

The idea that animals will defend a certain territory they use for nesting, food gathering, or mating was first recognized by an obscure English birder named Henry Eliot Howard. In a lifetime of observing many different species of bird, he noticed that males of particular species established themselves in certain locations, driving other males of the same species away when they wandered in. He also noticed that females of that species tended to be attracted to these areas, and mating and nesting took place within them. Since that discovery, the phenomenon of territoriality has been recorded in many species. By establishing a *territory* (a term introduced into the vocabulary of animal behavior by Howard in the 1920s), a mating pair enhances its ability to find sufficient food and to carry out whatever other activities are needed to raise their young undisturbed.

Territoriality takes different forms in different species. In some, it involves nesting and mating activities as well as food gathering, while in others it involves only mating and nesting (with food gathering taking place in communal areas), and in still other species territories are associated only with mating. The term *stamping ground* (as in "I'm glad to be back in my old stamping ground") comes from the behavior of some ungulates—such as elk and some species of antelope—whose males stake out territories during the mating season, stamping to signal to females that they have done so.

Territorial boundaries may not be apparent to us humans, but they are clearly recognized by the animals who establish them. Typically, males defend their boundaries against incursions by other males, though the territorial combat is seldom to the death. In fact, the general rule from birds to fish seems to be that once a territory is established, it is very difficult for one male to invade the territory of another.

Two male prairie chickens at the boundary of their respective territories. The bird on the right has inflated his throat pouches as a warning signal to his rival.

Anthropic Principle

There is some connection between the existence of life in the universe and the fundamental laws of physics that govern the universe's behavior

As our knowledge of the cosmos has increased, so too has our understanding of matter on the very smallest of scales. It has become apparent that, had the universe been put together just a little differently, we could not be here to contemplate it. It is as though the universe has been made for us—a Garden of Eden of the grandest possible design.

If the force of gravity were stronger than it is, then the expansion of the universe (*see* BIG BANG) would have ended long ago, the enhanced force of gravity pulling everything back to the starting point before stars and planets had a chance to form, and before life could develop. If, on the other hand, the force of gravity were weaker, then objects the size of stars and planets wouldn't have formed at all. Of all the values the force of gravity could have, only a restricted range of values are consistent with the appearance of life in the universe.

You can make the same sort of argument about the values of other fundamental properties of the universe. If the charge on the electron were much larger, for example, the repulsive force between protons in the atomic nucleus would be strong enough to blow it apart. If the charge were smaller, electrons wouldn't become bound to the nucleus. In either case, there would be no atoms, no chemistry, no life. If the strong force that holds the nucleus together were weaker, many of the nuclei essential to the formation of the chemical elements (*see* BIG BANG) would be unstable, whereas if it were stronger, the nuclear reactions that power stars (and provide energy for life on neighboring planets) would be affected.

In fact, of all of the ways a universe could be put together, and of all the values the various fundamental constants could have, only a small range of values are consistent with the development of life. This simple fact is the basis for what is called the anthropic principle. It was first elucidated by astrophysicist Robert H. Dicke (1916–97), and given its name in 1973 by cosmologist Brandon Carter (1942–) as an extension of the cosmological principle (*see* COPERNICAN PRINCIPLE). Carter distinguished two forms of the principle, which he called weak and strong.

The *weak anthropic principle* states simply that the universe is arranged in such a way that life is possible. It notes that the question "Why is the universe the way it is?" should really be asked in the form "Why is the universe the way it is, *given that an intelligent being is asking the question*?" In other words, the fact that questions about the values of the fundamental forces are being asked already implies that life has developed. In a universe with a larger or smaller gravitational constant, life wouldn't have developed and the question would never have been asked.

In this form, the anthropic principle isn't implying any deep reason why the constants of nature have to be as they are. There may be billions

Many Universes?

If the universe, by definition, is the totality of everything that exists, how can there be many universes? One answer is provided by what is called the many-worlds interpretation of quantum mechanics, first put forward to explain an otherwise baffling outcome of the double-slit interference experiment. When a single photon is aimed at the two slits, it seems to pass through both of them, because a series of single photons aimed at the slits will build up a pattern of interference fringes on a screen beyond the slits. According to some theorists, the only satisfactory interpretation is that every time an "interaction" happens, the universe splits into two close copies of itself. That would mean that a huge number of universes somehow co-exist, completely independent of one another on the macroscopic scale, but still capable of interacting with one another on the quantum scale. English astronomer Martin Rees (1942–) coined the term *multiverse* to describe this multiplicity of worlds.

The multiverse concept gives us a natural explanation of the weak anthropic principle. We might well wonder why "the" universe should have given rise to our existence. But with an almost infinite supply of universes to choose from, there will be a huge number in which life is possible, so we should not be in the least surprised that one of them, like baby bear's porridge, is just right for us.

of other universes with all sorts of different values for these constants (*see* sidebar), but the question will be asked only in universes where the values of the constants are such that life is possible.

Here's an analogy that may help you think about this notion: If you flip a coin ten times in a row, the odds are 1024 to 1 against getting all heads. But if you do a ten-flip cycle 1024 times, the odds are that one of those cycles will come up all heads, just by chance. There's no point in looking at that cycle and asking why it came up all heads—it's just a matter of chance that it was that cycle and not some other. In the same way, if there is a multitude of universes, it's odds-on that in some of them the fundamental constants will have values in the range needed to support life, and it is only in those universes that the question will be asked.

Some scientists want to make more of the compatibility of the universe with life, and this leads to the *strong anthropic principle*: the universe must be arranged so that life is possible. In this version of the principle, we go beyond the weak version to claim that life is, in fact, inevitable. Proponents of this view argue that there is some (as yet unknown) law that requires the fundamental constants of the universe to be as they are. Some take this argument to the limit, claiming that the universe must develop not only life, but consciousness as well.

Most scientists find the weak statement of the anthropic principle unobjectionable, since it seems to be a simple exercise in logic (some would say a tautology—"we're here because we're here"). The strong principle, however, does not enjoy this level of support as there seems to be no way to justify its conclusions. For the record, I find myself siding with the majority on both of these issues.

Antibiotics, resistance to

Over time, it is possible for populations (but not individual organisms) to acquire resistance to the effects of chemicals such as antibiotics and pesticides

We have all heard the scare stories about bacteria that are no longer susceptible to antibiotics—indeed, many of us may already have had the experience of being switched from one antibiotic to another when the first one proved to be ineffective. That populations of bacteria should develop resistance to antibiotics is not some strange revenge of Mother Nature on the human species, but a simple consequence of the theory of EVOLUTION.

One of the underlying principles of evolution is that there will be differences between members of a given population. Today, we understand that those differences are carried in the genes. Furthermore, the life of any organism, whether it be bacterium or human, is basically a series of chemical reactions between molecules. An antibiotic works because it inhibits a chemical reaction that is essential to the organism's life. For example, PENICILLIN blocks molecules that build new cell walls for the bacteria it is used against. Think of their action as analogous to gum adhering to a key and preventing the key from opening a lock. (Penicillin has no effect on the cells of humans or other animals because our cells do not have the same outer coverings as those of bacteria.)

In a given population of bacteria, there will be variations in the shape of the molecules that are targeted by antibiotics. Some will be more susceptible to being "gummed up," others less. Just by chance, a small number will have molecules that are affected less by an antibiotic such as penicillin. When the population of bacteria is exposed to the drug, this small proportion will not be killed outright. Over many generations, natural selection will work to favor those bacteria whose genes code for molecules that are less susceptible to the drug. The end product of this process? A population that is impervious to it.

It is important to note that this effect manifests itself over many generations. No single individual bacterium can acquire immunity—the shape of the molecules is fixed by the genes, and does not normally change over the organism's lifetime.

A classic experiment on antibiotic resistance was carried out in 1952 by the American geneticists Joshua and Esther Lederberg. Colonies of bacteria were started in a petri dish with ordinary nutrients, then parts of each colony were transferred to another dish whose medium contained penicillin. Most colonies died after transfer, but eventually they found one that survived. They then went back to the parent colony of the survivor on the original dish and transferred it to a dish containing penicillin. This colony survived too, showing that it had resistance to penicillin, even though it had never been exposed to the drug. This experiment confirmed the idea that resistance is a random event in a population and is selected by the environment.

The same sorts of effects can be seen in higher organisms such as insects, which "evolve away" from pesticides, and plants, which do the same when exposed to herbicides. This happens in a similar way to that in which bacteria acquire their resistance to antibiotics. By chance, some organisms in a population will have some chemical quirk in their nature that makes it possible for them to overcome the effects of a pesticide or herbicide. It may be, for example, that they possess molecules that lock onto the pesticide or herbicide molecules and prevent them from working, or that they possess chemicals that flush the pesticide or herbicide out of cells before much damage is done. Organisms that have this trait will be selected for by the normal processes of natural selection, and eventually the entire population will share the trait.

Knowing what they did about natural selection, scientists should have expected target organisms to behave in this way. That they didn't probably tells us more about human nature than about the effects of antibiotics and other chemicals. Nevertheless, this acquired resistance doesn't cancel out all the good that has been done by antibiotics and other chemicals. It just means that the war against disease cannot be won by a single battle. In fact, a better military analogy for this process would be an arms race, in which one side gains a temporary advantage which the other side learns to counter. The first side then develops a counter to the counter, at which point the other side counters the counter-counter … An arms race never ends—and, I suspect, neither will our battle against the evolutionary potential of microbes. All that is necessary is that we stay far enough ahead to keep diseases under control. That will be victory enough.

Antiparticles

For every known particle, it is possible to find an antiparticle—a particle that has the same mass but is opposite in every other way

Carl Anderson pictured in 1931 in the Guggenheim Laboratory with the cloud chamber in which antimatter, predicted by Paul Dirac, was first seen.

The subatomic world was very simple in the 1920s, after the introduction of QUANTUM MECHANICS. Just two particles, the proton and the neutron, made up the nuclei of atoms (though the existence of neutrons wouldn't be verified experimentally until the early 1930s) while one particle, the electron, had an existence outside the nucleus, orbiting it. From these three particles the entire complexity of the world could be built.

Alas, such simplicity was not to last. Scientists in mountaintop laboratories around the world were exposing equipment to the bombardment of cosmic rays (*see* ELEMENTARY PARTICLES), and beginning to discover all sorts of things that were not part of the familiar triad. In particular, they discovered a weird breed of objects called *antiparticles*.

Antiparticles create a kind of mirror-image world to our own. An antiparticle has exactly the same mass as its normal equivalent, but is opposite in every other way. The electron has a negative electrical charge, for example, while its antiparticle, called the *positron* ("positive electron"), has a positive charge. The proton has a positive charge, the antiproton has a negative charge, and so on. When a particle and its antiparticle come together, they undergo a process of annihilation in which the two particles disappear and their mass is converted into a spray of photons and other particles.

The existence of antiparticles was first predicted by Paul Dirac in a paper published in 1930. The easiest way to visualize how Dirac's theory works is to imagine a level field. If you dig a hole in the field, there will be two things there—a hole and a pile of dirt. Think of the pile of dirt as the normal particle, and the hole, or the "absence of a pile of dirt", as the antiparticle. If you shovel the dirt back into the hole, both hole and pile disappear—the equivalent of the process of annihilation—and you are left with a level field again.

While all of this theoretical work was going on, a young experimental physicist at the California Institute of Technology, Carl Anderson (1905–91), was putting together some experimental apparatus in a laboratory atop Pike's Peak in Colorado to study cosmic rays. Working with Robert Millikan (*see* MILLIKAN OIL-DROP EXPERIMENT), he devised a system in which cosmic rays would strike a target located between the poles of a powerful magnet. The target was located inside a device where passing particles left behind a string of

tiny droplets forming a long thin cloud along the particle's track. The cloud could be photographed to record the particle's path.

With this apparatus, called a *cloud chamber,* Anderson was able both to record the particles produced by the collisions of cosmic rays with the target and, by measuring how their tracks were deflected by the magnetic field, to determine their electrical charge. By 1932, he had recorded a number of events that showed the production of particles having the same mass as the electron but which were deflected by the magnetic field in the opposite direction to an electron—and therefore had to have a positive electrical charge. This was the first detection of an antiparticle. Anderson published his result in 1932, and received a share of the 1936 Nobel Prize for Physics. He is probably the only person in history who got a Nobel prize before he got a permanent university position!

Although this scenario may well seem to be a perfect example of the prediction–verification process of the scientific method described in the Introduction, the historical reality is a little more complicated. Apparently, Anderson was unaware of Dirac's paper until well after he had his data in hand. It appears that this scenario was more a case of simultaneous discovery than anything else.

After the positron, the anti-world equivalents of other particles were eventually produced in particle accelerators. Today, they are produced routinely for experiments, and are not regarded as being in any way unusual.

PAUL ADRIEN MAURICE DIRAC (1902–84) English theoretical physicist. He was born in Bristol. Dirac's father, a Swiss national who taught French in his adopted country, apparently refused to talk to his son unless the boy addressed him in French —a requirement that may account for Dirac's rather famous reticence later in life. He studied electrical engineering at what is now Bristol University, graduating in 1921. He went on to read mathematics and physics at Cambridge, where he received his Ph.D. in 1926 and where, six years later, he became Lucasian Professor of Mathematics, holding this prestigious post for nearly forty years. Even before he obtained his doctorate, Dirac published important papers on quantum mechanics. In 1928, he published a paper in which he accounted for the behavior of the electron by combining, for the first time, elements of relativity with elements of quantum mechanics. It was this paper that predicted the existence of antiparticles and which eventually led to Dirac's Nobel Prize for Physics, which he shared with Erwin Schrödinger in 1933.

Archimedes' Principle

The buoyant force on a floating object is equal to the weight of the fluid it displaces

"*Eureka!*" This is supposed to be what the Greek philosopher Archimedes shouted when he discovered the principle of buoyancy. (It means "I have found it!") According to legend, he had been asked by King Heiron II of Syracuse to find out whether or not a crown was made of pure gold, but without damaging the crown. Archimedes could weigh the crown, of course, but that itself couldn't reveal its composition. What he needed was a way of measuring the crown's volume so that he could find the density of the material from which it was made.

According to the story, Archimedes was getting into his bathtub one day when he noticed that, as he lowered himself into the water, the water level rose. He realized that his own volume was displacing the water in the tub, and that the crown would do the same. He shouted out in delight and, in some versions of the story, triumphantly raced through the streets to the palace without bothering to put on his clothes.

What is true is that Archimedes had discovered the *principle of buoyancy*. If a solid object is put into a fluid, it will displace an amount of fluid equal to its own volume. The pressure that was once exerted on the missing fluid is now exerted on the solid object, creating an upward force. If the upward force exceeds the downward pull of gravity, then the object will float; otherwise it will sink. In effect, an object will float if its density is less than that of the fluid.

Archimedes' principle can be understood in terms of KINETIC THE-ORY. In the undisturbed fluid, pressure is exerted by the collisions of molecules. When the fluid is displaced by the solid object, the molecules in the fluid will collide with the body, exerting the same pressure as they did before the object was placed there. For a completely submerged object, the pressure will be less on its top than on its bottom—the molecules of the fluid will be hitting the bottom of the object with a greater force than those hitting the top. This is the molecular origin of the upward buoyant force.

This picture of buoyancy explains how a ship made of steel can float, even though the steel by itself would sink if it were placed in water. The point is that the volume of fluid displaced is equal to the volume of the steel plus the volume of the air in the hull of the ship. Together, the density of the two is less than that of water, so the force exerted by the upward collisions of the water molecules overcome the downward pull of gravity, and the ship floats.

ARCHIMEDES OF SYRACUSE (c. 287–212 BC) Greek mathematician and scientist. Little is known of his life. In addition to proving many mathematical theorems, he is credited with the invention of various mechanical devices. According to legend, he was killed by a Roman soldier during the sack of Syracuse when he refused to leave a mathematical proof on which he was working.

Argon, discovery of

The noble gas argon was discovered as the result of a small discrepancy between two measurements

In 1892, the British scientist John Strutt, now better known as Lord Rayleigh (*see* RAYLEIGH CRITERION), was engaged in one of those exercises that are part of the unglamorous but necessary routine of experimental science. He was investigating the optical and chemical properties of the atmosphere, and his goal was to measure the weight of a liter of nitrogen to an accuracy never before reached in the sciences.

His results, however, were puzzling. When he got the nitrogen by removing every other known substance (such as oxygen) from the air, he got one answer, but when he used nitrogen obtained from a chemical reaction (by passing ammonia over red-hot copper) he got a different answer. The nitrogen from the air seemed to weigh 0.5% more the nitrogen obtained chemically—a small but troubling discrepancy. After satisfying himself that the discrepancy did not arise from experimental error, Rayleigh wrote a letter, published in the journal *Nature,* asking if anyone had any ideas.

At this point Sir William Ramsay (1852–1916), then at University College, London, contacted Rayleigh. Ramsay thought there might be an as yet undetected gas in the atmosphere, and proposed using the best chemical techniques available to isolate it. In the experiment that was carried out, an electrical spark was passed through a container containing clean air and water, causing the nitrogen to combine with the oxygen and go into solution. At the end, when both these gases had been exhausted, there was still a tiny bubble left. When the spark was passed through this gas, hitherto unknown spectral lines were seen (*see* SPECTROSCOPY), signaling the discovery of a new element. Rayleigh and Ramsay published their result in 1894, naming the new gas *argon,* from the Greek for "lazy." The two received Nobel prizes for this work in 1904. Oddly enough, they didn't share a prize, as they would today, but each received a prize in a separate field—Rayleigh in physics, and Ramsay in chemistry.

Actually, there was something of a conflict. At the time, many scientists considered that they "owned" particular areas of study, and it wasn't clear whether Rayleigh had given Ramsay permission to work on this problem. Fortunately, both men were sophisticated enough to realize that there was plenty of credit to go around, and, by publishing their results jointly, they effectively eliminated what could have been a nasty priority battle.

Atomic Theory

Matter is composed of atoms

The word "atom" comes from the Greek, and translates roughly as "that which cannot be divided." The Greek philosopher Democritus (who, in the delightful language of classicists, "flourished" in the 5th century BC) is generally given credit for being the first to argue that matter, seemingly smooth and continuous, is actually made up of constituents so small as to be invisible. We know almost nothing of Democritus' life, and none of his original writings survive. We know of him primarily through quotations in works of other authors, most notably Aristotle.

Democritus' argument, recast in modern language, was very simple. Imagine, he said, that you had the world's sharpest knife. You take any piece of material you have to hand, and you start to cut it. Whatever you have cut in half, then you cut one of those halves in half, then one of those halves in half, and so on. Eventually, he argued (largely on philosophical grounds), you will be left with a piece that you cannot divide any further. This piece will be an *atom*.

In Democritus' scheme, atoms were eternal and unchanging as well as indivisible. Changes in the universe took place because of changes in the relationships between atoms, but the atoms themselves were unchanging. In this way, he finessed an old argument in Greek philosophy about whether the real world could encompass change, or whether change was only apparent.

About the only thing that survives today from the old Greek atomic theory is the word "atom." We now recognize that atoms are made up of particles more fundamental still (*see* ELEMENTARY PARTICLES). It is easy to see that the Greek theory bears little resemblance to modern scientific work—for one thing, it has no grounding at all in observation or experiment. All Democritus did, like all the ancient "natural philosophers," was to think about the natural world.

Democritus' work has not gone unrecognized in modern times. His face was on the face of the last Greek 10 drachma coin (now ousted by the euro), and there was a schematic representation of an atom on the back. I am indebted to my friend Hans Von Bayer for pointing out that the atom shown on the coin had three electrons, and hence must be lithium. Democritus was known as the "laughing philosopher" (he apparently had a most unphilosophical sense of humor). What better atom for his coin than lithium, an element used to treat depression?

The notion that matter was composed of atoms languished as an obscure piece of philosophy until early in the 19th century. By then, the foundations of the modern science of chemistry were in place. Scientists studying chemical reactions had found that many substances that could be broken up by chemical reactions—water, for example, could be separated into hydrogen and oxygen. There were, however, some materials (like oxygen and hydrogen) that could not be broken down by chemical means. These materials were called *chemical elements*. By the turn of the 19th century, almost 30 elements were known (at the time of writing, counting those created in laboratories, there are rather more than 110—*see* PERIODIC TABLE). Furthermore, it was known that in chemical reactions the amounts of these elements were always the same for a given reaction. Thus, water was always formed by eight parts of oxygen (by weight) to one part of hydrogen (*see* AVOGADRO'S LAW).

The man who made sense of these facts and regularities was John Dalton (whose name is immortalized in DALTON'S LAW). His chemical researches were concentrated on the behavior of gases (*see* BOYLE'S LAW, CHARLES'S LAW, IDEAL GAS LAW), but his interests were more wide-ranging than that. Starting in 1808, he began the publication of a two-volume work titled *New System of Chemical Philosophy,* a book that was to change chemistry forever. In it Dalton suggested that the way to under-stand the new results from experimental chemistry was to realize that to each chemical element there corresponded one kind of atom, and that it was the coming together of these different kinds of atoms in different proportions that created most of the substances in the world. Water, for example, was made from two atoms of hydrogen combined with one atom of oxygen (the familiar H_2O). The fact that all atoms of a given species are identical explained the fixed proportions found in chemical reactions. The two atoms of hydrogen in water, for example, will always be the same, no matter where the water is found, and they will always have the same relationship to their partner oxygen atom.

For Dalton, as for Democritus, the atoms were indivisible. Indeed, there are drawings in Dalton's notes and books in which atoms are depicted as something like cannonballs. The main plank of his work—that chemical elements are associated with a given type of atom—forms the basis of modern chemistry. This is true regardless of the fact that we now understand that atoms are composite structures (*see* RUTHERFORD EXPERIMENT) composed of a heavy, positively charged nucleus with light, negatively charged ELECTRONS in orbit around it. Throw in the complications of QUANTUM MECHANICS (*see also* BOHR ATOM and SCHRÖDINGER'S EQUATION) and you can bring the atom right into the 21st century.

Not bad for a concept that originated in philosophical arguments some 2500 years ago!

Before its currency was replaced by the euro, Greece honored Democritus, who originated the concept of the atom, on its 10 drachma coin.

Aufbau Principle

The lowest orbits in atoms will fill with electrons before higher orbits do

The hydrogen atom is simple enough—a single electron in orbit around a nucleus consisting of a single proton. The aufbau principle (in German, *Aufbau* means "building up") outlines a simple procedure by which we can start with hydrogen and move on to more complex atoms. The basic idea is that QUANTUM MECHANICS tells us how the electron orbits around a nucleus work, and the PAULI EXCLUSION PRINCIPLE tells us that no two electrons can occupy the same state. These two taken together tell us that there is a maximum number of electrons that can reside in each orbit in the atom.

The aufbau principle works like this: We can imagine ourselves building up the structures of atoms by starting with hydrogen and then adding protons (and the requisite number of neutrons) to the nucleus one at a time, to make larger atoms. Each time we add a positively charged proton to the nucleus, we have to add an extra electron. As we go from small to large atoms, the spaces available in each electron orbital fill up, so we have to put the new electron into the first slot in the next available orbital (*see* BOHR ATOM). The process is a little like watching a bricklayer construct a brick wall. One course of bricks is laid down, brick by brick, and then the next row is started.

Although it wasn't realized at the time, the aufbau principle actually depends on the Pauli exclusion principle. This principle states that no two electrons can occupy the same state, which means that electrons in atomic orbits are a little like cars in a parking lot—once a space is filled, other electrons have to look elsewhere for a space of their own. In the same way, as we build up complex atoms, once spaces in inner orbits are filled then electrons can only be added to higher orbits.

One extra piece of information we need in order to see how the aufbau principle works is that electrons possess a quality called *spin*—think of them as spinning upon an axis in the same way that the Earth rotates about a line through its poles. It is always possible to pair electrons—one spinning clockwise, the other spinning counterclockwise.

Up one level of complexity from hydrogen is the helium atom, which normally has two protons and two neutrons in its nucleus. The second electron we need to add can be paired with the electron already present in the hydrogen atom—we just have to give it the opposite spin. Thus, according to the aufbau principle, both electrons in a helium atom should be in the lowest available orbital, but have opposite spins. That prediction has been verified experimentally.

The next element, lithium, always has three protons (and normally four neutrons) in its nucleus, and the atom must therefore have three electrons. The lowest electron shell has been filled, however, so the extra electron has to go into the next level up. For beryllium (four protons), the extra electron will be paired with the third electron we needed to make up lithium.

In this way, we can proceed to fill up the second shell, which has space for four pairs of electrons. (The element with ten electrons—two in the lowest shell, eight in the next shell up—is neon.) We can then move on to the third electron shell: The atom with just one electron in this shell is sodium, and when the shell is full we have an argon atom. After this, quantum mechanics predicts that the orbitals get a little more complicated—there are, for example, spaces for nine pairs of electrons in the third and fourth shells, and even more higher up. Nevertheless, the same basic principle holds. Each orbit contains room for only so many electron pairs, and once they are filled the aufbau principle tells us that we have to move on.

The aufbau principle also explains the regularities in the chemical properties of the elements which were discovered by Dmitri Mendeleyev and marshaled into order in the PERIODIC TABLE of the elements.

Avogadro's Law

Equal volumes of gas at the same temperature and pressure contain equal numbers of molecules

When you watch a piece of wood burning, what you are seeing is a chemical reaction in which carbon in the wood combines with oxygen in the air to produce carbon dioxide (CO_2). Now, a single atom of carbon has the same mass as 12 hydrogen atoms, while two oxygen atoms have the same mass as 32 hydrogen atoms. So the ratio of the masses of carbon and oxygen taking part in the reaction is always 12 : 32 (or to put it as simply as possible, 3 : 8). It really makes no difference what units we use—12 grams of carbon will combine with 32 grams of oxygen, 12 tons of carbon with 32 tons of oxygen, and so on. What matters in chemical reactions is the relative number of atoms of each element taking part. When you watch that campfire burning in the night, you can be sure that for every atom of carbon taken from the wood, two oxygen atoms will be taken from the air, and that the ratio of masses will be 12 : 32.

If this is so, then it follows that there must be as many atoms of carbon in 12 grams of carbon as there are atoms of oxygen in 16 grams of oxygen. Chemists call this amount of atoms a *mole*. Basically, a mole is a mass of a substance for which the number of grams is equal to the relative atomic mass of the material expressed in grams. From this discussion, it follows that the number of atoms or molecules in a mole of any substance has to be the same. The mole is an example of a measure of amount, something we use all the time. We speak, for example, of a *pair* of socks, a *dozen* rolls, a *sixpack* of beer. And just as there is always the same number of socks in a pair and the same number of bottles or cans in a sixpack, the number of atoms or molecules in a mole of a substance is always the same.

But how did scientists come to this way of thinking about materials? It is, after all, easier to count socks than to count atoms. To answer this question, we have to go back to the Italian chemist Amedeo Avogadro. He knew that when chemical reactions take place between gases, the volume ratios of those gases are the same as the molecular ratios of the gases. For example, when three hydrogen molecules (H_2) combine with a nitrogen molecule (N_2) to produce two molecules of ammonia (NH_3), the volume of hydrogen taking part in the reaction is three times the volume of nitrogen. Avogadro realized that this meant that the number of molecules in the two volumes had to be in the ratio 3 : 1, or that equal volumes of gas contain equal numbers of atoms or molecules—what is now known as Avogadro's law. He did not know what that number actually was; we now know that it is 6×10^{23}, and we call it *Avogadro's number* (or the *Avogadro constant*). It is represented by the symbol N.

Avogadro's work was ignored by mainstream scientists for several decades. Most historians blame this curious state of affairs on the fact that Avogadro worked in Turin, far from the centers of scientific research in Germany, France, and England. It was, in fact, only after he traveled to Germany to present his work that it was recognized as part of the main corpus of science.

Establishing the value of N turned out to be a difficult task. It wasn't accomplished until early in the 20th century, when the French physicist Jean Perrin (1870–1942) found several different ways to measure this number, all of which gave the same result. In the best-known method, Perrin used Albert Einstein's analysis of BROWNIAN MOTION. This is the continuous random motion of a small body (such as a pollen grain) caused by statistical fluctuations in the number of atoms or molecules colliding with it over time. The movement of the pollen grain depends on how often it is hit and, therefore, on the density of atoms in the material surrounding it.

LORENZO ROMANO AMEDEO CARLO AVOGADRO (1776–1856) Italian physicist and chemist. He was born in Turin into a family of minor Italian nobility and became an ecclesiastical lawyer. His private studies of mathematics and natural philosophy, begun in 1800, eventually led to his appointment to a professorship at the College of Vercelli six years later. Avogadro eventually became the professor of mathematical physics at Turin (a chair which he lost when it was suppressed for political reasons in 1821, but to which he was reappointed in 1834). He was extremely modest and worked largely in isolation, and his work was not recognized by his peers for most of his life.

Balance of Nature

Scientists no longer believe that ecosystems, left to themselves, will attain a fixed, stable state of high productivity

The myth of the balance of nature is lodged firmly in the public consciousness. It says that, in the absence of human intervention, natural systems will inevitably evolve into stable, permanent, and highly interdependent states in which everything is finely tuned. A good deal of popular (but not scientific) writing on ecology plays on this theme to portray nature as fragile and endangered, in constant peril of destruction at the hands of humans, whose activities threaten this delicate balance.

For example, it is well known that after a disturbance such as a forest fire in a northern-hemisphere temperate zone, the recovery of vegetation over time follows a well-defined law of ECOLOGICAL SUCCESSION. First we see weeds, then pioneer species like pine, and finally hardwoods such as oak and maple. It used to be thought that this succession would ultimately lead to something called a *climax forest,* a stable ecosystem containing the maximum possible amount of organic matter, with maximum storage capacity for chemical elements necessary to life, and exhibiting the greatest possible biological diversity. But this is not how forest progression works, as we can see from an example. As individual trees grow, they store the most material in their early stages of growth. It is also at these stages that their capacity for storing chemical elements is at a maximum. A mature forest, on the other hand, will probably experience a net loss of material as trees age and die.

Furthermore, as the biological progression goes on, geological and other forces act to change the environment. Fires, floods, and changes in rainfall are examples of processes that change the environment in which a forest grows. The plants, of course, will respond to these changes. The ecosystem, in other words, is always aiming at a moving target. The "balance of nature" depends on the environment, and the environment is always changing. Nature is thus better thought of as being in a state of constant flux—always going somewhere but never arriving. Human intervention is just another way of changing the environment, and hence the direction of development of the ecosystem.

Band Theory of Solids

The electrical properties of a solid depend on how electrons from atoms arrange themselves into bands when the solid forms

As we know from the model known as the BOHR ATOM, electrons in atoms are found in orbitals at specific distances from the nucleus. Different orbitals have different energies. When atoms combine to form solids, these electron orbitals are shifted slightly. There are two ways to think about this shifting: You can note that when an electron is in a solid it will experience electrical forces of attraction and repulsion from electrons and nuclei in other atoms, as well as from those in its own atom. Alternately, you can note that because the PAULI EXCLUSION PRINCIPLE forbids any two electrons from being in the same state, there has to be a small difference in energy levels.

In any case, you can think of a solid being formed in terms of atoms coming together, and each energy level in the atom being smeared out slightly into a group of closely spaced energy levels known as an *energy band*. All the electrons in a given energy level have approximately the same energy. For energy levels close to the nucleus, electrons aren't free to move about the solid because, even though it is energetically possible for electrons to move from atom to atom in this band, all the available states are already filled by electrons from other atoms.

The most important band is the one which is created from the energy levels of valence electrons (the outermost electrons, which are those that take part in chemical reactions and give an element its *valency,* or combining power), known as the *valence band*. Unlike the inner bands, this band is usually not filled with electrons. In an unfilled band there are energy levels available that allow electrons to move from atom to atom inside the solid. Electrons in an unfilled valence band can move pretty much at will throughout the material—they can, in fact, be marshaled into an electrical current. (For this reason, the lowest band in which electrons can move freely is referred to as the *conduction band,* which may or may not be the valence band.)

We can use band theory to explain the ELECTRICAL PROPERTIES OF MATTER. If the valence band of a solid is completely filled, and there is a large energy gap before the next unfilled band, then it is extremely unlikely that an electron will jump up. Electrons will not be able to move within the valence band because it is filled, and they cannot get into the conduction band because they don't have enough energy. They cannot move, and the solid will be an *insulator* that doesn't conduct electricity.

A *conductor,* on the other hand, is a material with a partially filled valence band—one in which electrons are free to move. A *semiconductor* is like an insulator in that the valence band is filled and there is an energy gap between it and the conduction band. In a semiconductor, however, the energy gap is small, so that electrons can get enough energy from normal thermal motion to make the jump.

BCS Theory of Super-conductivity

Superconductivity results from the formation of Cooper pairs—two electrons acting as a single particle

Superconductivity is a strange, counterintuitive phenomenon. When an electrical current passes along a normal wire, heat is generated when the moving electrons that comprise the current collide with the heavy atoms that comprise the metal of which the wire is made. In each such collision, on average, the electron slows down a little bit and the atom is made to vibrate a little more. We perceive this additional kinetic energy of the atoms in the wire as heat—this is the energy you see being emitted when your toaster wires glow red. This collision mechanism guarantees that every time an electrical current flows through a wire, it will lose energy. The wire has a *resistance* that drains energy from the current.

In 1911 the Dutch physicist Heike Kamerlingh Onnes (1853–1926) made a startling discovery. He found that when he put a wire into a bath of liquid helium, just 4 degrees above absolute zero (–273°C or –460°F), the wire was capable of carrying electrical current without any loss of energy. By some unknown mechanism, nature seemed to be able to turn off the electron–atom collisions when the temperature was low enough. The phenomenon was given the name *superconductivity*.

Why nature behaved in this way remained a mystery until 1957, when three physicists—John Bardeen, Leon Cooper (1930–), and John Robert Schrieffer (1931–)—came up with an explanation. Their work, now called the BCS theory in their honor, remains the standard explanation of superconductivity at very low temperatures.

At low temperatures, the heavy atoms in the metal don't vibrate very much, and can be thought of as being essentially stationary. Because the atoms have given up electrons to form the metal (*see* CHEMICAL BONDS), they now carry a positive charge. When a moving electron associated with an electrical current passes between two such atoms, the atoms are attracted to the negatively charged electron. They respond sluggishly, however, so by the time the two atoms have moved toward where the electron used to be (and toward each other), the electron is long gone. For a while, though, before they begin their sluggish way back, the two atoms create a slight concentration of positive charge, capable of exerting an attractive force on electrons. By this process, two electrons can become associated with each other, and the two can move through the metal more easily than either one could alone. This phenomenon is a little like the technique of "drafting," when cyclists ride near each other to cut down wind resistance. The two electrons connected in this way are called a *Cooper pair*.

At low enough temperatures, all the electrons in the current form into Cooper pairs. Think of each pair as a strand of spaghetti, each end of the strand being one of the electrons. Like a plate of spaghetti, the pairs are intertwined and locked together. In essence, the formation of Cooper pairs locks all of the electrons in the metal into a single intertwined mass. When this happens, the electrons can no longer slow down when they encounter an atom, and therefore cannot transfer energy to it. In order to derive energy from the collision, the atom would have to slow down the entire mass of electrons, which it cannot do. The electrons therefore move through the metal without giving up any energy to the atoms that reside there.

So, as long as the temperature is low enough for the vibration of the atoms not to upset the delicate balance of the Cooper pairs, electrons will be able to flow through the material without transferring energy. The material will, in other words, be a superconductor. For their discovery of this mechanism, Bardeen, Cooper, and Schreiffer shared the 1972 Nobel Prize for Physics.

Superconductors have gone from being a laboratory curiosity to a worldwide, multibillion-dollar industry. The reason for this is that electrical current flowing in a wire can produce a magnetic field (*see* FARADAY'S LAWS OF INDUCTION). Since superconducting wires will carry current without loss for as long as they are kept sufficiently cold, they are ideal for making electromagnets. You will have been close to a superconducting electromagnet if in the course of a medical investigation you have been examined using a magnetic resonance imaging (MRI) machine. MRI scanners provide the magnetic field that allows physicians to image soft tissues in the human body.

Superconductors whose operation depends on the BCS mechanism can remain superconducting up to temperatures of about $20\,\mathrm{K}$ (20 degrees above absolute zero). For a long time it was thought that this

Looking into a magnetic resonance imaging scanner, known more familiarly by its initials, MRI. The patient lies inside a powerful superconducting magnet. Changes in the magnetic field associated with atoms in the patient's body are processed to give a cross-sectional image which doctors can use for medical diagnosis.

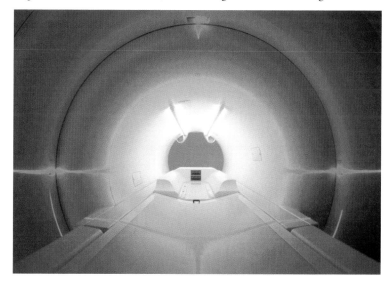

marked the limit of superconductivity. Then, in 1986, Georg Bednorz (1950–) and Alexander Müller (1927–) at IBM's laboratory in Switzerland discovered a new kind of material that remained superconducting at a temperature of 30 K. Today, materials of this kind have been developed to the point where they will remain superconducting at temperatures of 160 K or more (that's a little colder than –100°C). There is as yet no generally accepted theory to explain this class of *high-temperature superconductors,* but it is clear that they cannot be explained by the BCS theory. At the moment, there are no practical applications of these new superconductors (the materials tend to be very brittle and hard to work), but there is an intense research effort to change this state of affairs.

JOHN BARDEEN (1908–91) American physicist, one of a small number of people to have received two Nobel prizes. He was born in Madison, Wisconsin, into an academic family (his father was a professor of anatomy), and he was educated at the universities of Wisconsin and Princeton. He interrupted his graduate education for several years to work at Gulf Oil, helping to develop seismic methods for detecting oil deposits. After World War II, which he spent with the Naval Ordnance Laboratory in Washington, DC, he moved to Bell Telephone Laboratories, where he played a crucial role in the development of the transistor. For this work he shared the 1956 Nobel Prize for Physics. He then joined the faculty of the University of Illinois, where he worked on developing the BCS theory, for which he shared a second Nobel Prize for Physics, in 1972.

Beauty Criterion

Scientific theories are judged on aesthetic as well as on pragmatic criteria …

… in other words, a scientific theory is to be judged as a work of art as well as by whether it explains nature. On one level this is hardly surprising—every working scientist has heard, and probably engaged in, this kind of talk at some point in his or her career. To the general public, though, it often comes as a shock to discover that scientists aren't always the hardheaded rationalists they are usually portrayed to be, but are as appreciative of beauty and elegance as everyone else.

There are many examples of the operation of this criterion in science. The general theory of RELATIVITY, for example, was accepted by scientists very quickly because of its elegance, even though it took decades for experimenters to come up with a handful of tests of its predictions. That said, I do want to stress that, while beauty and elegance may predispose scientists to accept certain theories, they can't overcome experimental evidence to the contrary. Had relativity not passed its experimental tests, it would have been scrapped or modified, beautiful or not. How the beauty criterion operates is by raising or lowering the bar on the burden of proof in having a theory accepted, not by being a deciding factor itself.

Because of its qualitative nature, the beauty criterion is not as sharply defined as the other concepts outlined in this book. For example, there is no definition of what "beauty" means in connection with scientific theories. Nevertheless, there are some generally accepted benchmarks. The more general a theory is, for example, the greater the chance that it will be considered beautiful. The less arbitrary or jury-rigged material there is in a theory, the less it will seem to have been contrived to fit a particular set of data, and the more elegant it seems. And simplicity certainly contributes to elegance (*see* OCCAM'S RAZOR). In these respects, at least, there seems to be some universal consensus in the sciences.

But I wonder. Is it possible to state criteria for scientific beauty that are more objective than those applied to, say, paintings or music? As I read through material on this subject, I find myself disagreeing over and over again with other writers on the question of whether a particular scientific idea is or is not beautiful. There are those, for example, who find a flat universe—one in which spacetime looks like a grid on a tabletop—to be beautiful. I don't see that as either beautiful or ugly. Others think that a universe with a cosmological constant that causes acceleration is beautiful, but my friend Rocky Kolb, a prominent astrophysicist, refers to such a universe as "unspeakably ugly." In science as in art, beauty is often in the eye of the beholder.

Bell's Theorem

It is possible to determine experimentally whether or not there are hidden variables in quantum mechanics

"God does not play dice with the universe."

With these words, Albert Einstein threw down the gauntlet to his colleagues who were developing the new science of quantum mechanics. In his mind, the discovery of HEISENBERG'S UNCERTAINTY PRINCIPLE and SCHRÖDINGER'S EQUATION gave the subatomic world a kind of unwholesome indeterminacy. He was sure that God would not have been so unkind as to have made the world of the electron so different from the familiar Newtonian world of billiard balls. In fact, he spent a good part of his later years playing devil's advocate, thinking up clever paradoxes to confound the creators of the new science. But in doing so he played a crucial role in forcing people to think deeply—always a good idea when a new field is being developed.

It is ironic that Einstein has become known as an opponent of quantum mechanics, because he was actually one of the pioneers in the field. His 1921 Nobel Prize for Physics was awarded not for the theory of relativity, but for explaining the PHOTOELECTRIC EFFECT in terms of the new quantum ideas that were sweeping though the world of early 20th-century science.

What Einstein most objected to was the need to describe events at the subatomic level in terms of probabilities and wave functions (*see* QUANTUM MECHANICS), rather than in terms of more familiar quantities such as position and velocity. This is what lay behind his "playing dice" remark. He acknowledged that if you describe electrons in terms of position and velocity then you wind up with the uncertainty principle—the notion that you cannot simultaneously know both the position and velocity of a particle. But, he argued, there must be other variables, as yet undiscovered, that would return the quantum world to wholesome determinacy. God only appeared to be playing dice, in other words, because we were approaching His creation in the wrong way. Over the years, this came to be known as the *hidden variable approach* to quantum mechanics. The basic idea was that electrons "really" had a definite state, like a Newtonian billiard ball, but that we were forced to the probabilistic quantum mechanical description because we weren't describing that state correctly.

You can visualize a hidden variable theory in this way: The physical basis of the uncertainty principle is that the only way we can make a measurement on a quantum object such as an electron is to have it interact with a similar quantum object in such a way that the original object is altered by the interaction. But there might be some other way of making the measurement, using as yet unknown probes. These probes (think of them as "sub-electrons") could then be made to interact with a quantum object without changing it, and the uncertainty principle wouldn't apply to measurements made in this way. Despite the fact that there was no evidence whatsoever to support these kinds of theories, they enjoyed a shadowy existence on the fringes of mainstream quantum mechanics—mostly, I believe, because it is so difficult to let go of our comfortable Newtonian view of the universe.

In 1964, John Bell produced a startling new theoretical result. He showed that there was a certain type of experiment (I'll describe the details in a moment) in which the outcome depended on whether you describe quantum objects with a probabilistic wave function or assumed that they "really" have a definite position and momentum, like a Newtonian billiard ball. Bell's theorem, as it is now called, showed that if certain ratios were measured in these experiments, a distinction could be made between any hidden variable theory and conventional quantum mechanics. It might be, for example, that a hidden variable theory required a particular ratio to be $1:2$, whereas quantum mechanics required it to be greater than $3:4$. Measure these ratios, Bell told the physics community, and you can decide once and for all whether God is playing dice with the universe.

(I should mention that I was a graduate student at Stanford about the time that Bell developed his theorem. With his red beard and heavy Irish accent, he was a hard man not to notice. I remember standing in a corridor at the Stanford Linear Accelerator Center one day when he came out of his office in a state of high excitement and announced that he had found something really interesting. Although I have absolutely no proof of it, I like to think that I was present on the very day that Bell came up with his theorem.)

The kind of experiment Bell suggested was easy to describe, but at the time it was thought to be beyond the capability of experimental science to perform. The experiments would work like this: Some process within an atom was triggered in such a way that the atom emitted two particles back to back. For example, an atom might be stimulated with a laser to emit two photons. The two particles would then be allowed to travel away from their source, and then their spin would be measured. From these measurements the numbers called for in Bell's theorem (technically, they are called *correlation coefficients*) can be calculated.

The biggest surprise came after Bell published his theorem, calling for a series of seemingly impossible experiments. Within a decade, experimentalists had not only mastered the techniques needed to perform them, but had amassed a large body of data as well. The only way I can think of to convey the enormity of this advance is to say that at the time it seemed as likely as a million proverbial monkeys pounding away on typewriters and re-creating a Shakespeare play.

The results of the experiments were unambiguous. The correct way to describe the particle travelling between its source and the detector was with a probabilistic wave function. There was no room for hidden variables. The old hope that somehow quantum mechanics would turn out to be less puzzling had to be abandoned. This is the only time in the history of science when a brilliant theorist made a prediction, brilliant experimentalists pushed the boundaries of their art to test it, the reigning theory in science was resoundingly confirmed, and everyone was miserable!

However, it was not all wasted effort. In a surprising recent development, scientists and engineers are starting to find practical applications for effects associated with Bell's theorem. The two particles emitted in the experiment are said to be "entangled" because their wave functions are related to each other when they start their journey. The idea is to use this phenomenon to create an absolutely secure encryption process, based on the fact that anyone trying to read a message encrypted using entangled photons will inevitably (the uncertainty principle again) destroy the entanglement between the particles.

So it looks as though Einstein was wrong: God really does play dice with the universe. Perhaps Einstein should have listened to his old friend and colleague Niels Bohr who, exasperated by one more repetition of the "dice" dictum, is supposed to have snapped, "Albert, stop telling God what to do!"

JOHN STEWART BELL (1928–91) Northern Irish physicist. He was born in Belfast into a poor family, and in 1949 he got his first degree at Queen's University there, where he had first worked as a lab assistant. After working at the Atomic Energy Research Establishment at Harwell, in 1960 Bell moved to the European Centre for Nuclear Research (CERN) in Geneva, Switzerland, where he remained for the rest of his working life. His wife, Mary Bell, was also a physicist at CERN.

Bernoulli Effect

The faster a fluid moves, the lower its pressure

Have you ever wondered how an airplane weighing many hundreds of tons can take off from a runway? You can start to see how this is possible by looking closely at the shape of an aircraft wing the next time you're at an airport. You'll notice that the wing cross-sections are concave, so that the upper surface is actually longer than the lower one. It is this subtle design feature, something you probably wouldn't notice unless you were looking for it, that allows the plane to get off the ground.

The basic idea is this: The airstream splits when it encounters the leading edge of the wing, part going over the upper surface and part going under the lower surface. For the two streams to match up when they meet just after leaving the trailing edge of the wing, the air going over the top has to move faster than the air going underneath. Enter an effect first discovered by Daniel Bernoulli, one of a cantankerous, feuding family of Swiss scientific geniuses.* Daniel was the son of Johann Bernoulli, a prominent mathematics professor at Groningen. (Johann later went to Basel to take over the chair of Greek, but moved over to the chair of mathematics when its previous occupant, his brother Jakob, died. Daniel's book *Hydrodynamica* was published in 1738, about the same time as his father published *Hydraulica,* a book whose date had been set back to 1732 to establish precedence. What a family!)

By combining NEWTON'S LAWS OF MOTION and the conservation of energy (*see* the first law of THERMODYNAMICS), Daniel Bernoulli was able to show that the pressure exerted by a fluid drops as the speed of the fluid increases (a fluid can be either a liquid or a gas). Because of this effect, the upward pressure of the slower-moving air under the wing is greater than the downward pressure of the faster-moving air over the wing. This means that as a plane gathers speed, there is an increasing

The Bernoulli effect is what makes it possible for birds and airplanes to fly. The natural and artificial wings have very similar cross-sections: The shape creates a difference in airflow between the upper and lower surfaces, which is what produces lift.

* No fewer than eight members of the Bernoulli family have entries in the authoritative *Dictionary of Scientific Biography.*

upward force, called *lift,* on the wings caused by this imbalance in pressure. When the pressure difference is high enough, this lift becomes greater than the downward force of gravity, and the plane is literally pushed upward into the air. Lift also keeps the plane aloft during flight.

There is something else related to the Bernoulli effect you may have noticed if you are a frequent flier: When it leaves your home airport, your plane will take off in different directions on different days, and it will do the same when it lands. The choice of direction isn't arbitrary—it depends on the direction the wind is blowing. If an airplane is moving into the prevailing wind, the speed of the air over its wings will be equal to the speed of the airplane relative to the ground plus the speed of the wind itself. There will be a corresponding increase in the lift associated with the Bernoulli effect if the plane is headed into the wind. So, for a given expenditure of energy to get the plane rolling at a specific speed, a plane moving into the wind gets an added bonus in terms of the lift generated by airflow over its wings.

Another place where you may encounter the Bernoulli effect is in front of a fireplace on a stormy evening. When there is a particularly strong gust of wind, the flames shoot up into the chimney. What's happening is this: As the wind passing over the top of the chimney speeds up, the pressure at that point drops. The higher air pressure inside the house, then, literally pushes the flames up the chimney. You may have noticed spiral vanes around the outside of chimneys on, for example, industrial plants. They are there for the same purpose: By channeling any wind around and over the chimney-top, waste gases are exhausted most efficiently.

My commonest first-hand encounter with the Bernoulli effect is rather unusual. I customarily do in-line skating as part of my exercise program, and my favorite venue in the Washington, DC, area is a paved track alongside the Potomac river. I usually start my workout on a part of the track near National Airport, and I have learned to watch the way the planes are moving as I approach the parking lot. If they're landing one way, I'll have the wind at my back on the return leg, a situation I much prefer. For me, then, the work of Daniel Bernoulli is of more than academic interest—it tells me how tired I'll be at the end of my daily workout.

DANIEL BERNOULLI (1700–82) Swiss mathematician and scientist. He was born at Groningen, the Netherlands, into a dynasty of Swiss mathematicians and intellectuals. Educated first as a physician, he turned to mathematics and in 1725 went to St Petersburg as professor of mathematics. Returning to Switzerland seven years later, he held university chairs in botany, physiology, and, finally, physics. His studies of hydrodynamics resulted in the statement of what is now called Bernoulli's principle, and foreshadowed the KINETIC THEORY of gases.

Big Bang

The universe began in a hot, dense state about 15 billion years ago and has been expanding and cooling ever since

Astronomers use the term "big bang" in two different but related senses. On the one hand, it refers to the event that marked the origin of the universe about 15 billion years ago; on the other hand, it describes the entire scenario of expansion and cooling that has taken place since then.

The concept of the big bang begins with the discovery of HUBBLE'S LAW in the 1920s. This law captures in a simple equation the results of observations that showed the universe to be expanding, with galaxies moving farther and farther apart from one another. It's not too hard, therefore, to imagine "running the film backward" to arrive at the origin of the universe in a highly compressed state, billions of years ago. This dynamic picture of the universe has been bolstered by two important pieces of evidence:

The cosmic microwave background

In 1964, two American physicists, Arno Penzias and Robert Wilson, found that the universe is filled with radiation in the microwave range. Subsequent measurements established that these microwaves are part of a classical BLACK-BODY RADIATION pattern that is characteristic of an object at a temperature of about –270°C (3 K), just three degrees above absolute zero.

A simple analogy will help you to interpret this result. Imagine sitting by an open fire and watching the coals. While the fire is roaring, the coals may look yellow. As the fire becomes less fierce, the color of the coals fades to orange, then to a dull red. When the fire begins to die, the coals stop giving off visible light, but if you put your hand out toward them you will still feel heat, which means that they are still giving off infrared radiation. When an object cools down, then, the radiation it emits (*see* STEFAN–BOLTZMANN LAW) moves to longer and longer wavelengths. What Penzias and Wilson detected were the "coals" of a universe that had been cooling for 15 billion years, and whose radiation had shifted in wavelength into the microwave range.

Historically, this discovery established the big bang as the cosmologists' theory of choice. Other models of the universe (e.g., the STEADY-STATE THEORY) can explain the Hubble expansion, but none can also explain the presence of the cosmic microwave background.

The abundance of light elements

The EARLY UNIVERSE was very hot. If a proton and a neutron came together in an attempt to form an elementary nucleus, a subsequent collision with another particle would soon have torn them apart. It turns out that the universe has to cool for about three minutes after the big bang before collisions are gentle enough to allow simple nuclei to stay together. This marked the opening of a window of opportunity in the history of the early universe during which light nuclei could be made. Any nucleus made before three minutes would be broken up, while those made afterward would survive.

However, this early building of nuclei (called *nucleosynthesis*) took place in an expanding universe. Shortly after the first three minutes, the expansion had carried the particles so far away from one another that the collisions that can form nuclei happened extremely infrequently. This marked the closing of the window. During that brief period, collisions produced deuterium (an isotope of hydrogen whose nucleus contains a proton and a neutron), helium-3 (two protons and a neutron), helium-4 (two protons and two neutrons), and a little lithium-7 (three protons and four neutrons). Every element heavier than these would be made later, in a star (*see* STELLAR EVOLUTION).

The big bang theory tells us the temperature of the universe and the collision rates of particles in it. It can therefore predict the proportions of these light nuclei that were created at the very beginning of the universe. When we compare these predictions with measurements of the abundances of light elements (adjusted for the production of these elements within stars), we find an amazingly good agreement between theory and observation. To my mind, this is the strongest piece of evidence for the big bang.

In addition to these two bodies of evidence, recent work (*see* INFLATION-ARY UNIVERSE) has shown that the melding of big bang cosmology with the modern theory of ELEMENTARY PARTICLES resolves many outstanding issues about the structure of the universe. Of course, some problems remain: We do not understand the very beginning of the universe, nor is it clear that our current laws of physics applied at this time. But there is now ample and persuasive evidence to support the theory of the big bang.

ARNO ALLAN PENZIAS (1933–) and ROBERT WOODROW WILSON (1936–) American physicists. Penzias was born in Munich, and his family emigrated to America in 1940. Wilson was born in Houston. Both men joined Bell Laboratories, in Holmdel, New Jersey, in the early 1960s. In 1963 they were given the task of tracking down the origin of the radio "noise" that was interfering with communications. After eliminating many possible sources (including pigeon droppings in their antenna), they were left with a persistent signal that seemed to originate beyond the Milky Way. It turned out to be the cosmic background radiation that had been predicted and even searched for by theoretical astrophysicists, including Robert Dicke, Jim Peebles, and George Gamow. For their discovery, Penzias and Wilson received the 1978 Nobel Prize for Physics.

Biot–Savart Law

The magnetic field at a point due to a current element is proportional to the strength of the current, inversely proportional to the square of the distance to the point, and lies in a direction perpendicular both to the current and the direction to the point

One of the great intellectual currents of 19th-century science was the series of discoveries that established that two seemingly unrelated phenomena—electricity and magnetism—were just two sides of the same coin. One of the key pieces of this puzzle was the realization that moving electrical charges (what we call an electrical current) could produce a magnetic field. The discovery was made by the Danish scientist Hans Christian Oersted (*see* OERSTED DISCOVERY) and quantified by the French scientist André-Marie Ampère (*see* AMPÈRE'S LAW). The Biot–Savart law represents a kind of generalization of this work, and constitutes the most complete statement of the relation between electrical currents and the magnetic fields they produce.

Jean-Baptiste Biot was a prominent and adventurous scientist, a professor of physics in Paris and a member of the French Academy. After Oersted's discovery, he teamed up with his colleague Félix Savart to study the newfound connection between electrical current and magnetic fields.

Unlike Ampère, who studied magnetic fields indirectly by measuring the forces exerted on a pair of wires carrying an electrical current, Biot and Savart measured magnetic fields directly by using small compass needles. Their law is best understood by imagining a current-carrying wire being cut up into small segments (each segment is called a *current element*). If the segments are short enough, it makes no difference if the wire is straight or curved—the segment itself can be treated as if it were a straight line. If we want to know the magnetic field *B* a distance *r* from the segment, the Biot–Savart law tells us that it will be proportional to

$$IL/r^2$$

where *I* is the strength of the current, and *L* is the length of the current element.

I said above that the Biot–Savart law gives us the most general form of the relationship between current and field. What I mean is this: You can take any arrangement of current-carrying wires, no matter how complex and how asymmetric, and break each wire up into current elements. Each element will then make a contribution to the magnetic field at the point for which you are making the calculation. All you then have to do is add up the contributions from all of the elements in all the wires to find the total magnetic field. (This adding-up process involves the techniques of calculus, and isn't always simple.) Thus, Ampère's law, which relates the current passing through a surface to the magnetic field around the edge of the surface, can be derived as a special case of the Biot–Savart law.

The one point I haven't discussed is the direction of the magnetic field predicted by the Biot–Savart law. It involves something known as the *right-hand rule*, the scourge of generations of physics and engineering students. The rule goes like this: Point the index finger of your right hand in the direction of the current in the current element, and the middle finger of your right hand in the direction of the point where you are calculating the field. With these two fingers stretched at right angles, and your thumb at right angles to both of them (another stretch), your thumb will be pointing in the direction of the magnetic field.

As mentioned above, the full mathematical expression of the Biot–Savart law requires calculus, and is what is called an integral equation. It is in fact a general solution of the fourth of MAXWELL'S EQUATIONS.

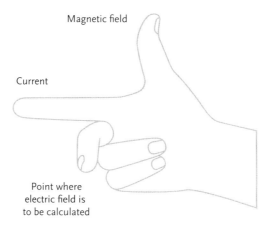

Magnetic field

Current

Point where
electric field is
to be calculated

A time-honored way of remembering the direction of the magnetic field produced by an electric current.

JEAN-BAPTISTE BIOT (1774–1862) French physicist. Biot, born in Paris, grew up during the French Revolution, and first joined the army. He then became one of the first students at the École Polytechnique in that city and went on to follow an academic career in the sciences, eventually becoming professor of physics at the Collège de France in Paris. In 1803 he was sent by the Ministry of the Interior to investigate a reported fall of meteorites at l'Aigle, and was able to show that they really did fall from the sky, which until that time had been doubted. The next year, with Joseph-Louis Gay-Lussac (1778–1850), he made the first balloon flight for the purpose of scientific experimentation, to study the Earth's magnetic field at high altitudes. His most famous work involved the Biot–Savart law, and he also made great advances in understanding the polarization of light.

FÉLIX SAVART (1791–1841) French physician and physicist. He was born in Mézières. Savart was educated as a physician, but joined the Collège de France as a professor of acoustics. He studied the working of musical instruments, especially the violin, by studying patterns in a layer of sand spread on a flat plate caused by the vibrations of sound waves. He also invented *Savart's wheel* for measuring musical tones, and the *Savart quartz* plate for investigating the polarization of light. His most famous work was the collaboration that led to the Biot–Savart law.

Black Holes

Black holes represent singularities in the space-time continuum

Of all the hypothetical entities that have been thrown up by scientific theories, black holes are surely the weirdest. Although their existence was proposed well over a century before Albert Einstein published the general theory of RELATIVITY, it is only recently that persuasive evidence for their existence has been found. In fact, I can remember the professor in my relativity class in graduate school arguing that, though they may have been predicted by general relativity, black holes couldn't possibly form in the real world.

Let's start by talking about how general relativity deals with the phenomenon of gravitation. NEWTON'S LAW OF GRAVITATION tells us that there is an attractive force between any two objects in the universe. The Earth therefore circles the Sun because there is an attractive force between the two. General relativity gives us an alternate way of looking at the Earth–Sun system. In this theory, the presence of a mass like the Sun warps the spacetime around it. One way to picture this is to imagine a stretched rubber sheet onto which a heavy weight (like a bowling ball) is placed. The sheet will deform, creating a depression around the weight. In just the same way, the mass of the Sun deforms spacetime in its vicinity.

In this picture, the Earth rolls around that depression the way a marble would roll around the depression in the rubber sheet (if we allow ourselves to forget that the marble will lose energy by friction, and spiral down). What we perceive as gravity in everyday experience becomes, in general relativity, an effect of the altered geometry of spacetime rather than a force in the Newtonian sense. At the moment, general relativity is our best theory of gravitation.

Now imagine what would happen in this picture if we started to increase the mass of the bowling ball without, at the same time, making it bigger. As the ball gets more and more massive, the pressure on the rubber gets higher and higher and depression in the sheet gets deeper and deeper until, eventually, the sides of the depression touch each other above the ball. In the real universe, if the mass of an object becomes sufficiently great and concentrated it will wrap the fabric of spacetime around itself and cease to be connected to the rest of the universe. It will have become a black hole.

An artist's impression of a supermassive black hole at the center of a galaxy. Surrounding the hole is an accretion disk of inwardly spiraling material. Huge jets of gas are shot out on either side of the black hole by the intense heat.

The essential feature of a black hole is that once something falls into it, it can never get back out. This applies to light, which explains the name "black hole"—something that absorbs all the light that falls on it will appear black. According to general relativity, when any object gets within a certain distance of the center of a black hole, a distance called the *Schwarzschild radius,* it can never get out again. (It was the German astronomer Karl Schwartzschild (1873–1916) who, in the last year of his life, used Einstein's equations of general relativity to calculate the gravitational field around a mass of zero volume.) The Schwarzschild radius for the Sun is about 2 miles (3 km), so to make a black hole out of the Sun we would have to compress its mass into a volume smaller than a small city!

Inside the Schwarzschild radius the theory predicts something even stranger: The black hole's matter concentrates in a point of infinite density at the center, something mathematicians call a *singularity*. At infinite density, matter—quite literally—occupies no space. Whether this really happens inside a black hole, of course, is something we can't find out directly, since nothing that went into the black hole to investigate could ever get out again.

Even though we cannot "see" a black hole in the conventional sense, we can detect its presence through the gravitational influence it exerts on matter around it. Various kinds of black holes are believed to exist in nature. Astronomers are confident that two types really do exist:

Supermassive black holes

In the center of the Milky Way and other galaxies is a huge black hole with a mass millions of times that of the Sun. These supermassive black holes, as they are known, were detected by observing the motion of gases near the centers of galaxies. These gases seem to be in orbit around a heavy object, and a simple application of NEWTON'S LAWS OF MOTION tells us that the attracting object has to be extremely compact and extremely massive. Only a black hole could produce the motion we see. In fact, astronomers have documented dozens of these massive black holes at galactic centers, and suspect that every galaxy has one.

Stellar-mass black holes

According to our current understanding of STELLAR EVOLUTION, when a star more than about 30 times as massive as the Sun dies in a fiery supernova, the remnant left behind will be a black hole. For a star isolated in space, it would be very difficult to detect this remnant, since there is nothing close enough to register its gravitational attraction. If, however, the star were part of a binary system (two stars orbiting each other), then the black hole would exert a force on its companion. Astronomers know of about a half dozen candidates for this sort of system, although the evidence isn't completely airtight for any of them.

In a binary system in which one star is a black hole, matter from the other star can be pulled onto the black hole. The matter will spiral in toward the black hole, bunching up as it gets near the Schwarzschild radius. The matter will be compressed and heated by collisions, and will eventually get hot enough to give off X-rays. Astronomers observing the system will see a flickering in the emitted X-rays, and from this and other data can estimate the mass of the object attracting the material. If that mass is greater than the CHANDRASEKHAR LIMIT of 1.4 solar masses, the object cannot be a white dwarf. In most of these so-called X-ray binaries, the attracting object is a neutron star. In just a handful of known cases, it is a black hole.

Other types of black hole are much more speculative and theoretical, with no experimental evidence at all for their existence. First, there are *mini black holes,* in which the mass of a mountain is compressed into the volume of a proton, suggested by the English cosmologist Stephen Hawking (*see* CHRONOLOGY PROTECTION CONJECTURE) to have formed shortly after the big bang. Hawking suggested that the explosion of mini black holes could explain the enigmatic, ultra-high-energy phenomena known as gamma-ray bursts. Second, some theories of elementary particles predict that on the smallest scales, the universe should be littered with a kind of black-hole foam. These black holes are only about 10^{-33} cm across—many billions of times smaller than a proton. At the moment there is no prospect that we will be able to find experimental evidence for their existence, much less investigate their properties.

JOHN MICHELL (1724–93) English geologist and clergyman. Little is known of his early life. He was elected to the Royal Society in 1760 on the strength of his study of the earthquake that devastated Lisbon in 1755, which he correctly attributed to shock waves emanating from a disturbance beneath the Atlantic seabed. In astronomy, he established that most double stars—two that appear very close together—must be true binary systems, physically associated, for there were too many of them to be accounted for by chance alignments. But Michell's most remarkable insight was to predict the existence of what he called "dark stars," so massive that light could not escape from them. He even pointed out that such an object could be detected if it were part of a binary system.

Black-Body Radiation

Radiation is emitted at all wavelengths from a black body—an object which absorbs perfectly at all wavelengths when it is heated

Toward the end of the 19th century, scientists investigating the interaction of radiation (e.g., light) and atoms began to run into serious problems—problems that eventually led to the development of QUANTUM MECHANICS. To understand the first of these problems, imagine a large rectangular box with a small hole in one side and mirrors on all the interior walls. If you were to shine a burst of light into the hole, it would bounce around inside forever (in principle, at least—if the hole were infinitesimally small), and never reemerge through the hole. An object into which light disappears is black, so this apparatus is an approximation to what is called a *black body*. (A true black body—like many concepts in physics—is a hypothetical object, but a sphere with a tiny hole in it, maintained at a uniform temperature, comes pretty close.)

You have probably seen some pretty good approximations to black bodies. In log fires it sometimes happens that several logs settle in such a way that there is a largely enclosed space between them. Inside this space, heat (which is infrared radiation) and light bounce around for long periods before they escape—if you peer into such a space you will see the orange-yellow glow of intense heat. Radiation is trapped temporarily between the logs in just the same way that light is trapped in the box described above.

The box gives us a way of thinking about how light interacts with atoms. The important thing to recognize is that once light enters the box, it will be absorbed by an atom, reemitted, then absorbed by another atom and reemitted again until a state of equilibrium is reached. In this equilibrium, the light and atoms will have come to a state in which, whenever light of a particular frequency is absorbed by an atom in one place, light of the same frequency will be emitted by an atom in another place. Thus, the amount of light at each frequency will stay the same, even though waves may be absorbed or emitted from different atoms.

This much of the behavior of a black body is easy to understand. The problem for classical physicists ("classical" in this sense means before the advent of quantum mechanics) came when they started to calculate the amount of energy stored in the radiation in a black box at equilibrium. They quickly discovered two things:

— the number of waves in the box increased with the frequency (i.e., the higher the frequency range they looked at, the more waves the classical theory predicted ought to be seen); and
— the higher the frequency of the waves in the black body, the more energy was stored.

These two findings taken together led to an impossible conclusion: The energy stored in the radiation in a black body should be infinite!

This flouting of the laws of classical physics was dubbed the *ultraviolet catastrophe,* because high-frequency radiation is in the ultraviolet range.

Order was restored by the German physicist Max Planck (of PLANCK CONSTANT fame), who showed that the problem disappeared if atoms emitted light only at certain well-defined frequencies. (Einstein would later generalize this idea to the notion that light also came in well-defined bundles of energy, called *photons.*) In this scheme, many of the frequencies of light that could exist in the box in the classical picture couldn't appear because they couldn't be emitted by the atoms of the box, and many of the frequencies emitted by the atoms couldn't fit into the equilibrium radiation in the box. By pruning away the allowed frequencies of radiation, Planck averted the ultraviolet catastrophe and moved science toward an understanding of the subatomic world. He also calculated the characteristic distribution of frequencies that characterizes a black body in which the atoms and radiation are in equilibrium.

This distribution achieved widespread publicity almost a century after Planck published it when cosmologists showed that the microwave radiation that pervades the universe—the so-called cosmic microwave background radiation (*see* BIG BANG)—followed the black-body curve exactly. The entire universe can be regarded as the ultimate black body, pervaded by radiation with a distribution characteristic of a body at about three degrees above absolute zero.

Bohr Atom

Electrons in atoms can only be in allowed orbits

When John Dalton first discussed the modern ATOMIC THEORY, he pictured atoms as indivisible, something like microscopic billiard balls. Over the course of the 19th century, however, this view became less and less tenable. A turning point was the discovery of the ELECTRON by J.J. Thomson in 1897, which showed that there were discrete particles inside the nucleus—prima facie evidence against indivisibility. The final nail in the coffin was the discovery of the atomic nucleus (*see* RUTHERFORD EXPERIMENT) in 1911. The upshot of these discoveries was that the atom is not only divisible, but has a discernible structure, with a massive, positively charged nucleus at its center and light, negatively charged electrons in orbit around it.

But there were problems with this simple picture of the atom. For one thing, according to the laws of nature as they were then understood, such an atom wouldn't be able to exist for more than a fleeting moment—a statement that we can safely say is not supported by experience. The argument goes like this: According to NEWTON'S LAWS OF MOTION, an electron in orbit is actually accelerating. Therefore, according to MAXWELL'S EQUATIONS, it should emit electromagnetic radiation; and because of the conservation of energy (*see* the first law of THERMODYNAMICS) the orbit should decay and the electron should fall into the nucleus. It's a standard problem for first-year graduate students in physics to show that this chain of reasoning means that no atom could exist for more than a second. Obviously, something was wrong with the simple picture of the atom, because atoms have obviously been around for billions of years.

The man who resolved this difficulty and sent scientists on the road to understanding the atom was a young Danish post-doctoral fellow named Niels Bohr. Bohr took as his point of entry the new notions of QUANTUM MECHANICS, the idea that at the subatomic level everything comes in tiny bundles called *quanta*. The German physicist Max Planck had used the notion that atoms emit light in bundles of radiation (later called *photons* by Einstein) to solve the longstanding problem of BLACK-BODY RADIATION. Albert Einstein had used the notion of photons to explain the PHOTOELECTRIC EFFECT. Both Planck and Einstein were awarded Nobel prizes for their work.

Bohr took the quantum theory a step further and applied it to the orbits of electrons in atoms. In technical language, he assumed that the angular momentum of the electron (*see* STERN–GERLACH EXPERIMENT) was quantized. He was able to show that the electrons could not be in orbits just any distance from the nucleus, but were restricted to certain orbits, now called *allowed orbits*. Electrons in these orbits can't emit electromagnetic radiation in arbitrary amounts, because they would then most likely have to move to a lower, non-allowed orbit. Like airplanes stuck on a runway because their destination airport is closed, the electrons have to stay where they are because they have nowhere else to go.

Electrons can, however, move from one allowed orbit to another. Like much of what happens in the quantum world, this process isn't something we can easily visualize. The electron simply disappears from one orbit and appears in the other without ever crossing the space in between. The event is called a *quantum leap* or *quantum jump*—terms that have entered the popular lexicon to indicate sudden improvement ("a quantum leap in wristwatch technology!!!"). If the electron jumps to a lower orbit, it has to lose energy in the process, so it emits a photon of a specific energy and a specific wavelength. We perceive these photons as specific colors—the blue of copper in a fire, the yellow of a sodium vapor street lamp. By the same token, an electron taking a quantum leap to a higher orbit requires an input of energy. Typically, the electron will absorb a photon to move up.

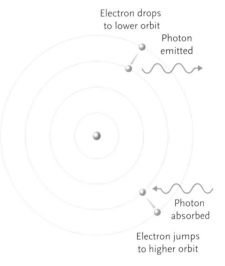

Electron drops
to lower orbit

Photon
emitted

Photon
absorbed

Electron jumps
to higher orbit

The Bohr picture of the atom, then, has electrons moving up and down, using the allowed orbits in the way we would use the rungs of a ladder to climb up and down. Each jump is accompanied by either the emission or the absorption of a bundle of electromagnetic energy we call a photon.

In Bohr's model of the atom, an electron jumps to a higher orbit when it absorbs a photon, and falls back to a lower orbit when it emits a photon.

Over time, the ad hoc nature of the Bohr conjecture gave way to the systematic formulation of quantum mechanics, and particularly the understanding of the wave–particle nature of elementary particles (*see* COMPLEMENTARITY PRINCIPLE). Today, instead of thinking of electrons as microscopic planets circling the nucleus, we now see them as probability waves sloshing around the orbits like water in some kind of doughnut-shaped tidal pool, governed by SCHRÖDINGER'S EQUATION. Physicists routinely calculate the details of these waves for complex atoms, then apply those calculations to help them understand atomic behavior. Nevertheless, the basic picture of the modern quantum mechanical atom was painted back in 1913, when Niels Bohr had his great insight.

Bohr Explanation

It works whether or not you believe in it

Niels Bohr was one of the pioneers of 20th-century physics, the founder of the Copenhagen school of quantum mechanics, and, among other honors, the recipient of the Nobel Prize for Physics in 1922. More than that, he was a father figure and mentor to an entire generation of theoretical physicists in the United States and Europe, a man respected even by those who disagreed with him.

The story goes that Bohr would often invite students and colleagues to visit him at his summerhouse on one of the islands off the coast of Denmark. One day, a young physicist who was going through the stage of hyper-rationality that afflicts many of us in our youth noticed a horseshoe nailed above the cabin door.

"Surely, Professor Bohr," he said in a state of high indignation, "you don't believe in all that silliness about a horseshoe bringing good luck?"

"No, no, of course not," replied Bohr with a gentle smile, "but I understand that it works whether you believe in it or not."

NIELS HENDRIK DAVID BOHR (1885–1962) Danish physicist. Bohr was one of the founders of quantum mechanics. Born to the family of a noted professor of physiology in Copenhagen, he quickly showed promise in science. His master's thesis at the University of Copenhagen, an experimental study of the surface tension of liquid drops, is still a standard reference in the field of hydrodynamics. It won Bohr a gold medal from the Danish Academy of Sciences and marked him as a rising star in that country's academic life. Switching to theoretical physics (a field he would pursue for the rest of his life), Bohr began thinking about the problems that were bedeviling physicists in the early part of the 20th century—problems having to do with the world of the atom. For his doctoral thesis, he studied the behavior of electrons in metals.

In 1911, Bohr went to England on what we would now call a post-doctoral fellowship to work in the laboratory of J.J. Thomson, who had discovered the electron. After a short stay (apparently Thomson had lost interest in questions of atomic structure), Bohr moved to Manchester, where he joined a group working under Ernest Rutherford, who had just discovered the existence of atomic nuclei (*see* RUTHERFORD EXPERIMENT). There, in the space of a few

months in 1912, the Dane worked out the BOHR ATOM, a model which is the basis for much of our current understanding of the subatomic world.

Although physicists were at first reluctant to accept Bohr's revolutionary ideas —which engendered a sense of scandal in some conservative German universities —the model quickly gained the approval of experimentalists as it resolved some longstanding difficulties in the understanding of atomic spectra (*see* SPECTROSCOPY). Bohr was offered a lectureship at Manchester, then a professorship in Copenhagen. Three years after his return to his native city, the Danish government built a laboratory for him. This was the Institute for Theoretical Physics, which became the center for the development of quantum mechanics in the decades that followed. All of the great names in quantum mechanics worked there with Bohr, and the so-called Copenhagen interpretation formed the central plank of quantum mechanics for over half a century.

For his work, Bohr received the 1922 Nobel Prize for Physics. The relatively short time between the introduction of the theory and the award is a sure indication of the work's fundamental significance. Not one to rest on his laurels, in the 1930s Bohr led his institute into the new field of nuclear physics, developing theoretical models that explained aspects of the fission of uranium and the working of both the nuclear reactor and the atomic bomb. During World War II, he was smuggled out of Nazi-occupied Denmark and moved to America, where he contributed to the Manhattan Project.

After the war, he became a leading spokesman for an "open world," which he considered the only feasible way for human society to deal with nuclear weapons. His son Aage Niels Bohr (1922–) also won a Nobel Prize for Physics (1975) for work on the structure of the atomic nucleus.

Niels Bohr, one of the founding fathers of our modern view of the subatomic world, developed the first quantum-mechanical model of the atom. He is pictured here at Princeton University in 1948.

Boltzmann Constant

The Boltzmann constant is the bridge between temperature and the energy of atoms

Ludwig Boltzmann was one of several scientists who developed the KINETIC THEORY of gases: our modern picture of the connection between the motion of atoms and molecules on the one hand and macroscopic properties like temperature and pressure on the other. In the picture they created, the pressure exerted by a gas is caused by its molecules colliding with its container, and the temperature of the gas is related to the speed (or, more precisely, the kinetic energy) of those same molecules. The faster the molecules move, the higher the temperature.

The Boltzmann constant provides a way of moving between these two worlds—the microscopic world of the atom and the macroscopic world of the thermometer. The key equation that creates this bridge is

$$\tfrac{1}{2}mv^2 = kT$$

where m and v are respectively the mass and the average velocity of the molecules of a gas, T is the temperature of the gas (measured in degrees kelvin), and k is a quantity known as the Boltzmann constant. The bridge between the two worlds is manifest in this equation, which links atomic properties, on the left side, with what we call *bulk properties*—ones we can measure on the human scale—on the right. The link, the constant k, has the value of 1.38×10^{-23} joules per degree kelvin.

The field of physics that relates the macro- and micro-worlds is *statistical mechanics*. There is hardly an equation in this field in which the Boltzmann constant does not appear. One of them, which the German discovered himself and is known simply as *Boltzmann's equation*, is this:

$$S = k \log p + b$$

where S is a quantity called entropy (*see* the second law of THERMODYNAMICS), p is a quantity called the statistical weight (an important ingredient of the statistical approach), and b is another constant.

Throughout his life, Boltzmann was a champion of the atomic theory and engaged in vigorous debates with scientists who felt that atoms weren't real but merely a convenient mathematical construct. His sense that this argument was going against him in the early 20th century exacerbated his depression, which led to his suicide. Boltzmann's equation is engraved on his tombstone.

LUDWIG EDWARD BOLTZMANN (1844–1906) Austrian physicist. He was born in Vienna into the family of a civil servant. At the University of Vienna he studied with Josef Stefan (*see* STEFAN–BOLTZMANN LAW), receiving his doctorate in 1866. Boltzmann then followed an academic career that saw him professor of physics and mathematics to the universities of Graz, Vienna, Munich, and Leipzig. He was a leading proponent of the reality of atoms, and his many theoretical discoveries illuminated how the atomic constituents of matter affect its behavior.

Boyle's Law

If a gas is held at constant temperature, its volume is inversely proportional to its pressure

Robert Boyle was one of those gentleman scientists associated with a past in which the pursuit of science was the prerogative of those with wealth and leisure time. He did most of his research in what today we would call chemistry, although he would undoubtedly have referred to himself as a natural philosopher. Apparently his interest in the behavior of gases began when he saw plans for an early model of an air pump. After building a much improved version of the pump for himself, he undertook to study how the high pressures produced by his instrument affected gases. As well as being a gifted experimentalist, he espoused the (then rather new) view that science must be based on empirical study, rather than on philosophy.

Basically, Boyle's law says that if you squeeze a gas you will make it contract. Think of putting your hands around a balloon and pushing inward. Because there is a lot of space between the molecules in the gas, there is no force to counter the inward pressure, and the volume of the gas in the balloon decreases. This is one way in which a gas differs from a liquid. In water, for example, the molecules are packed tightly together, like marbles in a bag. Consequently, it is very difficult to compress water through the use of outside pressure. (If you don't believe this, try pushing down on a tightly fitting plug in the neck of a container of water.) Boyle's law, along with CHARLES'S LAW, is one of the key ingredients of the IDEAL GAS LAW.

ROBERT BOYLE (1627–91) Anglo-Irish physicist and chemist. He was born in Lismore Castle, Ireland, the 14th child of the Earl of Cork, a great Elizabethan adventurer. He was educated at Eton, where he was one of the first of the "gentlemen" students, and then traveled extensively on the Continent, continuing his education at Geneva. After his travels he began his private scientific research, first in London, and then, after 1658, in Oxford, assisted by Robert Hooke (*see* HOOKE'S LAW). Boyle was a founder-member of the Royal Society, which grew out of a group that first met in Oxford. He carried out many pioneering experiments in chemistry, including studies of ACIDS AND BASES, and is sometimes credited with developing the notion of the

chemical element. Away from science, he was a member of the East India Company and was active in attempting to spread Christianity in Britain's eastern colonies.

Bragg's Law

X-rays scattered from a crystal will display constructive interference when a certain mathematical condition is met

The atoms in a crystal are arranged in a regular lattice structure, rather like oranges stacked on a fruit stall. One of the tasks of physicists is to unravel the structure of crystals. A common technique for doing this depends on a law discovered by the Australian-born physicist Sir William Lawrence Bragg, who worked closely with his father.

When an X-ray beam hits the atoms in a crystal, each atom acts as a center for reflected Huygens wavelets (*see* HUYGENS' PRINCIPLE). Sets of parallel planes pass through the atoms in the crystal. (Think of each of these sets of planes as passing through the atoms at a different angle— one diagonally across a square of atoms, one from one atom to the opposite corner one square away, another to the opposite corner two squares away, and so on.) The reflected wavelets from atoms in neighboring planes will not, in general, reinforce each other, but if they arrive at the observation point exactly one wavelength apart, they will do so. The condition for this to happen is that

$$2d \sin \theta = n\lambda$$

where d is the spacing between the planes, θ is the angle through which the X-rays have scattered, λ is the wavelength of the X-rays, and n is any integer (i.e., any whole number). If $n = 1$ then the wavelets from neighboring atoms are one wavelength apart, if $n = 2$ they are two wavelengths apart, and so on.

This condition, which is known as Bragg's law, tells you that for a given X-ray wavelength you will see intense scattered X-rays at certain angles, and that those angles are related to the distances between planes of atoms.

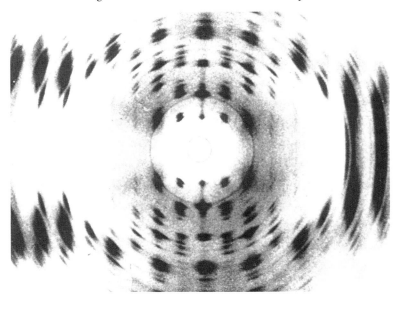

The technique of X-ray diffraction is not restricted to crystalline substances. This is an X-ray diffraction image of a DNA molecule, whose double-helix structure gives rise to repeated patterns.

Each set of planes will produce bright X-ray spots when the Bragg condition is met.

So, when an X-ray beam is sent into a crystal, it will emerge as a scattered beam consisting of some pattern of bright spots. By working backward from the pattern of spots, scientists can calculate the distances between the different sets of planes, and thus the distances between the atoms in the crystal. This procedure is called *X-ray diffraction*. This procedure has become particularly important in biotechnology, because X-ray diffraction is one of the main tools that scientists use to discover the structure of the MOLECULES OF LIFE.

WILLIAM HENRY BRAGG (1862–1942) and **WILLIAM LAWRENCE BRAGG** (1890–1971) English physicists, the only father-and-son team to be jointly awarded a Nobel prize. William Henry was born in Westwood, England, was educated at Cambridge, and held professorships at various English and Australian universities. He began experimental investigations of how newly discovered radioactive particles interact with matter. His most important research, carried out with his son, was into the way that X-rays scatter from crystals, for which they were awarded the 1915 Nobel Prize for Physics. William Henry went on to become head of the Royal Institution and president of the Royal Society. William Lawrence, born in Adelaide, Australia, spent much of his career building the science of X-ray crystallography, which he and his father founded.

Brewster's Law

Light reflected at a specific angle is polarized parallel to the reflecting surface

Light, like other forms of electromagnetic radiation, is composed of oscillating electric and magnetic fields oriented at right angles to each other. The direction of the electric field specifies the direction in which an electric charge will move as the wave goes by. The *polarization* of the wave is defined as the direction in which this electric field points.

Light waves can be *linearly polarized* (in which case the electric field is always pointing along a fixed line), *circularly polarized* (in which case the electric field rotates like the hands of a clock), or *elliptically polarized* (in which case the electric field both rotates and changes magnitude). Brewster's law deals with the way that polarization changes when a light wave is reflected from a surface. The law states that at a specific angle, which varies from one substance to another, reflected light will always be polarized parallel to the reflecting surface. The angle, called the *Brewster angle*, is defined as the angle whose tangent is equal to the refractive index of the reflecting substance. Even away from this angle, much of the scattered light will be polarized the same way.

The refractive index of a substance is defined to be the ratio of the velocity of light in a vacuum to the velocity of light in the substance. A typical piece of glass, for example, might have a refractive index of 1.5. This means that light, which travels at 186,000 miles per second (3×10^8 meters per second) in a vacuum, travels only two-thirds as fast in the glass. For this substance, the Brewster angle would be about 57°.

The place where you are most likely to come into contact with Brewster's law is in the use of polarized sunglasses. On a sunny day, reflected light will be polarized mainly in a horizontal direction, parallel to the ground. Polarized sunglasses are designed to transmit only light polarized in a vertical direction, so that they keep out a lot of the light reflected from objects around you—light which you would otherwise perceive as glare.

DAVID BREWSTER (1781–1868) Scottish physicist. He was born in Jedburgh. He studied at Edinburgh University to enter the church, becoming a minister and one of the founders of the Free Church of Scotland, but abandoned this early calling for a career studying and popularizing science. He specialized in optics, in particular spectroscopy and polarization, and discovered the law which now bears his name. He achieved a kind of immortality by inventing the kaleidoscope, in 1816. Brewster was made principal of Edinburgh University in 1859.

Brownian Motion

Small particles suspended in a fluid undergo random motion caused by the impacts of molecules

The latter part of the 19th century saw a serious debate between physicists about the nature of atoms. On the one side were scientists who, like the redoubtable Ernst Mach (known for his research on SHOCK WAVES), maintained that atoms were simply useful mathematical fictions, without any physical reality. On the other side were the likes of Ludwig Boltzmann (*see* BOLTZMANN CONSTANT), who championed the reality of atoms. Neither side in the debate realized that the experimental result that could resolve the issue had already been obtained by the botanist Robert Brown.

In the summer of 1827, Brown was using a microscope to study pollen grains. He examined very fine pollen grains of the plant *Clarkia pulchella* suspended in water, and saw that they were executing a random, jittering motion. He determined that the motion didn't result from currents in the water or from evaporation, but was at a loss to explain where it did come from. With characteristic thoroughness, he showed that this effect (now referred to as Brownian motion) was shown by all sorts of pollen grains, mineral powders, and other fine materials.

It was Albert Einstein, in 1905, who finally understood that this seemingly obscure experimental observation was the key to the debate on the reality of atoms. His explanation went like this: The pollen grain is constantly being bombarded by water molecules. On average, these collisions occur in equal numbers on all sides of the grain. For a small grain, however, we can expect that at any given moment, just by chance, more molecules will strike it on one side than on the other. In response, the grain will start to move in the direction of the side experiencing fewer collisions. A moment later, just by chance, more molecules will hit a different side of the grain and it will start off in a new direction. Using the rules of statistics and the KINETIC THEORY of gases, Einstein showed that the observed Brownian motion could be explained by these sorts of random fluctuations in collisions. (Incidentally, his paper explaining Brownian motion appeared in the same volume of the German journal *Annalen der Physik* as his paper proposing the theory of RELATIVITY and his paper on the PHOTOELECTRIC EFFECT. It was the last of these that won him the 1921 Nobel Prize for Physics.)

In 1908, the French physicist Jean-Baptiste Perrin (1870–1942), in a brilliant series of experiments, confirmed Einstein's explanation of Brownian motion. It became clear that the observed motion of the pollen grains was due to random collisions of molecules. Since "useful mathematical fictions" can't cause real motion, the debate on the reality of atoms was settled once and for all. As a side benefit, Einstein's analysis allowed Perrin to estimate the number of atoms or molecules colliding with the pollen grain in a specific period of time, and hence to estimate the number of atoms or molecules in the surrounding water. The basic idea here is that at any moment the acceleration of the pollen grain (*see* NEWTON'S LAWS OF MOTION) will depend on the number of molecular collisions, and hence on the number of molecules actually present in the fluid. This number, of course, is what had been called *Avogadro's number* (*see* AVOGADRO'S LAW), one of the fundamental constants in the universe.

ROBERT BROWN (1773–1858) Scottish botanist. He was born in Montrose, the son of a Scottish clergyman. He studied medicine at Edinburgh University and served as an army surgeon, but in 1798, after meeting Joseph Banks (1743–1820), the foremost botanist of his time, Brown turned to botany, a subject in which he would achieve the same prominence as his mentor. He sailed as a naturalist on exploratory voyages to Australia, and eventually became head of the botany department at the British Museum. He made many important advances in identifying, classifying, and elucidating the morphology of plants, but is best known for his discovery of Brownian motion.

Carbon Cycle

Carbon is continually cycled through the Earth's biosphere by a variety of chemical and other processes

The carbon atom is the basis of all life on Earth. Every molecule in every living thing is built around a skeleton of carbon atoms linked together. Over the course of time, carbon atoms are cycled through many different parts of the *biosphere,* the narrow shell around the Earth that supports life. Tracing the movement of carbon reveals a dynamic picture of life on our planet.

The largest reservoirs of carbon on Earth are the carbon dioxide (CO_2) contained in the atmosphere and dissolved in the oceans. Let's start with a molecule of carbon dioxide in the atmosphere. This molecule can be taken up by a plant and, through the process of PHOTOSYNTHESIS, the carbon atom can be incorporated into the plant's structure. Once this has happened, there are several possibilities:

— the carbon can stay where it is until the plant dies, at which point *decomposers* such as fungi and termites will use the molecule for food and eventually return the carbon to the atmosphere as CO_2;

— the plant can be eaten by a herbivore, in which case the carbon will either be returned to the atmosphere (by the herbivore's respiration while it is alive, and by decomposition after it dies) or be ingested by a carnivore (after which the same two outcomes apply); or

— the plant can die and be buried, eventually turning into a fossil fuel like coal.

Alternately, that original molecule of CO_2 can go into seawater as a dissolved gas. Again, there are several options:

— it can simply move back into the atmosphere (this kind of back-and-forth exchange between oceans and atmosphere goes on all the time); or

— it can be incorporated into plants or animals in the ocean, in which case it may end up in sediment on the ocean floor and, eventually, in limestone (*see* ROCK CYCLE) or, after spending time in the sediment, may return to the ocean.

Incorporation into sediments and fossil fuels, then, provides a way for carbon to be removed from the atmosphere. Throughout the history of the Earth, carbon removed in this way has been replaced by CO_2 injected into the atmosphere by volcanic eruptions and other geothermal processes. This natural replenishment has been joined in modern times by the burning of fossil fuels by human beings. Because CO_2 contributes to the GREENHOUSE EFFECT, understanding the carbon cycle has become an important goal for atmospheric scientists.

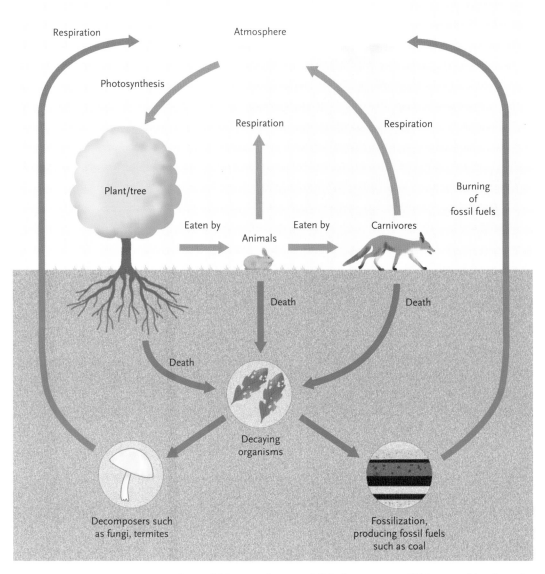

Respiration

Atmosphere

Photosynthesis

Respiration

Respiration

Plant/tree

Burning of fossil fuels

Eaten by

Eaten by

Carnivores

Animals

Death

Death

Death

Decaying organisms

Decomposers such as fungi, termites

Fossilization, producing fossil fuels such as coal

Carbon is continuously cycled through the Earth's biosphere around a network of interlinking pathways. To the natural processes can now be added the effects of burning fossil fuels.

Part of this quest is to establish the amount of CO_2 stored in plant tissue (in a newly planted forest, for example)—an example of what scientists call a *carbon sink*. As governments attempt to reach an international agreement on limiting emissions of CO_2, the question of how to balance a nation's sinks against its carbon emissions has become a major bone of contention among the industrialized nations. However, scientists doubt that reforestation by itself can counter the build-up of atmospheric CO_2.

Carnot's Principle

There is an idealized engine whose efficiency is the highest possible for any engine operating between two specific temperatures

Some of the most important machines don't actually exist, or at least they do so only as mental constructs. Each of these hypothetical machines, among which the *Carnot engine* holds an important place, illustrates an important point. (Even the castle in the sky that is the PERPETUAL-MOTION machine serves to show that you can't get something for nothing.) The Carnot engine, a concept developed by Sadi Carnot, underlines an important consequence of the second law of THERMODYNAMICS.

The main working part of the Carnot engine can be thought of as a piston in a gas-filled cylinder. As befits the theoretical nature of the engine, the piston operates without friction and suffers from no imperfection. The piston can move back and forth between two reservoirs of heat, one at a high temperature, the other at a low temperature. (Think of the high temperature as being produced by a burning mixture of gasoline and air, and the low temperature as associated with the surrounding atmosphere.) The engine then goes through a cycle as follows:

1 The cylinder is first in contact with the high-temperature reservoir, and the gas in the cylinder is allowed to expand, always staying at the same temperature and pressure. During this stage, heat flows into the gas from the high-temperature reservoir.
2 The cylinder is then wrapped in insulation, so that no heat can enter or escape, and the gas is allowed to continue to expand until it reaches the temperature of the low-temperature reservoir.
3 The insulation is removed, and the gas in the cylinder, now in contact with the low-temperature reservoir, contracts, transferring some heat to the low-temperature reservoir as it does so.
4 When the contraction has reached the proper point, the cylinder is again wrapped in insulation and the gas is compressed by moving the piston up. This continues until the temperature of the gas has risen to that of the high-temperature reservoir; the insulation is then removed and the cycle is repeated from stage 1.

This engine has many of the characteristics of real engines: it runs on a cycle (in this case known, not surprisingly, as the *Carnot cycle*), it extracts energy from a high-temperature process (such as the burning of a fuel), and it dumps some of that energy into the surrounding environment. In the process, it does some work (in the case of the Carnot engine, by pushing the piston down). The efficiency of the Carnot engine, the *Carnot efficiency,* is defined to be the work it does divided by the energy (in the form of heat) it removes from the hot reservoir. It is easy to prove that this efficiency, E, is given by the formula

$$E = 1 - (T_c/T_h)$$

where T_c and T_h are the temperatures (in degrees kelvin) of the cold and hot reservoirs, respectively. This efficiency obviously has to be less than 1 (or 100%).

Carnot's great insight, which is what we now call Carnot's principle, was that no engine operating between two temperatures can be more efficient that a Carnot engine. If it did, it would violate the second law of THERMODYNAMICS, because the engine would be effectively extracting heat from the low-temperature reservoir and passing it to the high-temperature reservoir. (In fact, the second law derives from Carnot's studies.) The above equation therefore places an upper limit on the efficiency attainable by any real engine operating in the real world. Engineers can approach this limit, but they can't surpass it. Thus, it turns out that Carnot's imaginary engine plays a major role in the real, greasy, noisy world of industry—another example of the usefulness of basic research.

NICOLAS LÉONARD SADI CARNOT (1796–1832) French physicist and military engineer. He was born in Paris, the son of Lazare Nicolas Marguerite Carnot (1753–1823), a prominent general, politician, and mathematician. Sadi studied at the École Polytechnique, graduating in 1814 and then taking time out to fight with a student volunteer detachment for Napoleon. He spent the next years working as a military engineer, but in 1819, after the fall of Napoleon, he left the military to study science, economics, and the arts. He was also interested in the many new industrial developments going on at the time. In the course of investigating the principles of steam engines he became one of the pioneers of the science of thermodynamics and developed the concept of the Carnot engine. His ideas were set out in his classic book, *Réflexions sur la puissance mortrice du feu* ("Reflections on the Motive Power of Heat"). After a short return to the military as a captain, he died in Paris at the age of just 36 from cholera, following an attack of scarlet fever.

Catalysts and Enzymes

A substance that helps a chemical reaction to proceed, but is not itself changed in the process, is called a catalyst or (if the reaction is biochemical) an enzyme

The rate at which some chemical reactions proceed can be greatly increased by the presence of a substance that participates in the reaction but is not consumed in the process. The best way to understand this is to think of the role of a real estate broker. The broker brings together people who want to buy a particular property with the person selling it, thereby facilitating the sale and transfer. The broker, however, does not actually buy or sell anything in the transaction. In a similar fashion, a catalyst or enzyme facilitates a reaction between two substances but is left at the end of the reaction in its original form.

Probably the most familiar catalyst is in your car, in the catalytic converter. This is a fine mesh made of the metals palladium and platinum through which the exhaust of the car's engine passes. The metals catalyze a number of chemical interactions. First, they absorb carbon monoxide (CO), nitrogen monoxide (NO), and oxygen, and each NO molecule is broken into its constituent atoms. The CO is combined with an oxygen atom to produce carbon dioxide, while nitrogen atoms combine to form nitrogen molecules. At the same time, the extra oxygen allows hydrocarbons that were not burned in the car's cylinders to burn completely into carbon dioxide and water. In this way, an exhaust stream that contains carbon monoxide (a lethal poison) and substances that lead to ACID RAIN, as well unburned fragments of the original molecules in the gasoline, is converted into a relatively innocuous mix of carbon dioxide, nitrogen, and water.

To understand the action of enzymes, we need to know that when complex organic molecules interact with one another, it is not enough for them simply to come together. The reaction depends on particular atoms in each molecule approaching each other in just the right orientation so that they can form CHEMICAL BONDS, rather like a key being presented to a lock the right way up. Thus, three-dimensional geometry plays an extremely important role in the chemistry that goes on in living systems (*see* MOLECULES OF LIFE).

In biochemistry, it is extremely unlikely that two complex molecules, left to themselves, will just happen to be at the right orientation to each other for an interaction to take place. What is needed for that reaction to happen at an appreciable rate is the action of a type of molecule known as an enzyme. An enzyme attracts two other molecules to itself and holds them in the right orientation so that the molecules can interact. Once the interaction has taken place, the enzyme is free to go off and repeat the process with another set of molecules. All the enzymes found in living systems are PROTEINS, which can assume many complex shapes. As enzymes, they govern the rate at which chemical reactions proceed, and are coded for in DNA.

Cell Theory

(1) All living things are made from one or more cells, (2) the chemical reactions of living things take place inside the cells, (3) all cells come from other cells, and (4) cells contain the hereditary information that is passed from one generation to the next

The first person to see a cell was the English scientist Robert Hooke (of HOOKE'S LAW fame), who was curious about why cork was so buoyant. In 1663 he was looking at a thin slice of cork under an early microscope. He found it to be divided into tiny compartments that reminded him of the small rooms occupied by monks in a monastery, so he named them *cells*. The Dutch instrument-maker Anton van Leeuwenhoek (1632–1723) was the first to see what he called "animalcules"—living organisms moving in a drop of water, which he observed in 1674. Thus, by the early 18th century, scientists knew of the existence of cells in living things.

It wasn't until 1838, however, that Matthias Schleiden, after spending years making detailed studies of plant tissue, suggested that all plants are composed of cells. The next year he was joined by Theodor Schwann in proposing that all living things are made of cells. This idea is the basis of modern cell theory. In 1858, the theory was extended by the German pathologist Rudolph Virchow (1821–1902), who said, "Where a cell exists, there must have been a pre-existing cell"—in other words, that life can arise only from other life. The fourth tenet above was added when MENDEL'S LAWS of heredity were rediscovered and turned scientists' attention to questions of inheritance. Today, we recognize that hereditary material is contained within the DNA in cells (*see* MOLECULAR BIOLOGY).

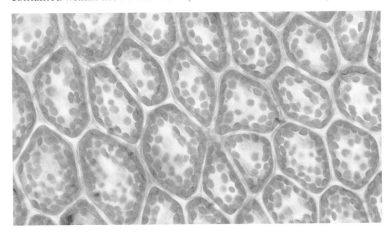

A magnified view of a section through a leaf, showing the cells (the polygon-like structures) of which it is composed.

MATTHIAS JAKOB SCHLEIDEN (1804–81) and **THEODOR SCHWANN** (1810–82) Schleiden, a German botanist, was born in Hamburg, the son of a prominent physician. He trained as an attorney but abandoned the law to study botany, eventually becoming a professor at the University of Jena. Unlike other botanists of the time, who were content to study plant classification, he studied the growth and structure of plants through the microscope. The German physiologist Theodor Schwann was born in Neuss. He began his career by studying for the priesthood, but soon transferred to medicine. Obtaining his MD in Berlin, he made a number of discoveries in the field of biochemistry. In later life, as a professor at the University of Liège, he became something of a religious mystic.

Centrifugal Force

Observers in a rotating system will experience a force which tends to move them away from the axis of rotation

You will probably have had the experience of riding in a car that made a sharp turn. You felt as though you were being thrown to the outside of the curve. You might even have invoked NEWTON'S LAWS OF MOTION and maintained that your motion toward the car door had to result from the action of a force. This apparent force is usually called the centrifugal ("center fleeing") force. It is the centrifugal force that makes so many fairground rides exciting, because it plasters riders up against the side of rotating equipment. (Incidentally, the word, which comes from the Latin *centrum* and *fugus,* was first used by Isaac Newton in 1689.)

An observer watching you from the outside, however, would describe events differently. To such an observer, you were simply continuing to move in a straight line, as any object would if not acted upon by an external force, and the car curved away from you. To this observer, in other words, it's not that you are being pushed against the door—it is that the door is being pushed against you.

There is nothing inherently contradictory between these two views. They lead to exactly the same description of events and exactly the same equation describing those events. The interpretation of those events is all that differs between observers. In this, centrifugal force resembles the Coriolis force (*see* CORIOLIS EFFECT) that also acts in rotating systems.

Because not all observers see a force acting, physicists often refer to the centrifugal force as a *fictitious force* or a *pseudoforce,* but I find these terms somewhat misleading. There is, after all, nothing fictitious about a force you can feel pushing you against the side of the car. The reality of the situation, though, is that you are still trying to move in a straight line and the car is turning away from that straight line and so pushing against you.

To illustrate the equivalence of the two descriptions of the centrifugal force, we need to do some simple math. An object moving around a circle at constant speed is accelerating, because it is always changing direction. The acceleration needed to produce this change is v^2/r, where v is the speed and r is the radius of the circle. The size of the centrifugal force felt by someone in a rotating system is just mv^2/r.

Now consider something moving in a circular path, such as you inside the car, a ball being twirled around your head on a string, or the Earth in orbit around the Sun. In each case there is a force tending to pull the object in, exerted, respectively, by the side of the car, the string holding the ball, and the gravitational attraction of the Sun. Let us call this force F. From the point of view of someone in the rotating system, it appears that the object doesn't move. This means that the inward force F is balanced by the outward centrifugal force

$$F = mv^2/r$$

From the point of view of someone outside the rotating system, however, the object (you, the ball, the Earth) is actually being accelerated by the force that is acting. Newton's second law of motion tells us that the relation between the force and the acceleration is then $F = ma$. Using the acceleration for an object moving in a circle, this becomes

$$F = ma = mv^2/r$$

But this, of course, is precisely the equation that would be written down by an observer in the rotating system. So the two observers arrive at the same equation, even though they describe the terms differently.

This illustrates a very important aspect of the science of mechanics. Observers in different frames of reference may give very different descriptions of the events they see. All observers, however, agree on the basic laws of nature (as in this case, when they agree on the equation). This is the basic idea behind the theory of RELATIVITY.

Chandrasekhar Limit

A white dwarf star can be no more than 1.4 times as massive as the Sun

Like everything else in the universe, stars are born, live out their lives, and eventually die (*see* STELLAR EVOLUTION). Depending on the mass of the star, its end can be the fiery explosion of a supernova or the sedate decline of the white dwarf.

A star's life is a constant battle against the inward pull of gravity. Right now, for example, nuclear reactions in the core of the Sun are releasing energy and raising the temperature of the matter there, which behaves as though it were an ideal gas. According to the IDEAL GAS LAW, if the temperature increases then the pressure increases as well, so the effect of the nuclear reactions is to raise the pressure at the core of the Sun. This pressure is what holds up the Sun's outer layers, which would otherwise rapidly collapse under the force of gravity.

Eventually (in about 6½ billion years) the Sun will run out of fuel for its nuclear furnace, and gravity, which has been held off for 11 billion years, will take over. It will start to collapse until something else can take the place of nuclear-generated pressure. For stars like the Sun, that something else is provided by the electrons in the star. Electrons obey the PAULI EXCLUSION PRINCIPLE, which means that no two electrons can occupy the same state. Think of this as a requirement that electrons need their "elbow room" and can't be crammed too closely together.

When a star of the Sun's mass collapses down to roughly the Earth's size, the electrons cannot be forced any closer together and the collapse stops. The star no longer generates energy, but may still shine from its accumulated heat. A star like this is called a *white dwarf,* and we know of many of them in the sky. In effect, the white dwarf is supported against gravity by a kind of pressure exerted by the electrons. Astrophysicists call this *degeneracy pressure.* (More massive stars go through a different life cycle, ending in a cataclysmic explosion known as a supernova.)

In the early 1930s, a young Indian theoretician, Subrahmanyan Chandrasekhar, worked out the theory of white dwarfs. His most important result was that if the star is too massive, the gravitational force it exerts is so large that it overcomes the electron

Subrahmanyan Chandrasekhar as a young man. He was responsible for our understanding of the white dwarf—the end state of stars like the Sun.

degeneracy pressure, and the collapse will continue. The largest stars that can be held up against gravity by electron degeneracy pressure are about 1.4 times as massive as the Sun, a number that has become known as the Chandrasekhar limit.

Artist's impression of the Chandra X-ray Observatory in orbit. Chandrasekhar himself made fundamental contributions to the theory of black holes and to other phenomena that the observatory, launched in 1999, was designed to study.

SUBRAHMANYAN CHANDRASEKHAR
(1910–95) Indian-American astrophysicist. He was born in Lahore (now part of Pakistan), where his father was a senior civil servant in the British Indian government. Chandrasekhar attended the University of Madras and then went to Cambridge to finish his education. In 1937 he joined the faculty of the University of Chicago, where he remained until his death. He contributed to many fields of theoretical physics and astrophysics, and for his discovery of the Chandrasekhar limit he received the 1983 Nobel Prize for Physics. He had a courtly manner, almost always wore a conservative black suit, and spent much time encouraging young theorists. NASA's current orbiting X-ray observatory is named Chandra in his honor.

Changes of State

It requires energy to change matter from the solid to the liquid state (melting), from the liquid to the gaseous state (evaporation), or from the solid to the gaseous state (sublimation). Energy is released when there is a change of state in the opposite direction

Matter normally comes in three states—solid, liquid, and gas (*see* STATES OF MATTER). Each state corresponds to a different arrangement of atoms and a different amount of energy tied up in keeping the atoms in that arrangement. When the arrangement is changed (e.g., when a solid melts), energy has to be provided to break the bonds that keep the atoms in one state so that they can be rearranged to form another.

If you start with a solid, you can imagine the atoms locked in a kind of rigid lattice, with the relations between them fixed. The atoms will be vibrating around their average positions—the higher the temperature, the greater the vibration. As the temperature of the solid increases, these vibrations become more and more violent until, eventually, the atoms tear themselves loose and begin to move around freely. At this point the material is in the process of changing from a solid to a liquid, the process we call *melting*. The energy required to break the bonds that hold the atoms together in the solid is called the *heat of fusion*.

The way that the temperature of a solid changes as it moves through the melting point is also interesting. Below the melting point, adding heat to the solid causes the atoms to vibrate more rapidly so that each bit of added heat makes the temperature go up. When the solid reaches the melting point, however, it stays at that particular temperature until enough energy has been added to break the bonds that hold the solid together. Adding energy at this point does not increase the temperature at all, but merely breaks more bonds. This is why an ice cube in a summer drink will stay at the same temperature until it is completely melted. While it is melting, it is draining energy away from the drink (and therefore lowering its temperature) in order to break the bonds that hold it together.

The amount of heat needed to melt or boil a substance is called the *latent heat of fusion* or *latent heat of vaporization,* respectively. It can be very large: it requires 420,000 joules of heat energy to raise 1 kilogram of water from 0°C to 100°C, but 2,260,000 joules to convert that same kilogram of water at 100°C to a kilogram of steam at 100°C.

Once the solid has been completely converted into a liquid, adding heat will once again begin to raise the temperature. In a fluid, molecules are close together and are weakly attracted to one another, but they are not held rigidly in position. Adding energy to the fluid increases the velocity of the molecules, until the point is reached at which the molecules are moving fast enough to overcome their weak attraction. It is at this point that the liquid begins to convert into a vapor—a process we call *boiling* or *vaporization*.

As with melting, a fluid will stay at its boiling point until all of its molecules have entered the gaseous state. It requires a certain amount of energy to break the relatively weak bonds that hold the fluid together, and this energy is called the *heat of vaporization*.

All these processes take place in reverse as we begin to cool a gas. At first, removing energy from a gas will lower its temperature, but when you reach the boiling point the gas will begin to change into a liquid, which we call *condensation*. Energy is then released as the attractive forces between the molecules begin to take over. This energy is sometimes called the *heat of condensation,* but it is equivalent to the heat of vaporization. You get as much energy out of the system when it goes from gas to liquid, in other words, as you had to put in to get it from liquid to gas.

The fact that the heat of condensation can be fairly high explains why a burn with live steam can be so painful. Not only is your skin absorbing the heat needed to cool the gas, but it is also absorbing the heat needed to convert the steam into liquid water.

When a liquid is cooled down to its melting point (which in this direction is called the *freezing point*), it again gives up energy as it changes into a solid. Here again, the heat given up when this happens is the same as the amount of heat that is required to convert the material from a solid to a liquid.

There is one more type of phase transition: the process by which a solid is converted directly into a gas, or vice versa. This is called *sublimation*. The commonest everyday experience of sublimation occurs if wet clothes are hung out on a line in below-freezing temperatures. The water in the clothes will turn to ice, and the ice will then convert directly into a gas, without ever becoming liquid. Similarly, the dry ice used at rock concerts to produce swirls of mist is solid carbon dioxide subliming directly to gaseous carbon dioxide (and freezing out the water vapor in the air to produce the mist). The energy required to convert a solid into a gas is called the *heat of sublimation*.

Charles's Law

If a gas is held at constant pressure, the volume is proportional to the temperature

A contemporary etching of the launching of a Montgolfier balloon in 1783.

Jacques Alexandre César Charles came to science as a result of his fascination with hot-air ballooning, which was then just becoming possible. I am told by today's balloonists that modern balloon design is essentially the same as that developed by Charles two centuries ago. Small surprise, then, that Charles's research interest as a scientist was the properties of gases. He formulated his eponymous law in 1787, after experimenting with oxygen, nitrogen, hydrogen, and carbon dioxide.

The best way to think about Charles's law is to think of a gas as a collection of molecules whizzing around, colliding with one another. The pressure exerted by the gas is caused by collisions of those molecules with the walls of the container—the more collisions, the higher the pressure. For example, molecular collisions from the air around you exert a pressure, about 14.7 pounds for every square inch of your body (in metric units, 101,325 pascals, also known as 1 bar, especially in the context of meteorology). To understand Charles's law, think of air inside a balloon.

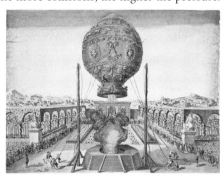

For a given temperature, the air inside the balloon will expand or contract until it, too, exerts exactly this much pressure, 14.7 pounds per square inch, and balances the pressure of the air. In other words, for every collision by a molecule in the air which tends to push the air in, there is a collision from the inside pushing the balloon out. If you lower the temperature of the air in the balloon (e.g., by putting it into a big enough refrigerator), the molecules will move more slowly, so that each collision with the container walls contributes less. The pressure of the outside air will then push in, shrinking the balloon, and hence shrinking the volume of the gas inside, until the increased density makes up for the lower temperature, and equilibrium is once again established. Charles's law is one of several that lead to the IDEAL GAS LAW, which relates the pressure, volume, and temperature of a gas, and the amount of substance present.

JACQUES ALEXANDRE CÉSAR CHARLES
(1746–1823) French physicist and chemist. He was born in Beaugency, and worked first as an official in the Ministry of Finance in Paris. He became interested in hot-air ballooning and developed the modern design for these balloons. Charles was the first to use hydrogen (which is lighter than air) in a balloon, and piloted hydrogen balloons to set new height and distance records: He achieved a height of about 10,000 feet (3000 meters), and once managed to travel 27 miles (43 km). It was his ballooning exploits that led him to study the properties of gases.

Chemical Bonds

Atoms can attach themselves to one another by giving up or accepting electrons, by sharing pairs of electrons between neighboring atoms, by sharing electrons with many other atoms, or through the effects of polarization

The electrons in atoms can be considered as occupying a series of nested shells (*see* BOHR ATOM), so that the electrons that are likely to be affected by the presence of another atom are those in the outermost shell, known as the *valence shell*. When the electrons in two atoms have arranged themselves so that a force is generated that holds the two atoms together, we say that a chemical bond has formed. We can identify several types of chemical bonds.

Ionic bond

The total energy of atoms is often minimized when the outermost electron shells are completely filled. A sodium atom, for example, which has one electron outside of a filled shell, is likely to give that extra electron away. Conversely, a chlorine atom, which needs one electron to fill its outermost shell, can accept an electron to fill its outermost shell. When a sodium atom is near a chlorine atom, the sodium will give up its outermost electron and the chlorine will accept it. In the process, the sodium atom becomes a positively charged sodium ion (because it has lost a negative charge) and the chlorine atom becomes a negatively charged chloride ion (because it has acquired an extra electron). COULOMB'S LAW says that there will then be an electrical attraction between the two ions, and this force creates the chemical bond that holds them together. (*See also* OCTET RULE.)

Incidentally, it is one of the wonders of chemistry that a violently reactive substance like sodium and a highly poisonous gas like chlorine combine to give us ordinary table salt, a regular item in many people's diet.

Covalent bond

Some atoms, most notably carbon, form bonds in a different way. When two of these atoms approach each other closely enough, they begin a process that you can think of as a continuous mutual exchange of electrons. It's as though each atom tosses one of its outer electrons to the other atom, then catches the other's electron and tosses it back in a continuous game of catch. According to the laws of QUANTUM MECHANICS, this exchange generates an attractive force that holds the atoms together.

The key point is that an atom such as carbon, which has four electrons in its outermost shell, can complete its valence shell of eight electrons by forming covalent bonds with four other atoms. This gives carbon atoms the ability to form long-chain molecules of the type found in living systems. In fact, some scientists (myself included) argue that because of this feature of the carbon atom, *all* life in the universe must be carbon based, like life on Earth.

Metallic bond

In metals, a third kind of chemical bond is formed. Each atom in a metal gives up an electron or two, in essence sharing these electrons with all of their fellow metal atoms. These loose electrons form a kind of background sea in which the ponderous positive metal ions are situated, rather like a three-dimensional lattice of marbles in a sticky volume of molasses. If you push one of the marbles, it will move slightly, but will maintain its position relative to the others. In the same way, metal atoms can be disturbed by an external mechanical force, but will be held together by the sea of electrons. This is why a metal will dent when you hit it with a hammer, but probably won't break. It is the electron sea that makes metals good conductors of electricity (*see* FREE ELECTRON THEORY OF CONDUCTION).

Chemical bonds have to do with how the electrons in atoms arrange themselves with respect to other electrons and the nucleus, and with the electrical attraction between positive and negative charges.

Hydrogen bond

Strictly speaking, this isn't a chemical bond in the way that the other three types are, but rather an attraction between individual molecules. Many molecules, although they are electrically neutral overall (i.e., they have as many negative electrons in their atoms as there are positive

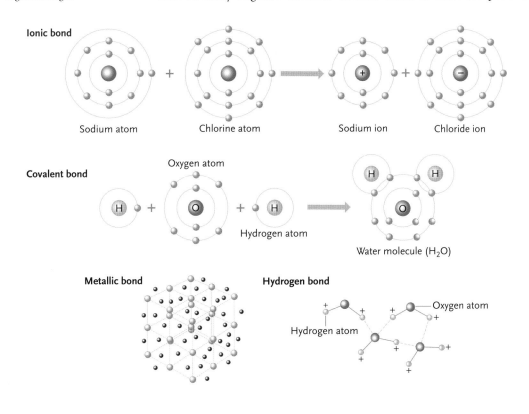

Ionic bond

Sodium atom   Chlorine atom   Sodium ion   Chloride ion

Covalent bond

Oxygen atom

H + O + H

Hydrogen atom

Water molecule (H₂O)

Metallic bond   **Hydrogen bond**

Oxygen atom

Hydrogen atom

protons in their nuclei), but are *polarized*. By this we mean that some parts of the molecules have a net negative charge, while other parts have a net positive charge. The net charge of the molecules is zero, of course, but the positive and negative charges are not evenly distributed.

Imagine that a *polar molecule,* as it is called, is approaching another molecule, so that its negatively charged region is closest to the target molecule. The electrical force exerted by this negative region will be greater than the force exerted by the positive region, because the positive region is farther away. This electrical force will drive electrons in the target molecule away from the contact point, leaving that point with a slight positive charge. The result is an attractive force between the two molecules, which creates a bond.

The commonest polar molecule is water, in which the negative charge tends to congregate around the oxygen atom, leaving the hydrogen atoms with a small positive charge. It is because of this polarization that water is such a good solvent. Its molecules are capable of creating bonds stronger than those that hold the target molecules to one another. When such bonds are created through the action of positively charged hydrogen atoms—as they usually are—chemists call them hydrogen bonds. Because there are so many hydrogen atoms in the MOLECULES OF LIFE, hydrogen bonds are quite common. The bonds that hold the two helices of the DNA molecule together, for example, are hydrogen bonds.

Cherenkov Radiation

When a particle moves through a material faster than the speed of light in that material, a characteristic type of radiation is emitted

PAVEL ALEKSEYEVICH CHERENKOV
(1904–90) Russian physicist. Cherenkov was born in Novaya Chigla, in Voronezh province, to a peasant family. He attended the University of Voronezh, then joined the Lebedev Institute of Physics in Moscow, a branch of the USSR Academy of Science, where he remained for the rest of his career. After the work that led to the discovery of Cherenkov radiation, he went on to study cosmic rays and to work on the design of large particle accelerators.

When light moves through a transparent material such as glass, it moves more slowly than it does in vacuum. Just as an air traveler crossing a continent by making intermediate stops goes more slowly than one on a direct flight, light rays slow down when they interact with atoms, and cannot travel as quickly as they can in a vacuum. RELATIVITY tells us that no material object can travel as fast as light does in a vacuum, but that doesn't prevent material objects, such as high-energy subatomic particles, from traveling faster than this reduced velocity of light within the transparent medium. In glass or water, for example, light typically travels at 60–70% of its speed in vacuum, and it is indeed possible for high-speed particles (e.g., protons and electrons) to travel faster than light in that medium.

In 1934, Pavel Cherenkov was studying how water absorbed radiation from radioactive sources. He found that a characteristic type of radiation (which now bears his name) was emitted by fast-moving electrons. This radiation is actually the optical equivalent of the sonic boom created by supersonic airplanes. We can picture what happens by imagining Huygens wavelets (*see* HUYGENS' PRINCIPLE) expanding outward at the speed of light in the form of circles, each centered at successive points along the path of the particle. If the particle is moving faster than light in the medium, it will outrun the wavelets. The crests where these wavelets reinforce one another then form the wavefronts of the Cherenkov radiation.

The radiation is emitted as a cone around the path of the particle. The angle at the top of the cone depends on the velocity of the particle and the speed of light in the medium. That is what makes Cherenkov radiation so useful in elementary-particle physics, because detecting the angle of the radiation cone allows physicists to measure the speed of the particle. The result, coupled with measurements from other instruments, allows physicists to identify the particles in their apparatus. In modern experiments, Cherenkov detectors are stacked with other instruments in huge, multistorey detectors. An example is the Super Kamiokande detector at Kamioka in Japan, which contains 50,000 tons of water and has 11,000 photomultipliers. You can also see Cherenkov radiation directly in small nuclear-research reactors, which are often placed at the bottom of a pool of water for shielding. The reactor core is surrounded by an eerie bluish glow, which is the Cherenkov radiation given off by fast particles streaming out from the nuclear reactions.

Because the analysis of this radiation played such an important role in the fledgling science of nuclear physics, Cherenkov shared the prize in 1958 with his colleagues Igor Tamm (1895–1971) and Ilya Frank (1908–90). It was Tamm and Frank who, in 1937, had shown that the radiation from Cherenkov's electrons was emitted because the electrons were traveling faster than the speed of light in water.

Chronology Protection Conjecture

There is an as yet undiscovered law of nature that forbids time travel

Time travel has always been a puzzling subject. On the one hand, if time really is just another dimension, as the principle of RELATIVITY seems to imply, then going backward in time (instead of forward, as we normally do) should be no more difficult that moving left instead of right. On the other hand, the minute you start thinking about sending objects, information, or (most interestingly) people into the past, you run into logical paradoxes.

Consider, for example, the *grandfather paradox*: You acquire a time travel machine and use it to go back in time. The purpose of your journey is simple: You wish to prevent your grandfather from meeting your grandmother (in more bloodthirsty versions of the paradox, you actually shoot your grandfather). If you do so, you won't be born, which means you couldn't have acquired the machine and engaged in time travel. But if you didn't, then your grandfather and grandmother would have met, in which case you would have been born, acquired the machine, and gone back to prevent the meeting, in which case … If we can indeed alter the past, we are faced with a logical contradiction.

General relativity predicts that time-like paths become space-like paths near a sufficiently large, dense, and rapidly rotating cylinder, so such an object could become a time machine. The practicalities of spinning a cylinder with the mass of a galaxy and a density near to that of a black hole at a speed approaching the speed of light are formidable, and certainly beyond any conceivable human technology. There is no prospect, then, of a time-travel machine in the foreseeable future.

The paradoxical nature of the phenomenon was probably what led astrophysicist Stephen Hawking to his chronology protection conjecture. In order for nature not to be paradoxical, he says, there must be some law that prevents time travel. The best argument for the existence of such a law was stated by Hawking: If time travel is possible, where are all the tourists?

STEPHEN WILLIAM HAWKING (1942–) English theoretical physicist. He was born in Oxford, where he took his first degree, and obtained his Ph.D. from Cambridge. While at Cambridge he was diagnosed with a rare form of motor neuron disease. Despite the progression of this disease, which later confined him to a wheelchair and obliged him to communicate with a speech synthesizer and computer, he became one of the leading researchers of his age. Hawking has applied the ideas of quantum mechanics and relativity to BLACK HOLES and cosmology, and popularized his ideas in the bestselling *A Brief History of Time* (1988).

Circadian Rhythms

Organisms possess internal clocks

Virtually all living things, from algae to human beings, exhibit behavior that is tied to a time cycle, usually related to the length of the day. Many plant species, for example, open their leaves at dawn and close them at sunset, and anyone who has taken a long plane trip is aware of the phenomenon of jet lag, which results from a disruption of internal rhythms. In the mid-20th century, scientists debated whether these behaviors were a response to external stimuli or resulted from some sort of internal mechanism. Today, we know that they come from internal mechanisms that are loosely referred to as *biological clocks.*

Exactly how these biological clocks work in humans and other animals is still the subject of much research, but its existence is no longer in doubt. Experiments on fruit flies, for example, have shown that alteration of a single gene can produce flies with no internal clock, flies that could be described as insomniacs, and flies with sleep–wake cycles that are different from 24 hours.

It is well known that the human body is at its lowest ebb at around 3 or 4 a.m. (what one poet has called the "dark midnight of the soul"), and people are indeed more likely to die at this time than at any other. All sorts of physiological functions, from respiration to heartbeat, operate on these cycles. Jet lag results from the disruption of these cycles as the body tries to bring its internal clocks into synchronization with the hours of daylight in the new location. Since each of our physiological and mental functions comes back to equilibrium with a different response time, we feel out of sorts for a few days until everything is synchronized again.

Seasoned air travelers are well acquainted with the effects of circadian rhythms attendant on traveling across several time zones. Jet travel throws all the human circadian rhythms out of synchronization. The resultant feeling, called jet lag, persists until the cycles are able to reset themselves to the diurnal variation in the new time zone, a process that usually takes several days. The drug melatonin is often taken by travelers to help reset their sleep cycle, since it induces the body to sleep regardless of where in the cycle it is.

Clausius–Clapeyron Equation

The heat of vaporization of a substance increases as the temperature and vapor pressure increase

As we know from the KINETIC THEORY, atoms and molecules in substances are in constant motion. Occasionally, a molecule near the surface of a liquid will be moving fast enough to break loose. Thus, above any liquid there will be some molecules of the substance floating around. The pressure that would be exerted by these molecules if they were there alone is called the substance's vapor pressure. Sometimes we can sense this presence of a vapor above a liquid—the familiar feeling of humidity on the ocean is an example.

We also know that it takes energy to make a substance change its state from liquid to gas (*see* CHANGES OF STATE). The amount of energy it takes to do this is called the heat of vaporization. The Clausius–Clapeyron equation relates the heat of vaporization H, the vapor pressure p, and the temperature T of the substance. The equation says that

$$\ln p = H/RT + \text{constant}$$

where $\ln p$ is the natural logarithm of the vapor pressure, R is the RYDBERG CONSTANT, and the temperature T is measured in degrees kelvin.

The equation was first derived by Benoît Clapeyron, a specialist in the design and construction of steam engines. For him, the heat of vaporization was obviously a number of some interest. Clapeyron's work was largely ignored until Rudolf Clausius, the author of a version of the second law of THERMODYNAMICS, rediscovered it and brought it to the attention of the scientific community.

The commonest use of the equation is to provide an easy way of measuring the heats of vaporization of various substances. By measuring the vapor pressure at various temperatures and plotting a graph of $\ln p$ against $1/T$, a straight line is obtained. The slope of this line is the heat of vaporization.

BENOÎT PIERRE ÉMILE CLAPEYRON (1799–1864) French civil engineer. He was born in Paris and trained at the École Polytechnique and the École des Mines. He had a distinguished career building railroads, contributing to the design of railroad bridges and steam locomotives. To analyze the forces on a beam supported in more than two places, he came up with a "theorem of three moments." Clapeyron's achievements were eventually recognized by his election to the French Academy.

RUDOLF JULIUS EMMANUEL CLAUSIUS (1822–88) German physicist. He was born in Köslin (now Koszalin, Poland). As a boy he attended a private school run by his father. He went on to study at the University of Berlin, and got his doctorate from Halle in 1848. Though strongly attracted by history, he opted for a career in physics and mathematics, holding positions in Berlin, Zurich, Würzburg, and Bonn. Wounds received as a volunteer ambulance driver in the Franco-Prussian War limited his productivity later in life, but he did manage to establish the second law of THERMODYNAMICS in its modern form.

Cloning

It is possible to produce offspring of mammals that are genetically identical to living adults

In 1996, Dolly burst upon the world. Under the direction of Ian Wilmut, a sheep was born which was genetically identical to an adult sheep. In the normal course of affairs (*see* MENDEL'S LAWS) an individual grows from a single fertilized egg after receiving half of its genetic material from each parent. In cloning the genetic material is taken from the cell of one living individual. The procedure works like this: The nucleus (which contains the DNA) is removed from a single fertilized cell (a *zygote*). A nucleus is then removed from the cell of an adult member of the species and implanted in the enucleated zygote. This egg is then implanted into the uterus of a female member of the species and allowed to grow to term.

What makes cloning unusual, and what made Ian Wilmut and Dolly international sensations, has to do with the way the DNA in cells changes as an individual matures. At the start, all of the genes in a zygote are switched on—in other words, they can all work. As the individual ages, however, the cells start to specialize by switching off various genes, so that their effect is no longer detectable (in the language of geneticists, they can no longer be *expressed*). For example, every cell in your body contains the genes for manufacturing insulin within its DNA, but insulin is actually manufactured only in parts of the pancreas. In all other cells in your body (in your skin, for example, or the nerve cells in your brain) the gene for insulin is switched off.

Obviously, the DNA that is implanted into the fertilized egg has various "switches" turned off, the exact sequence being determined by which part of the adult body the cell is taken from. By a process we do not understand very well, the fertilized egg is able to reset all the switches on the DNA—to turn the "off" switches back to "on"—so that normal development can take place. This was at the heart of Wilmut's great discovery.

Not every attempt at cloning is successful. With Dolly, 273 eggs went through the procedure of DNA replacement, and only one of them resulted in a live adult. Since Dolly, many other mammals have been cloned—cows, mice, and pigs to name a few. With mice there have been generations of cloned animals—clones, clones of clones, clones of clones of clones, and so on.

The biggest issue in the debate on cloning is the potential application of aspects of this technology to human beings. On the one hand are those who feel that the moral dangers implicit in the new technology are so horrific that its application to humans should be banned. On the other hand, the technique offers a chance for many infertile couples to produce biologically related children, an outcome that many would regard as a moral good.

As the debate continues, there is one point about clones that has to be made. Technically speaking, a clone such as Dolly is nothing more than an individual whose DNA is identical to that of another individual. We have a great deal of experience in dealing with individuals with identical DNA—we call them twins. A clone is simply a twin born years or decades late—what has been called an "asynchronous twin." Just as we would never dream of expecting one twin to give up its heart for implantation into the other, the prospect of clones being raised for the harvesting of organs is a nightmare that will never come true. It has been my experience that if you substitute the word "twin" for "clone," much of the high-pitched debate on human cloning fades away.

For what it's worth, my own sense is that by around the year 2010, cloning will be regarded as no more morally reprehensible that in vitro fertilization or other modern fertility techniques. Because the cloning procedure is quite simple and uses standard techniques, I believe that cloned human beings will make their appearance soon (if they have not already done so by the time you read this).

IAN WILMUT (1944–) Scottish embryologist. He was born in Hampton Lucey, England. He graduated from the University of Nottingham in 1971, and obtained his Ph.D. from Cambridge in 1974 for research on methods of freezing boar semen. That year he joined the Roslin Institute near Edinburgh, where he continues to study the genetic engineering of livestock. He identified developmental and physiological causes of prenatal death in sheep and pigs, then began to investigate methods of improving livestock. Wilmut has said of the controversy over cloned animals, "I don't have any sleepless nights. I believe we are a moral species."

Ian Wilmut with Dolly— the first mammal to be cloned, in 1996. Dolly was created from cells taken from a Scottish Blackface ewe and a Finn Dorset ewe.

Coevolution

The evolution of one organism can depend on the evolution of another

According to the theory of EVOLUTION, organisms evolve over time in response to the pressures of their environment. Ordinarily, this means that climate, food supply, and the availability of water, for example, are factors in natural selection. However, the "environment," as the term is broadly understood, can include other organisms. Changes in one organism can then produce changes in another, which, in turn, produces changes in the first organism, and so on. This kind of organismal waltz through time is called coevolution.

For example, a plant may evolve a tougher coating on its leaves to foil insect predation. One of the insects that feed on the plant, in turn, may evolve specialized mouth parts to deal with the plant's new protection. The plant may then develop even more defenses (spines, for example) to keep the insect away, and the insect may develop counters to the new defense. The plant and insect are not responding to changes in their physical environment, but to the appearance of mutations in, respectively, their attacker and their food source.

Coevolution can lead to some truly spectacular results—results that have often puzzled evolutionary biologists. There are insects, for example, whose mouth parts are so specialized that they can take nectar from the flower of only one kind of plant. The plant, in turn, can be fertilized (i.e., have its pollen carried from plant to plant) only by that particular insect. In this case, the two have coevolved into something that, while not quite a single organism, is not two independent organisms either.

These sorts of systems are sometimes erroneously touted as evidence against the theory of EVOLUTION—as evidence for specialized creation of species. But as we can see, these seemingly miraculous pairings are straightforwardly explained in terms of natural selection.

This species of mantis mimics the plant on which it lives. This both protects it from predators and makes it easier for the insect to find its prey.

Cohesion–Tension Theory

The transport of water in plants depends crucially on hydrogen bonds that form between water molecules

Plants have an enormous need for water. Unlike animals, which retain much of the water they ingest, fully 90% of the water that enters a plant through its roots will be evaporated from its leaves in a process known as *transpiration*. This happens because pores in the leaves must remain open to allow carbon dioxide from the atmosphere to enter and take part in PHOTOSYNTHESIS. This prodigious need for water has led British ecologist John Harper to describe a plant as "a wick connecting the water reservoir of the soil with the atmosphere." Water from the soil enters plant roots by osmosis, and the question of how that water got to the top of tall trees was a puzzle for a while. The cohesion–tension theory is designed to answer that question.

It turns out that the properties of water molecules play an important role in the transport process. Each hydrogen molecule in water forms a hydrogen bond (*see* CHEMICAL BONDS) with oxygen atoms in neighboring molecules, creating a strong cohesive force that holds the water together. (This is the same force that gives rise to SURFACE TENSION.) Water leaks out of cells into openings in the leaf one molecule at a time, and as one molecule leaves a cell, another molecule moves in. Since all the molecules in the water are linked together by the hydrogen bonding, this movement creates a pull, or tension, in molecules throughout the plant's "plumbing," from the top to the bottom. In fact, you can imagine all of the water molecules as being linked together, from the one about to enter the leaf, down through the plant to the water in the surrounding soil— water that will eventually be sucked into the roots.

It is important to realize that this process is not to the same as what happens when you take a drink through a straw. There you are creating a partial vacuum in the straw and allowing air pressure to push the fluid upward. Such a process can lift water only about 32 feet (10 m) and many trees are taller than this. The cohesion–tension theory explains how to get beyond simple suction to move water to great heights.

Competitive Exclusion, principle of

If two competing species coexist in a stable environment, they do so as a result of niche differentiation. If there is no such differentiation, one species will be driven to extinction

The most important concept in ecology is that of the ecological niche. A *niche* describes the limits of environmental parameters within which a given species can grow and reproduce. For example, a niche for a plant might encompass the amount of rainfall, sunshine, and soil minerals needed for the plant to grow. The niche for an animal predator might include climatic factors, the availability of suitable prey, and the number of diseases endemic in an area.

The competitive exclusion principle says that if two species compete for the same niche, there are only two possible outcomes. Either the two species change slightly so that they each occupy a slightly different niche (*niche differentiation*), or one species is driven to extinction. Two plants in the same meadow, for example, may evolve so that one can get by on a little less sunlight, or in places where there is a little less phosphorus in the soil. Even though it may appear that there are still two species co-existing in the same niche (in violation of the principle), the niche has become sufficiently differentiated for it to be regarded as two niches.

There are many examples in ecology of species co-existing, but it is usually possible to find evidence for niche differentiation. Where such differentiation hasn't been established, it is not possible to know whether this is because of a violation of the principle or because the scientists have not looked at the right variables—it may be that they should be looking at the availability of potassium rather than phosphorus, for example. Because of the preponderance of evidence in favor of competition in other situations, and because there are theoretical reasons to suppose that competition does take place in nature, ecologists tend to accept the principle of competitive exclusion.

If different species of plant are to occupy the same territory, they must avoid competition by relying on different resources.

Complement-arity Principle

Quantum objects can be viewed as either particles or waves, and these two views complement each other

In our everyday world, there are two ways of moving energy from one point to another—by particle or by wave. If, for example, you want to knock a domino that's been balanced on end on a tabletop, you have two choices of how to deliver the energy. You can throw another domino at it (i.e., use a particle). Or you can set up a line of dominoes leading to it, and knock over the first one so that a wave moves down the line as each domino falls. In our everyday world there is no ambiguity about these two ways of moving energy—a basketball is a particle, sound is a wave, and that's it.

But in the world of QUANTUM MECHANICS things aren't so clear cut. In fact, it quickly became apparent from experiments that everything in the quantum world is different from what we're used to in our everyday, macroscopic world. Light, which we normally think of as a wave, can sometimes act as though it consists of a stream of particles (we call the particles *photons*), and particles such as electrons and protons can sometimes act as though they're waves.

Here's a simple experiment to show what I mean: Imagine a solid sheet of material with two horizontal slits cut in it. If you directed a beam of particles at the sheet you would expect to see those particles coming through only directly behind the slits. In particular, you would expect to see very few particles winding up on the centerline of a screen placed beyond the two slits.

If, though, you directed a wave toward the sheet, you would see something completely different. According to HUYGENS' PRINCIPLE, each slit would act as a source of secondary waves, and we would expect to see a maximum at the centerline of the screen. A sound wave, for example, would be loudest on the centerline, and a light wave would be brightest. The intensity of the wave would then vary between high and low as we move away from the centerline. Thus, a wave will appear strongest precisely where the particle number is lowest.

If a beam of electrons is directed at such an apparatus, the alternate high- and low-intensity bands characteristic of a wave will be seen on the screen. In this sense, the electron seems to behave as though it were a wave. On the other hand, if you shoot one electron at a time, it will show up at a specific location on the other side of the sheet—it behaves as though it were a particle. The same is true if you use a beam of photons instead of electrons—both seem to behave as particles in one situation, and as waves in another. (*See* DAVISSON–GERMER EXPERIMENT.)

Here's the clincher: If you direct electrons or photons toward the apparatus one at a time, watch them land individually, and then plot the number that arrive at various distances from the centerline of the screen beyond the slits, you find the familiar alternating high- and low-density lines characteristic of a wave. Quantum objects, in other words, seem to remember that they are waves even when they are acting as particles (and vice versa). This strange property of the quantum world is called *wave–particle duality*. Experimenters have even tried to trick quantum objects into revealing their "true" identity by changing the experimental apparatus from one that will detect particles into one what will detect waves while the object is en route from its source. It never works. Quantum objects seem to have wave–particle duality built into them.

The principle of complementarity is simply a statement of this fact. It tells us that when we do an experiment on a quantum object that measures its particle-like nature, we see it acting as a particle. Conversely, when we do an experiment that measures wave-like properties, we see it acting as a wave. The two views don't conflict with each other—in fact, as the name of the principle implies, they are complementary.

As I explain in the Introduction, I believe that philosophers of science have made much more out of wave–particle duality than is warranted. It's true that quantum objects don't behave in the same way as the things we're used to in our everyday world, but where is it written in stone that they're supposed to? Just as medieval philosophers tied themselves in knots trying to decide if projectile motion is "natural" or "violent," modern philosophers do the same over "wave" and "particle." In fact, electrons and photons are neither waves nor particles, but an altogether different kind of entity of which we have no direct experience in our everyday lives. If we insist on trying to force them into inappropriate mental categories, we can expect to run into paradoxes. But the point is that it's the mental categories, not the quantum objects, that are the problem.

Complex Adaptive Systems

In complex adaptive systems, unexpected properties may appear. These properties are, at present, very difficult to predict

The study of complexity is one of the great frontiers of modern science. A *complex system* is defined to be a system which has many independent agents, each of which can interact with the others. For example, a pile of sand can be thought of as a complex system since pressing on one grain increases the forces on all the other grains in the pile, and these grains, in turn, respond by distorting slightly so that they can exert forces back. A stock market is another example of a complex system, since buyers and sellers change their behavior in response to the behavior of other buyers and sellers. A system such as a stock market, where the agents change in response to the actions of other agents, is what is called a complex adaptive system.

It was not possible to study complex systems before the availability of high-speed digital computers. The systems are just too … well, too complex to be handled by ordinary mathematics. The most important result from computer studies of complex adaptive systems is the notion of an *emergent property*. Take that simple sand pile as an example. If you add sand grains to the pile, you eventually reach a point where a new type of behavior emerges. When you add the millionth grain (or however many it takes), you get an avalanche—a behavior that is completely different from the shifting of forces that has gone on before. With the millionth grain, in other words, we reach a point where "more" becomes "different."

An important feature of emergent properties is that they don't appear gradually. In other words, you don't get a millionth of an avalanche from one grain of sand, which can then be added together to make an avalanche from a million-grain pile. You get no avalanches at all until you get to that millionth grain, and then all of a sudden you get the avalanche. I think a very good case can be made that mental properties like consciousness and self-awareness are emergent properties of collections of neurons, somewhat analogous to avalanches being an emergent property of collections of sand grains. If this is so, it tells us that there comes a point in the evolution of nervous systems where "more" becomes "different" (*see* NERVE SIGNALS).

One of the great challenges facing science today is to develop the ability to predict the appearance of emergent properties from the properties of individual agents in a system. We cannot do this at present, and some scientists have suggested that we will never be able to. I believe that it is too early to give up on this problem just yet.

Compound Motion

The motion of an object in one dimension is independent of its motion in other dimensions

When I think about which one event I would pick to mark the birth of modern science, I often come to a date in 1537. The Duke of Milan had just acquired some cannon, the latest in military hardware, and had some questions about how to use his new toys. He summoned his chief engineer, a mathematician named Niccolò Tartaglia (*c.* 1500–57), and asked him a simple question: At what angle should we set the barrel to insure maximum range for the cannonball?

What happened next typified the new spirit of inquiry that was sweeping through Europe. Tartaglia didn't consult books written by ancient philosophers, nor did he retire to his study to ponder the question. Instead, he took some cannon to a field outside of Milan and started firing them. His result—that a barrel elevated at an angle of 45° gave the maximum range—would later be derived from NEWTON'S LAWS OF MOTION, but for Tartaglia it was simply a rule-of-thumb, seat-of-the-pants piece of engineering.

In fact, the problem of the trajectory of Tartaglia's cannonballs is an example of something that had puzzled philosophers for centuries. It was referred to as the "problem of projectile motion"—what happened to an object that was thrown or otherwise propelled through the air. Aristotle and other philosophers of ancient Greece had taught that there were two kinds of motion in the world, which they named "natural"—the motion an object would have if left to itself—and "violent"—motion imposed on an object by an outside agency. When he considered the trajectory of a spear (this problem seems to have had a long association with military themes), Aristotle argued that for the first part of its path the spear experienced violent motion because it had been thrown. Then, at some point along the trajectory, the violent motion gave way to natural motion. Since for Aristotle the natural tendency of every object was to move toward the center of the Earth—which for him was synonymous with the center of the universe—once natural motion took over the object would fall straight down.

For Aristotle, then, the trajectory would be triangular, slanting up during violent motion and coming straight down during natural. (Remember that, though this may sound absurd to us, going out and observing the flight of a projectile to determine its shape was in total contradiction to the modus operandi of natural philosophy in the world of ancient Greece.) In fact, a lot of the thought that went into this problem consisted of interminable arguments about when motion changed from natural to violent, whether the two could mix, and so on. Such an analysis, while it creates neat mental categories, offers little in the way of practical guidance for someone in Tartaglia's position. No wonder he went off to the fields outside Milan!

As with so much else about our knowledge of the motion of material objects, we owe the solution to the problem of projectile motion to Galileo. He had discovered the equations that describe ACCELERATED MOTION and, in particular, the equation for the motion of an object falling freely at the Earth's surface. His experiments had shown him that the distance d traveled by such an object in a time t is given by

$$d = \tfrac{1}{2}gt^2$$

where g, the acceleration due to gravity, is measured to be 32 feet per second (or 9.8 meters per second) each second.

A freely falling body, then, will travel 16 feet (4.9 meters) in one second, 64 feet (19.6 meters) in two seconds, 144 feet (44.1 meters) in three seconds, and so on. This fact explains the effectiveness of a traditional method for estimating the height of a cliff—you throw a rock over and time its fall, then use the equation to calculate the distance traveled by the rock, which is equal to the height of the cliff. It also explains an old student joke about a professor who gave his students a barometer and told them to measure the height of the physics building. Most students, as the professor expected, used the barometer to measure the difference in atmospheric pressure on the ground and on the roof, but one student took the barometer to the roof and threw it off. Different versions of the joke differ on the professor's reaction to this expensively unconventional (though perfectly legitimate) approach to the assignment.

Once Galileo had the problem of free fall solved, he needed one more piece of the puzzle to solve the problem of projectile motion. That piece was the idea of compound motion. In essence, he realized that the motion of a projectile can be split into two parts. In the vertical direction, the cannonball goes up and comes down, just like a ball thrown straight up in the air. In the horizontal direction, on the other hand, no forces are acting on the cannonball, so it will continue moving at whatever velocity it had at the beginning (neglecting air resistance, of course). Put simply, the horizontal motion doesn't affect the vertical motion, and the vertical motion doesn't affect the horizontal. The complicated problem of projectile motion, then, comes down to two simple (but separate) problems, each of which is easily solved.

Let's take a simple example to illustrate how it works. For simplicity, imagine shooting a cannonball out horizontally from the edge of a cliff. After 1 second, the cannonball will have fallen 16 feet, after two seconds it will have fallen a total of 64 feet, and so on. While this fall is going on, the cannonball is moving horizontally at whatever the initial muzzle velocity was. If, for example, it was 1000 feet per second, then after one second the cannonball will be 1000 feet out and 16 feet down.

The same basic technique can be used in the slightly more complicated case of firing a cannon from level ground with the barrel at an elevated angle. The vertical motion of the cannonball is the same as if it had been propelled upward. It will continue upward, slowing down because of the tug of gravity until it reaches the top of its arc, where its vertical motion is momentarily zero. It then begins to fall back to the ground, just as if it had been dropped from its maximum height. Using this analysis, Galileo was able to show that the curve traced out by a cannonball is a geometrical shape called a *parabola*—a far cry from Aristotle's triangle.

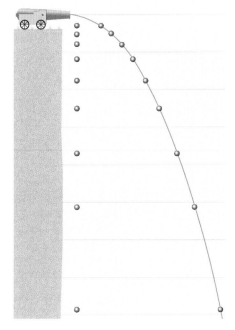

The trajectory of a cannonball results from two kinds of motion: a constant horizontal motion imparted by the cannon, and the downward accelerated motion caused by the Earth's gravitational pull.

One of the consequences of compound motion is somewhat counter-intuitive, and deserves special mention. If we dropped a cannonball off a cliff at the same time as we fired the cannon in our example, the principle of compound motion tells us that the vertical motion of both cannonballs will be the same. This means that they will both hit the ground at the same time, regardless of the fact that one travels a much longer distance than the other. (It may help you to note that the cannonball fired horizontally may travel farther, but it is also traveling faster.)

Galileo may have solved the problem of projectile motion, but that did nothing to resolve the ancient issue of natural *v.* violent motion. Even though we can describe precisely what the projectile does, we still don't have a clue about when the motion is natural and when it's not. This is because "natural" and "violent" are simply false categories —categories that appeal to human reason but have nothing to do with the real world. The medieval debate just illustrates how you end up down a cul-de-sac if you formulate problems in the wrong way.

Compton Effect

Photons scattering from electrons lose energy, and that energy loss depends on the angle through which the photon is scattered

The early decades of the 20th century were marked by a growing understanding that objects in the subatomic world have properties of both particles and waves (*see* COMPLEMENTARITY PRINCIPLE). Albert Einstein's explanation of the PHOTOELECTRIC EFFECT, which introduced the notion that all electromagnetic radiation, light included, comes in bundles called photons, was the beginning of this process. The discovery of the Compton effect by the American physicist Arthur Compton was another confirmation of the quantum nature of the photon.

The experiment that Compton did is simple to describe. A beam of electromagnetic radiation (Compton used X-rays) was directed at a crystal, and the energy and angle of the scattered beam was measured. In the classical picture of the interaction of radiation and matter (i.e., as it was understood before the application of the ideas of QUANTUM MECHANICS), the outgoing radiation has to have the same energy as the radiation coming in. What Compton found was different—the energy of the scattered wave was different from that of the incoming wave, and that energy difference seemed to depend on the angle through which the radiation was scattered, reaching a maximum at 90°. The only way to understand this result was to regard the interaction of the radiation with the atom as a collision between an incoming particle (a photon) and an electron. Like two billiard balls colliding, these two particles interact and bounce off each other. Because the electron is moving slowly it will, in general, pick up energy in the collisions—energy which the photon then loses.

The discovery of the Compton effect in 1923 changed the minds of most physicists who had still doubted the reality of the photon. The Compton effect is now used to help detect some astronomical gamma-ray sources by reducing the energy of the gamma-ray photon to the point where it may be detected by a conventional X-ray detector. Such an instrument was flown on the Compton Gamma Ray Observatory, launched by NASA in 1991.

ARTHUR HOLLY COMPTON (1892–1962) American physicist. He was the son of a philosophy professor at his birthplace of Wooster, Ohio. He received his doctorate in physics from Princeton in 1916 and spent his early career in private industry —he helped to develop the fluorescent light, for example. Returning to academic research, he spent most of his career at the University of Chicago, where he was made professor of physics in 1923. For his discovery and explanation of the Compton effect, he shared the 1927 Nobel Prize for Physics. Compton later became an important figure in American political life, and was involved in building the first controlled nuclear reactor as part of the wartime Manhattan Project.

Copernican Principle

The Earth does not occupy a special position in the cosmos

Whenever scientists don't have good data to go on, they try to make reasonable assumptions about how the world operates. If the assumption is made often enough, and sounds reasonable enough, it may get elevated to the rank of a "principle" even though there is no observational evidence to support it. This is never truer than when scientists discuss the existence of extraterrestrial life. We know only one solar system in detail—our own (though we know of the existence of others)—and we know of only one kind of life—the carbon-based life on our own planet. It is very difficult, to put it mildly, to draw conclusions from a single example, and that's where the Copernican principle comes in.

The principle, which can be put in a number of ways, says basically that there is nothing special about the conditions on our own planet, no reason why things here should be any different than they are anywhere else in the universe. (It is sometimes called the *principle of mediocrity*.) It rests on the fact that, as far as we can tell, the laws of nature are the same everywhere in the universe, and on the statistical argument that it is extremely unlikely that the Earth is in any way unique, given all the other stars and planets that exist.

A good deal of scientists' willingness to accept the principle has to do with history. Nicholas Copernicus, after all, showed that the Earth was not at the center of the universe, despite fiercely held philosophical and religious opinions to the contrary. This was the first step in making the Earth less of a special place. Later research showed that the Sun wasn't at

The Planisphaerium Copernicanum, *from a famous star atlas by Andreas Cellarius. The planets—including the Earth with the Moon orbiting it—are shown in their paths around the Sun. This is the view of the solar system devised by Copernicus (shown seated, lower right).*

the center of the Galaxy and that humans have evolved from other life forms. Each of these findings moved humanity further from a special place in the universe, and provided another example of the Copernican principle in action. Small wonder that, after several centuries of discoveries that supported it, the principle enjoys almost universal acceptance. Its ultimate modern form is the *cosmological principle*—the assertion, backed by observation, that on sufficiently large scales the universe is much the same in all directions.

Some critics point out that even though the Earth may not be at the geometrical center of the universe, it could still be special. It is, for example, the only place where we know life to exist, notwithstanding expectations that life might be found elsewhere in our own solar system. Conditions like the presence of a large moon and the nearly circular orbit of the Earth have been suggested as contributing to that special nature, and it has been argued that, without this or that special condition, life would never have developed on our planet. The FERMI PARADOX is cited in support of this view.

For the moment, most scientists seem content to accept the Copernican principle for lack of any reason to do otherwise. Whether this turns out to be justifiable will, of course, be decided by future observations. In the meantime, whether to accept the principle seems to be more a matter of taste than of science.

NICHOLAS COPERNICUS (1473–1543) Polish astronomer and church administrator who proposed that the planets, including the Earth, all orbit the Sun. He was born in Toruń and began his education at the University of Kraków. Like many ambitious and well-connected young men of his time, Copernicus chose to make a career in the Catholic Church. Through the good offices of his uncle, who was a bishop, he was elected as a canon in the cathedral chapter of Frauenberg, a position that guaranteed his financial independence for life. Before taking his position, he went to Italy to study both canon law and medicine at Bologna, Padua, and Ferrara.

His position involved administering many of the business affairs of the cathedral, and he was very much a man of affairs in his country. He apparently ran what we would call a free clinic in Frauenberg, and served on a royal commission to reform the Polish currency.

His interest in astronomy began during his stay in Italy and was strengthened by his observation of the Moon passing in front of the bright star Aldebaran in 1497. In Frauenberg, he had a small tower built for making astronomical observations. His major contribution, contained in his book *De revolutionibus orbium coelestium* ("On the Revolutions of the Heavenly Spheres"), was a model of the universe in which the Sun was stationary at the center and the Earth, along with the other planets, moved around it.

It is important to understand that Copernicus did not develop our modern view of the solar system at a single stroke. His planets, for example, circled the Sun on crystal spheres rolling within spheres, just as the planets used to circle the Earth in the old Greek models. But by moving the Earth away from the center of the universe, he opened the way for modern astronomy to develop—achievement enough for one individual.

Cope's Law

*Over time, species evolve
larger body sizes*

*According to Cope's law, animals
tend to become larger over time.
Cope reached this conclusion
from his studies of dinosaurs,
such as this iguanodon.*

Edward Cope was one of the last of his breed—an independently wealthy man who pursued science in his own way. He was instrumental in opening up the fossil fields of the American West, and is probably best known for his part in the "Dinosaur Wars"—a sometimes vicious competition with Othniel Marsh of Yale University. Both men uncovered many dinosaur fossils, and engaged in a competition to see who could make the most spectacular finds. Perhaps the most famous result of the war was the misidentification of the first brontosaurus fossil—a misidentification that later led to the renaming of brontosaurus as apatosaurus, the name originally proposed by Cope. (The mistake was formally announced in 1903, but it took a long time to percolate through to the public consciousness, and by then the name "brontosaurus" was known to pretty well everyone. Unless you're writing a paleontology paper, feel free to carry on using it!)

In any case, Cope's law, based on his study of dinosaurs, says that as evolutionary time goes by, a species will tend to develop a larger body size. The truth of this statement was accepted for almost a century. Recently, however, University of Chicago paleontologist Michael Foote (1963–) has called the law into question. Foote is one of a new breed of paleontologists, thoroughly at home with modern computer methods. Working with a massive database assembled from the fossil record of marine organisms called echinoderms, he has been able to show that over time scales of tens of millions of years there is no consistent trend in size. Some species in his database did, indeed, get larger, as required by Cope's law. Others, however, got smaller, while still others stayed the same. In other words, when you look at a lot of data, rather than just at dinosaurs, Cope's law doesn't seem to hold up.

Cope's law applied only to long-term changes in species stature, changes associated with significant change in their DNA. It was not meant to be applied to things like the increase in height of European people since the Middle Ages, which scientists believe has had to do with factors such as better nutrition and medical care. So when you look at a suit of jousting armor and wonder how the knights could have been so small, you are not seeing an example of Cope's law in action.

EDWARD DRINKER COPE (1840–97) American paleontologist. He was born in Philadelphia, Pennsylvania, and showed an interest in natural history from an early age. From 1864 he was professor of comparative zoology and botany at Haverford College, Pennsylvania, and eight years later he joined the U.S. Geological Survey. He later held chairs in geology and mineralogy (1889–95) and zoology and comparative anatomy at the University of Pennsylvania. Cope's career in paleontology was characterized by an intense rivalry with Othniel Charles Marsh (1831–99), professor of vertebrate paleontology at Yale. Both men are credited with the discovery of around a thousand extinct species.

Coriolis Effect

In a rotating frame of reference such as the surface of the Earth, the rotation causes objects to appear to an observer to move in a curved path. This is sometimes explained in terms of the existence of something called the Coriolis force

Imagine that someone standing at the north pole could throw a ball to someone else standing on the equator. While the ball was in flight, the rotation of the Earth would carry the catcher eastward, and if the thrower hadn't compensated for this motion in aiming the throw, the ball would fall to the catcher's west (left). From an observation point on the equator, in fact, it would appear that the ball started moving to the left as soon as it left the thrower's hand, and continued to do so until it landed.

NEWTON'S LAWS OF MOTION tell us that an object will deviate from motion in a straight line only if a force acts on it. The catcher on the equator would have to conclude, then, that a force was acting on the ball. If we could look at what is happening from a vantage point outside the Earth, however, there would not appear to be any force acting. The deviation is caused by the fact that the Earth rotates under the ball while the ball is moving in a straight line. Whether or not a force is perceived to act, in other words, seems to depend on where the observer is situated: what scientists call the observer's *frame of reference.*

This kind of thing arises whenever there is a rotating frame of reference, as provided by, for example, the Earth. Physicists often use the phrase *fictitious force* to describe the phenomenon, because the force isn't "really" there, but simply appears to observers in the rotating frame to be acting (another example is the CENTRIFUGAL FORCE). This causes no contradictions, because both observers will agree on the actual path taken by the ball and on the equations that describe that path. They simply disagree on the terms they use to describe the motion.

The fictitious force that operates in the example here is called the *Coriolis force,* named for Gaspard Coriolis. It is the Coriolis force that produces the swirling cloud patterns we associate with satellite pictures of storms. Air starts to flow in a straight line from areas of high pressure to areas of low pressure, but the Coriolis force deflects it and causes it to move in a spiral path. (Alternately, we could say that the Earth rotates underneath it so that it appears to move in a spiral to someone on the planet's surface.) A moment's reflection on the example of throwing a ball from pole to equator should show you that the direction of the Coriolis force is the opposite in the southern hemisphere to what it is in the northern. Storms in the southern hemisphere thus appear to rotate clockwise, while those in the northern hemisphere seem to rotate counterclockwise.

And this gives rise to one of the enduring bits of folklore associated with the Coriolis effect—the idea that water drains from a sink in opposite directions in the two hemispheres. (I can remember being one of a group of graduate students—including one Argentinian—who spent a fair amount of time in the Stanford Physics Department's men's room trying to verify this.) Actually, while it is true that the Coriolis force acts in opposite directions in the two hemispheres, the direction in which water runs out through a drain is only marginally affected by the effect. When water is poured into a container, currents flow for surprisingly long periods of time. Even though you can't normally see them with the naked eye, these currents keep the water swirling for hours after it has entered a bowl. Furthermore, the way the water is emptied from the container can set up similar currents. Either of these factors is more than enough to mask the effect of the Coriolis force. The direction of flow in northern- and southern-hemisphere sinks probably has more to do with the design of the plumbing than with the laws of nature.

How have we learned these fascinating facts? A group of physicists who apparently had nothing better to do actually performed experiments with a precisely machined basin equipped with a drain that could be moved straight down to release the water. They found that if they waited long enough and removed the plug carefully enough, they could see the Coriolis effect in operation. They even managed a few flushes in which the water started swirling one way and then, as the Coriolis force took over, reversed direction!

GASPARD GUSTAVE DE CORIOLIS (1792–1843) French physicist and engineer. He was born in Paris, where he was a student and later director of the prestigious École Polytechnique. (His directorship is remembered for his introduction of water coolers into classrooms there—coolers are still called "Corios" by the students at the École.) His main interest was in calculating the forces on moving machine parts, bringing theory to bear on the practical concerns of engineering. In the process, he came up with the modern definitions of work and kinetic energy.

Correspondence Principle

In the limit of large quantum numbers, quantum mechanics gives the same results as classical physics

The quantum world, dominated by HEISENBERG'S UNCERTAINTY PRINCIPLE and SCHRÖDINGER'S EQUATION, is very different from the world we're used to, the world dominated by NEWTON'S LAWS OF MOTION. Yet our macroscopic world is made from atoms, so there has to be a connection between the two. The correspondence principle was first enunciated by the Danish physicist Niels Bohr to address this problem. We can understand the principle by thinking about the BOHR ATOM.

In the Bohr atom, electrons can occupy only certain allowed orbitals. The orbitals are delineated by what is called the *principal quantum number*. The innermost orbital, the one closest to the nucleus, has a principal quantum number of 1, the next one out a principal quantum number of 2, and so on. Very large principal quantum numbers correspond to orbits far from the nucleus. In contrast, in a classical, Newtonian atom (i.e., a picture of the atom in which the effects of quantum mechanics are not taken into account) the electrons could be anywhere they want.

Now although the physical size of the orbitals increases steadily as the principal quantum number increases, the same is not true of the energies of the electrons in those orbitals. The increase in the energy gets smaller and smaller between successive principal quantum numbers, and in fact the value of the energy tends towards an upper limit known as the ionization energy. An electron with this energy would, theoretically, be in an orbital with an infinite radius, and so would be a free electron, leaving behind an ionized atom. Between this upper limit and the ground state, there is a region where the energies for the electrons allowed by quantum mechanics are so close together that they start to overlap. This arises because the value for each allowed energy (or size of the orbital) is not an exact number, but because of Heisenberg's uncertainty principle is slightly "blurred" (i.e., it has a small range of possible values). Thus the difference between the quantum mechanical atom, with allowed energies separated by finite amounts, and the classical atom, with electrons allowed to have any energy, simply vanishes. Far from the nucleus, in other words, the atom behaves as if it were a classical Newtonian system. This is an example of the correspondence principle in action.

The correspondence principle comes into play in the out-of-focus borderline where quantum mechanics merges into classical physics. It shows that there are no seams in nature, no sharp boundaries between theories. It also demonstrates what was said in the Introduction about the way that scientific theories develop. Quantum mechanics, for example, doesn't displace Newtonian mechanics, but incorporates it as a limit. Scientific theories grow by incorporating what is already known and adding to it, just as a tree adds new layers on the outside while preserving its heartwood.

Cosmic Triangle

Three independent measures of the amount of mass and energy in the universe together give a new picture of its structure in which most matter is invisible

Since the discovery of DARK MATTER, we have known that most of the stuff in the universe is not visible to us directly, and that the existence of dark matter must be inferred from its gravitational effects. One of these effects is the rate at which the universe is expanding (*see* BIG BANG, HUBBLE'S LAW), and that depends on its density. Until recently, the only way astronomers had of getting at the density of matter (both seen and unseen) in the universe was either to measure the expansion rate directly or simply to try to reckon how much material they could detect. The basic choice is between a *closed universe,* in which there is enough matter to reverse the expansion, and an *open universe,* in which the expansion goes on forever.

In the late 1990s, astronomers brought together three new methods of looking at the universe—three microscopes, if you like, that penetrate to the heart of the cosmos. Some of these methods are completely novel, while others are refinements of techniques that have been around for a while. Each provides its own picture of the universe, its own set of constraints on what the universe can be like. The amazing thing is that all three microscopes focus on the same picture—a picture of a universe quite unlike anything that had been imagined.

The first microscope measures X-ray emission from gases in the center of clusters of galaxies. The idea is that the more mass there is in a cluster, the more gravity there will be pulling in, and the hotter these gases will get. Since almost all of the matter of the universe (both ordinary matter and dark matter) is located in clusters of galaxies, this technique gives a way of measuring the total matter density of the universe. The results point to a universe with only about 20–30% of the mass needed to keep it closed. They point, in other words, to an open universe.

Detailed measurements of the cosmic microwave background (*see* BIG BANG) constitute the second microscope. The extreme uniformity of the background radiation points to a universe that is geometrically "flat," even though the X-ray data indicate that there is not enough mass to "flatten" it. The first two microscopes, in other words, seem to be giving contradictory results (if we assume that ordinary and dark matter is all there is in the universe), which is an indication either that we're not seeing the whole picture or that our basic assumptions and theories are wrong.

The third microscope is the most recent innovation. With it, we can make a direct measurement of the expansion rate of the universe. The idea is this: When we look at light from a galaxy 10 billion light years away, we are seeing the universe as it was 10 billion years ago. In particular, we are seeing how fast the universe was expanding 10 billion years ago. By comparing that to the current rate, we can tell how much the expansion has been slowed by the inward tug of gravity, and hence how much matter there is in the universe.

The difficulty with measuring what astronomers call the *deceleration parameter* is that it's very hard to know how far away a faint, distant galaxy really is. This is where the new measurement techniques come in. There are various types of supernova (the cataclysmic disruption in which a massive star ends its life). Those known as *Type Ia supernovae* occur in binary star systems in which one of the partners is a white dwarf that is pulling material from its companion onto itself. When the pirated material forms a blanket a few feet deep on the dwarf, the intense pressure and temperature trigger a nuclear reaction and the blanket flares off like a giant H-bomb. Because all Type Ia supernovae are the same kind of event, they all have about the same brightness, and that's the key to the third microscope. By finding a Type Ia supernova in a distant galaxy, figuring out how bright it really is, and then comparing that to the light that actually reaches Earth, astronomers can calculate how far away the supernova and its host galaxy are.

The results are startling: The expansion isn't slowing down at all, it's speeding up! This is impossible to understand if the only components of the universe are ordinary and dark matter, which would always be pulling in, slowing the expansion down. The only way to explain the result is by the presence of a new piece of the universe—a piece whose existence was hitherto well beyond the pale of mainstream scientific thought. It has to be a kind of antigravity, something that pushes matter apart (unlike gravity, which always pulls things together). Astronomers have taken to using the term *dark energy* to describe the origin of the antigravity. I like this term, because it emphasizes that we don't yet know what causes the effect.

The picture of the universe that the three microscopes give us, then, is one in which there is not enough ordinary matter to make the universe flat, but in which dark energy takes up the slack and keeps the expansion accelerating forever, even while the geometry of the universe is essentially flat. Seen this way, all three microscopes point to the same universe—as can be represented in a diagram called the cosmic triangle. The idea is that the hatched regions are the ones theoretically "allowed" by

A schematic version of the cosmic triangle, which uses three strands of evidence to home in on the nature of the universe.

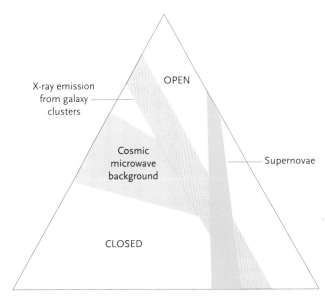

each of the three microscopes, and the real universe is located where the three regions intersect. The diagram shows that roughly one-third of the energy of the universe is in the form of matter (both ordinary and dark), and two-thirds is in the form of dark energy. Once again, we see that the solid stuff of our everyday lives doesn't amount to much in the universe at large.

One explanation of where the dark energy comes from (the one that has been accepted by most astronomers as a working hypothesis) is that it is an effect of the COSMOLOGICAL CONSTANT, a concept first introduced by Albert Einstein early in the 20th century. But there is now a rival to the cosmological constant as the explanation of dark energy. Called *quintessence,* it is a hypothetical new force whose properties remain to be worked out. Its name means "fifth essence," a reference to the four essences, or elements, of ancient Greek philosophy—earth, fire, air, and water. Whether either of these explanations will eventually prove to match the data will, of course, be decided by future measurements.

Cosmological Constant

The cosmological constant, if it exists, would explain the observed increase in the expansion rate of the universe

When Albert Einstein first conceived of the theory of general RELATIVITY, it was believed that the universe is static—that galaxies pretty much stayed in the same place. He noticed that because of the universal attraction of gravity, the universe described by his theory would actually contract— something he felt was impossible. To balance things out he introduced an extra term into his equations, a term that, in effect, provided a kind of antigravity, pushing galaxies outward just enough to counteract the inward tug of gravity. He called this extra term the cosmological constant and denoted it by the Greek letter Λ.

Even though the introduction of this term seemed to violate the BEAUTY CRITERION, it seemed to be required to keep everything in order. However, when the expansion of the universe (*see* HUBBLE'S LAW) was discovered, Einstein realized that it was no longer necessary to keep Λ in the equations, because in the Hubble picture the inward pull of gravity could act to slow down the expansion without the need for any counter-vailing force. Einstein promptly dropped the term from his equations. Later, he would refer to its inclusion as the biggest blunder he ever made.

For the better part of a century, the cosmological constant lived a kind of furtive existence in theoretical physics. A handful of intrepid theorists suggested that it could solve this or that puzzle, only to be ignored or ridiculed by their colleagues. Then, in the late 1990s, a rather surprising development brought Λ back to center stage.

The BIG BANG picture of the universe always implied a question about how the whole show will wind up. The outrushing galaxies will either be reined back by the force of gravity, resulting in a universal collapse some-times called the *Big Crunch*; or gravity will slow them down but not by enough to turn them around, resulting in an infinite expansion into darkness. The choice between these alternatives seemed to be all there was. Like Calvinists wondering whether they were predestined for heaven or hell, all cosmologists could do was try to discover which of these two fates awaited the universe.

One way is to measure the recession rate of very distant galaxies— galaxies many billions of light years from the Earth. Because it has taken billions of years for their light to get here, what we are actually measuring is the rate at which these galaxies were receding billions of years ago. If we compare this rate to the current one, as measured by observing galaxies closer to home, we can tell how much gravity has slowed the expansion, and hopefully learn the fate of the universe.

Measuring the velocity of recession (*see* DOPPLER EFFECT) is quite simple—it's just a matter of measuring the redshift of light from the galaxy, the displacement of lines in the galaxy's spectrum. Finding the distance to the galaxy is the problem. To do this, astronomers need what they call a *standard candle*—some object for which they know its intrinsic luminosity, how much energy it is emitting into space. By comparing this amount of energy emitted with the amount we actually receive on Earth, and hence finding how much the original energy has been diluted, we can establish the object's distance.

In the 1990s, astronomers began to develop a standard candle called a Type Ia supernova (*see* COSMIC TRIANGLE). When this technique began to be applied, an astonishing result emerged. The expansion of the universe isn't going to reverse itself—in fact, it isn't even slowing down. Against all expectations, the rate of expansion is actually increasing! There appears to be a force—some kind of antigravity—strong enough to overcome the force of gravity and push the galaxies ever farther apart. Once this took hold in the minds of cosmologists, there was a massive rehabilitation of Λ. Suddenly, theoreticians were falling all over themselves trying to incorporate Einstein's "biggest blunder" into their theories. Whether the return of the cosmological constant is permanent depends, of course, on the success of the new theories.

ALBERT EINSTEIN (1879–1955) German-Swiss-American theoretical physicist. He was born in Ulm, Germany. His father ran a small electrochemical factory in Munich, and it was there that Einstein began his formal education. After business difficulties forced the family to move to Italy, he completed his formal education in Zurich. (Here I should like to correct a common misconception about Einstein's youth. He was not a poor student, nor did he ever flunk mathematics. He did, however, have problems with the rigid German education system, as did many independent-minded students.)

In 1901, Einstein obtained a post at the Swiss Patent Office in Berne, and in that year he became a Swiss citizen, having developed a profound dislike for the rigidity and growing militarism of German institutions. In his seven years there he made most of his major contributions to science, including the explanation of the photoelectric effect, the explanations of Brownian motion, and the development of the theory of special relativity. In 1909, as his work was becoming known, he received the first of a series of academic appointments at Zurich, Prague, and, eventually, Berlin, where he was head of the Kaiser Wilhelm Institute for Physics. His marriage to his former fellow student Mileva Marić ended in divorce in 1919; he then married his cousin Elsa.

That year, Einstein was catapulted into international stardom when measurements of the bending of starlight by the Sun confirmed the predictions of general relativity. As a pacifist and a Zionist, however, he found the climate in Germany to be increasingly intolerant. In 1933, with Adolf Hitler on the rise, he left his native land and joined the Institute for Advanced Study at Princeton (he became an American citizen in 1940). During World War II, he wrote a letter to Franklin Roosevelt to advise him of the possibility of the development of nuclear weapons, although it is not clear what role this letter played in the program that led to the atomic bomb. After the war, he devoted much of his time to working for world peace.

An outspoken critic of quantum mechanics, Einstein's skepticism played an important role in forcing people such as Neils Bohr (a lifelong friend) to sharpen their ideas. He spent much time in his later years attempting, unsuccessfully, to unify gravity with the other forces of nature (*see* THEORIES OF EVERYTHING).

Coulomb's Law

The force between two electrical charges is proportional to the magnitude of the charges and inversely proportional to the square of the distance between them

The phenomenon of electrical attraction was well known to Greek scientists more than two millennia ago. They knew that if you rubbed a piece of amber with cat's fur or a piece of glass with silk, you could make objects exert forces on one another. If a small piece of cork were touched with the piece of amber, for example, the cork would skitter away from other pieces of cork that had been similarly touched, but would be attracted toward pieces that had been touched with the glass. This tendency of objects to attract or repel one another is a manifestation of what we now call *static electricity*. A modern example of attractive forces generated by static electricity is when items of clothing stick together when they are removed from a dryer, and the tendency of strands of hair to repel one another contributes to our "bad hair days."

The modern understanding of the phenomenon of electricity starts from the recognition that the two kinds of behavior (attraction and repulsion) observed by the Greeks arise because there are two kinds of electrical charge, positive and negative. In the atom, these charges reside in different places. The positive charges are located in the nucleus of the atom, while the negative charges are on the electrons that surround it. It was the American statesman and scientist Benjamin Franklin (1706–90) who first suggested that only one of the two charges actually moved in experiments like those described above. In modern language, he argued that removing electrons from a substance left it positively charged, because the electrons normally cancel the positive nuclear charge. Adding electrons, on the other hand, makes the object negatively charged.

Even though people had known about electricity for millennia, it wasn't until the 18th century that it was studied scientifically. (The people who carried out these early studies referred to themselves as "electricians.") Charles Augustin de Coulomb was one of the leaders of these studies. The law that now bears his name came out of his detailed studies of forces between electrically charged objects. In a typical experiment, he would suspend two small, electrically charged objects from fine threads, then bring them near each other. The effect of the electrical force would be to pull the objects toward each other (if they had opposite charges, one positive and one negative) or push them away from each other (if they had the same charge, both positive or both negative). The objects would move until the downward force of gravity balanced the electrical force. By measuring the angle of the thread, Coulomb could calculate the force the two charges exerted on each other.

We can summarize Coulomb's results with a single equation:

$$F = kQq/D^2$$

where Q and q are the magnitudes of the two charges, D is the distance between them, and k is a constant determined by experiment.

There are a couple of interesting points that can be made about Coulomb's law. The first is that its form is very similar to NEWTON'S LAW OF GRAVITATION. There is a reason for this similarity. In modern quantum theory, the electrical and gravitational forces are both generated by the exchange of massless particles—the *photon* in the case of the former, and a particle called the *graviton* in the case of the latter. Thus, despite their apparent differences, the two forces have a great deal in common.

The second point concerns the constant k. When the Scottish physicist James Clerk Maxwell (*see* MAXWELL'S EQUATIONS) wrote down the defining equations that governed electricity and magnetism, he found that this constant was related to the speed of light, normally denoted by the symbol c. Later, Albert Einstein showed that c played a fundamental role in his new theory of RELATIVITY. In this way, some of the most abstract yet potent ideas of modern science are descended from simple tabletop experiments done in the 18th century.

CHARLES AUGUSTIN DE COULOMB (1736–1806) French engineer and physicist. He was born in Angoulême into a prominent provincial family. Coulomb spent most of his life as a military engineer. After a term of service building forts and canals in France and in the Caribbean, he was posted to Paris, where his duties as a consultant left him time to begin a career in basic science. As well as his work on electrical charges and magnetism, he experimented with friction, and introduced the concept of the *thrust line*—still used in the design of buildings to calculate the off-vertical forces they must withstand from various parts of their structure, such as the roof. The modern unit of electrical charge, the coulomb, is named after him.

Curie Point

Ferromagnets can form only at temperatures below the Curie point

Most atoms have magnetic fields associated with them. In fact, atoms can be thought of as tiny magnets, each with a north pole and a south pole. These magnetic effects arise because moving electrons in an atom constitute a small electrical current, and electrical currents give rise to magnetic fields (*see* OERSTED DISCOVERY). When you add up the magnetic fields associated with each of the electrons in an atom, you get the total magnetic field of the atom.

In most substances the atomic magnets are arranged randomly, and their magnetic fields cancel one another out. But in a few substances, especially ones containing iron, nickel, and cobalt, the atoms arrange themselves so that the atomic magnets line up and reinforce one another. A piece of such a substance will be surrounded by a magnetic field. These substances can be made into what are called *permanent magnets* or, because they usually contain iron, *ferromagnets*.

We can understand how ferromagnets form by thinking about a piece of very hot iron. Because of the high temperature, the atoms are moving around vigorously, and there is no chance for the atomic magnets to align and reinforce one another. As the temperature is lowered, however, the random motion slows down until another force can take over. For iron (and a few other metals), there is a force at the atomic level that tends to make the magnetic dipoles of neighboring atoms line up with one another.

This atomic force is called an *exchange force,* and was first explained by Werner Heisenberg (*see* HEISENBERG'S UNCERTAINTY PRINCIPLE). Two electrons on neighboring atoms may exchange places. This links the two atoms and enables their magnetic fields to be parallel. The magnetic fields of the two atoms then reinforce one another instead of cancelling out. The same effect can occur over a region up to about one-twentieth of an inch (1 mm) across and containing up to 10^{16} atoms. Over this domain (*see* below) the atoms are aligned and there is a net magnetic field.

At high temperatures the effects of this force are masked by thermal motion, but at low temperatures the atomic magnetic fields can reinforce one another. The temperature where this transition takes place is called the Curie point of the metal, after the French physicist Pierre Curie.

Actually, the structure of a ferromagnet is rather more complicated than that. Typically, the atomic forces lead to the formation of regions called *domains,* some thousands of atoms across. Inside each domain the atomic magnets are lined up, but different domains will have their magnetic fields lined up differently. So normal iron isn't magnetic. Under certain conditions, however, the domains themselves will line up (as can happen if iron which has first been heated is then cooled in a strong magnetic field). That is what produces a ferromagnet, and it also explains why heating a magnet will usually destroy its magnetic properties.

MARIE SKLODOWSKA CURIE (1867–1934) Polish-French chemist. Marie Sklodowska was born in Warsaw to an intellectual family that had fallen on hard times in Russian-occupied Poland. As a schoolgirl she worked as a maid in her mother's boarding house, and later took on a series of governess posts with wealthy families to support her sister's medical education. In one of these positions, she and the youngest son of the family fell in love, only to have his parents break up the romance because of her social status. (Talk about missing a golden opportunity to upgrade your family's genetic capital!) When her sister finished her medical education in Paris, Marie left Poland and took up her studies in Paris.

Her outstanding performance in her *license* exams in physics and mathematics brought her to the attention of leading French scientists. She met Pierre Curie in 1894, and they were married the next year. This was the time when the entire field of what we now call radioactivity was opening up. Pierre and Marie Curie worked to isolate radioactive materials in ores taken from mines in Bohemia. This led to the discovery of several new elements, each of which was radioactive (*see* RADIOACTIVE DECAY). One of these, polonium, was named for Marie's native land. For this work, the Curies shared the 1903 Nobel Prize for Physics with Henri Becquerel (1852–1908), the discoverer of X-rays. Marie Curie was, in fact, the first scientist to use the term "radioactivity."

After Pierre's death in 1906, Marie Curie refused a pension and kept working. She confirmed that radioactive decay caused the transmutation of chemical elements, and founded the modern science of radiochemistry. For this work she was awarded the Nobel Prize for Chemistry in 1911—the first scientist to be honored with two prizes. (She was turned down for admission to the French Academy of Science in the same year— you'd think that after two Nobel prizes the boys would have gotten the message!)

During World War I, Marie Curie became heavily involved in the medical applications of her work, working on the front line with portable X-ray machines. In 1921, a public subscription among women in America raised enough money to buy her a gram of radium (about one-thirtieth of an ounce) for her research. As a part of a triumphal tour of the United States, she was presented with the key to the case containing the precious material by President Warren Harding.

Her later years were filled with work on important international initiatives in science and medicine. Her health declined in the early 1930s, the result of her long exposure to radioactive materials, and she died in a sanitarium in the French Alps.

PIERRE CURIE (1859–1906) French physicist. He was born in Paris into a prominent Alsatian family, and was educated at home. Initially enrolled as a pharmacy student in Paris, he quickly became involved in experimental studies of crystals with his brother Jacques, and eventually became director of experimental work at the École de Physique et Chimie. In 1895, the same year he married Marie Sklodowska, he published his doctoral thesis on the magnetic properties of materials (*see* CURIE'S LAW). With Marie, he carried out the work on radioactivity described above in difficult working conditions at the École. In 1904, he was appointed as a professor of physics at the Sorbonne, but was killed in a traffic accident before he could move to his new laboratory.

Curie's Law

The magnetization of a paramagnet is proportional to the magnetic field and inversely proportional to the temperature

Most atoms have magnetic fields associated with them (*see* CURIE POINT). In most substances the atomic magnets are arranged randomly, and their magnetic fields cancel one another out. In some substances, though, when an external magnetic field is applied to them the atomic magnets line up and reinforce the field. These materials, called *paramagnets,* do not normally display magnetic properties, but will do so in the presence of a magnetic field.

The French physicist Pierre Curie spent the early part of his career exploring the nature of magnetic materials and, indeed, gave us most of our current understanding of that field. He found that the extra magnetic field produced when atoms line up in a paramagnet is proportional to the applied magnetic field—that is, the stronger the external field, the more the atoms tend to line up. He also discovered that if the temperature of a paramagnet is raised, the extra magnetic field decreases. This happens because the increased thermal motion of the atoms starts to destroy the alignment as the temperature goes up. These results are summarized in Curie's law:

$$M = CB/T$$

where M is the extra magnetic field, or magnetization, of the material, B is the applied magnetic field, T is the temperature (in degrees kelvin), and C is a quantity known as Curie's constant. Curie's constant is always the same for a given material, but varies from one material to another.

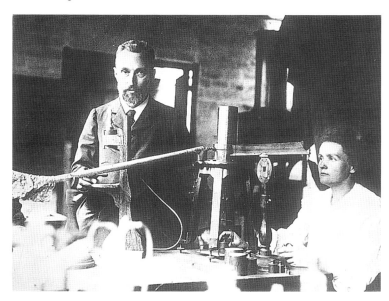

Marie and Pierre Curie in their laboratory, photographed about the year 1900. Conditions were primitive and dangerous. The room became so polluted with radioactive substances that everything in it glowed in the dark.

Dalton's Law

The pressure of a mixture of gases equals the sum of the partial pressures of each gas

John Dalton, the scientist behind the modern theory of atoms.

The air in the room where you're sitting contains several different gases, mainly nitrogen (around 80%) and oxygen (around 20%). The *partial pressure* of each of these gases is the pressure it would have if it alone filled the room. For example, if everything but the nitrogen were removed from the room, the pressure of what was left would be the partial pressure of nitrogen. Dalton's law says simply that the total pressure exerted by all the gases combined is the sum of the partial pressure of each gas taken separately. (Strictly speaking, the law applies only to IDEAL GASES, but it provides a close enough approximation to what happens with real gases.)

John Dalton formulated this law in 1801, though it follows directly from the KINETIC THEORY of gases, a model describing the behavior of gases that developed later in the 19th century. The pressure that a gas exerts on the walls of its container results from the collisions of the molecules of the gas with the walls. If we look at the gas with this in mind, it seems reasonable that the number of collisions that each kind of molecule makes with the walls should depend on how many of that kind of molecule there are in the gas. It follows, then, that the total pressure—which is just the sum of all the molecular collisions—should be obtained by adding up the number of collisions of each kind of molecule.

JOHN DALTON (1766–1844) English chemist and physicist. He was born in Eaglesfield, Cumbria, to a Quaker family. Dalton left school at the age of eleven, but then began to study meteorology. For two years he worked as a farm laborer, then became an assistant in a Quaker school. He received an informal education in science and mathematics from his colleagues, and eventually opened his own school to support himself. His most important contribution to science was the modern atomic theory. He was led to the idea of atoms through his interest in the weather and, hence, in the behavior of gases.

Dark Matter

Most of the matter in the universe can never be seen by us

When we conjure up a picture of a galaxy in our mind, we probably see a spiral of stars swimming against the blackness of space. If we had a powerful enough telescope, we could literally see the stars, of course, because they are emitting huge quantities of light and other radiation. We could also "see" dark regions in the galaxy—clouds of dust and gas that absorb light instead of emitting it.

Over the course of the 20th century, astronomers came to realize that this familiar image of a galaxy represents less than 10% of what's actually there. Over 90% of every galaxy is in the form of a mysterious something that has come to be called dark matter because it evidently neither emits nor absorbs light and is therefore invisible to us. (It has also been called *missing mass*—something of a misnomer, for it isn't missing at all.) The first intimations that this might be so go all the way back to 1933, when astronomer Fritz Zwicky (1898–1974) pointed out that there seemed to be more gravitational force holding together the members of the Coma Cluster of galaxies than could be accounted for by the matter we can see.

In the 1970s, Vera Rubin of the Carnegie Institution of Washington studied how galaxies rotate, especially out in their farthest fringes. Here, isolated atoms of hydrogen circle a galaxy like microscopic satellites. Studying their motion is a way of finding out how the matter in the galaxy is distributed. Consider our own solar system. Almost all of the mass is concentrated in the center, and the farther away a planet is from the Sun, the slower it moves. Jupiter, for example, takes over ten times as long as Earth to complete one orbit—it has a larger orbit and travels more slowly (*see* KEPLER'S LAWS). In the same way, if all of the matter in a spiral galaxy were concentrated in the spiral arms, where the stars are, then the farther away from the center the hydrogen is, the slower it should move. Rubin found that, as far away from the center as she could

measure, the hydrogen was moving at the same speed. It was as if the hydrogen were locked into a giant rotating sphere of invisible material.

We now know that dark matter is found throughout the universe, not only in individual galaxies but between them as well. What we *don't* know is what the dark matter actually is. Some of it could be

Galaxies come in clusters. This one, known as the Coma Cluster after the constellation in which it is located, is the nearest rich cluster, containing perhaps several thousand galaxies. Like all clusters, this one contains large amounts of dark matter.

ordinary stuff—Jupiter-sized objects too small to be seen by our telescopes. There is some evidence that at least some of the dark matter is in the form of "jupiters." It comes from studies in which stars in nearby galaxies are monitored to watch for sudden dips in brightness caused by dark objects in our Galaxy passing between us and them. Such dips are, in fact, observed, and the objects that cause them are called *massive compact halo objects*, or MACHOs ("halo" is the term for the outer reaches of a galaxy).

Most scientists believe that dark matter consists not only of ordinary stuff, but also of some as yet undetected types of particles. Called *weakly interacting massive particles*, or WIMPs, these particles do not interact with light or with other forms of electromagnetic radiation. The search for them is a modern echo of the old search for the luminiferous ether (*see* MICHELSON–MORLEY EXPERIMENT). The idea is that if the Milky Way is really encased in a sphere of WIMPs, then there must be a constant "dark matter wind" blowing through the Earth, just as a "wind" buffets the occupants of a moving car even on the calmest day. Occasionally, one of those particles will jostle an atom in a specially designed detector, registering a signal in sensitive monitoring instruments. There is some evidence from the laboratories conducting such experiments that there is a six-month variation in these signals, which is exactly what you would expect if the Earth were heading into the "wind" at one point in the year and had the wind behind it six months later.

WIMPS are examples of what is called *cold dark matter* because they are heavy and move slowly. They are thought to have played a crucial role in the formation of galaxies in the EARLY UNIVERSE. Some scientists also believe that at least some of the dark matter may be in the form of rapidly moving particles such as neutrinos. Such particles would be an example of *hot dark matter*. The basic problem with galaxy formation is that until atoms formed, approximately 300,000 years after the big bang, the universe was composed of a dense, opaque plasma. If ordinary matter started to clump together under the influence of gravity, the clumps would have absorbed radiation and been blown apart. After atoms formed, the matter in the universe became transparent and gravity could start to pull things together. Unfortunately, by that time the expansion of the universe would have carried all its material too far apart for galaxies to form. This used to be known as the *galaxy problem* because it seemed to pose a fundamental difficulty for the BIG BANG picture.

If there were dark matter particles mixed in with everything else soon after the big bang, however, there would have been nothing to prevent them from coming together. Because there were so many of them, they would have exerted significant gravitational forces, and, since they don't interact with radiation, there would have been nothing to blow clumps of dark matter apart. Thus, by the time atoms formed, dark matter had already collected into galaxies and clusters of galaxies, and these concentrations pulled in the newly liberated ordinary matter. Like leaves on a river, ordinary matter moves in response to forces exerted by something else. In other words, ordinary matter—including ourselves—is not only a small part of the galactic picture, it's also something of a Johnny-come-lately.

VERA COOPER RUBIN (1928–) American astronomer. She was born in Philadelphia and obtained her Ph.D. at Georgetown University, Washington, in 1954. Since then she has worked mostly at the Carnegie Institution, Washington. There she studied how galaxies rotate, concentrating on spiral galaxies and motions in their spiral arms. It was her discovery that the rotational speed of large gas clouds in the arms did not fall away with distance from the center, but instead increased, that provided the first persuasive evidence of dark matter in individual galaxies.

Davisson– Germer Experiment

The electron, usually thought of as a particle, exhibits behavior associated with waves

According to the principle of COMPLEMENTARITY, quantum mechanics allows us to think of things we normally call particles as waves, and vice versa. The electron, for example, had traditionally been thought of as a sort of miniature baseball, but in 1924 Louis de Broglie (*see* DE BROGLIE RELATION) argued that a particle with momentum p could also behave like a wave of wavelength λ, given by

$$\lambda = h/p$$

where h is the PLANCK CONSTANT.

The test for whether something is or is not a wave is whether it undergoes INTERFERENCE. In 1927 two American physicists, Clinton Davisson and Lester Germer, performed a classic experiment that demonstrated the wave nature of electrons. They used a heated filament in a vacuum chamber to produce a beam of accelerated electrons, then scattered that beam off a crystal. They found that the angles at which the electrons were scattered tended to cluster around 65°.

The explanation of this result is that what they were seeing was, in fact, a result of BRAGG'S LAW, which was first derived for the scattering of X-rays from crystals. Basically, each atom in the crystal acts as a source of Huygens wavelets (*see* HUYGENS' PRINCIPLE), and these waves exhibit interference by reinforcing one another at a certain angle. This was the angle where Davisson and Germer found the maximum number of scattered electrons. When the momentum of the electrons matched the wavelength calculated from the DE BROGLIE RELATION, they knew they had verified the prediction that electrons could exhibit wave properties.

After working at various universities and industrial laboratories, Clinton Davisson finished his career at the University of Virginia. When I was on the physics faculty there, I was assigned his old office. Prominently displayed on the wall was his yellowed 1954 edition of the PERIODIC TABLE of the elements. I left that chart in place when I moved on to my present position, regarding it as something of a cultural artifact. I hope it's still there!

CLINTON JOSEPH DAVISSON (1881–1958) American physicist. He was born in Bloomington, Illinois, and educated at the University of Chicago, and at Princeton, where he received his Ph.D. in 1911. After studying with J.J. Thomson (discoverer of the ELECTRON) at the Cavendish Laboratory in Cambridge, England, in 1917 he joined Western Electric Laboratories (now Lucent Technologies) in New York, where he first worked on the emission of electrons from metals; he remained there until 1946. His discovery, with Lester Halbert Germer (1896–1971), of the wave nature of electrons came after he shifted to studying what happened when crystal surfaces were bombarded with electrons. For this work Davisson shared the 1937 Nobel Prize for Physics with George Thomson (1892–1975), son of J.J., who independently discovered electron diffraction.

De Broglie Relation

The wavelength of a quantum particle is inversely proportional to its momentum

One of the facts of life in the subatomic world is that entities such as electrons and photons are not like objects we're used to in our everyday world. They seem to be neither particles nor waves, but something else that can display the properties of either depending on circumstances (*see* COMPLEMENTARITY). Now, it is one thing to make this sort of general statement, but quite another to say precisely how the wavelike and particle-like aspects of quantum objects are related to each other. This is exactly what the de Broglie relation does.

Louis de Broglie published his relationship as part of his doctoral thesis in 1924. It seems to be one of those crazy ideas that happened to be right and played an important role in the development of quantum mechanics. His subsequent career was much more conventional (he was a professor in Paris), and never again did he reach those dizzy heights.

Here's what the relationship is about: One way to characterize particles is according to their velocity. For various reasons, physicists prefer instead to use *momentum*, which is a particle's mass multiplied by its velocity, rather than velocity itself. When we think of a wave, on the other hand, we characterize it by its wavelength (the distance between successive crests) or its frequency (the number of crests that pass by a fixed point each second). If we call the momentum p and the wavelength λ, then the de Broglie relation says that

$$p = h/\lambda \quad \text{or} \quad \lambda = h/p$$

where h is the PLANCK CONSTANT.

What the relation says is this: If you wish to think of a quantum object as a particle, then you can assign it a momentum p. If you wish instead to think of that same quantum object as a wave, then you have to assign it the corresponding wavelength λ given by the above equation. In other words, the wave- and particle-like aspects of quantum objects aren't independent, but are related to each other through this simple equation.

The de Broglie relation helped to explain one of the great mysteries of early quantum mechanics. When Niels Bohr first proposed his model of the atom (*see* BOHR ATOM), it contained the notion of *allowed orbits,* in which electrons could orbit forever without losing energy. We can use the de Broglie relation to interpret this notion. If we regard the electron as a particle, then for any distance from the nucleus there will be one and only one velocity (or momentum, rather) that the electron can have if it is to remain at that distance.

If on the other hand, we regard the electron as a wave, a different set of considerations apply. In order for a wave to fit onto an orbit of a given radius, the circumference of that orbit has to be equal to an integral number of wavelengths. In other words, you have to be able to fit one, or two, or three (and so on) complete cycles into a length equal to the circumference of the electron's orbit. In this scheme, a non-integral number of wavelengths just won't fit on the orbit.

The point is that, given any orbit, we can always find a momentum (if we think of the electron as a particle) or a wavelength (if we think of the electron as a wave) that will allow the electron to stay in that orbit. For most orbits, however, the de Broglie relation says that an electron (considered as a particle) of a specific momentum will not have an associated wavelength (considering it as a wave) that will fit on the orbit. Conversely, if we think of the electron as a wave with a specific wavelength,

Prince Louis de Broglie, the man who explained wave–particle duality, wearing the Grand Cross of the Légion d'Honneur, which he was awarded in 1961.

then the momentum associated with that wavelength will not allow the electron to remain in orbit if you think of it as a particle. In other words, for most orbits, either the wave or the particle description will tell you that an electron cannot be that distance from the nucleus.

There are, however, a few orbits for which the two descriptions are compatible. In these orbits, the momentum the electron needs to keep it going (particle picture) corresponds exactly to a wavelength it needs to fit onto the circumference (wave picture). These are the allowed orbits of the Bohr atom. They are determined by the requirement that the wave- and particle-like aspects of the electron be consistent with each other.

Another way of saying this, which I find philosophically attractive, is to note that the Bohr orbits correspond to those states for which it makes no difference what mental categories humans apply to nature. The real world, in other words, seems to be arranged so that it doesn't matter what we think!

LOUIS VICTOR PIERRE RAYMOND DE BROGLIE (1892–1987) French physicist, born in Dieppe. In a profession dominated by upwardly mobile people from middle- and working-class backgrounds, de Broglie was a genuine blue blood—a French aristocrat, the second son of a Normand family that had produced a long line of soldiers, statesmen, and diplomats. In 1909 he entered the Sorbonne to study history, but a year later switched to physics. The de Broglie relation was part of his doctoral thesis in 1924, and for the discovery he received the 1929 Nobel Prize for Physics. He later became professor of theoretical physics at the Henri Poincaré Institute in Paris. In 1960, on the death of his elder brother, who had made the first studies of X-ray spectra, Louis became the 7th Duc de Broglie.

Determinism

Given the initial conditions of any system, it is possible to use the laws of nature to predict the final state of that system

One of the principal tenets of the scientific method is that the world is predictable—that for a given set of circumstances there is only one possible (and predictable) outcome. This philosophical doctrine is known as determinism. Perhaps the best example of a deterministic system comes from a combination of NEWTON'S LAWS OF MOTION and NEWTON'S LAW OF GRAVITATION. If you apply these laws to a single planet orbiting a star, and start the planet off from a certain place at a certain speed, you can predict where it will be at any moment in the future. This is what gave rise to the "clockwork universe"—an idea that had enormous impact not just on the development of science, but on the emergence of the intellectual movement known as the Enlightenment which reached its high point in the 18th century.

As an intellectual ideal, determinism has played (and continues to play) an important role in the sciences. It is important to understand, however, that in practice it is not always easy to predict how a system will end up (what scientists call the final state of a system), even when you know the initial conditions. For example, calculating the orbit of a single planet, as in the example above, is relatively straightforward. But introduce a few more planets into the picture, and things get much more complicated. Each planet exerts a gravitational force on all the other planets and, in turn, is influenced by them. The exact solution of such a many-body problem, as astronomers call it it, is nigh on impossible.

During the 19th century a prize was offered for the first person who could show whether or not the solar system is re-entrant. The question of re-entrance was this: If you could go forward in time as far as you liked, would you ever again see all the planets exactly where they are today, in the same relative configuration and moving with the same velocities? This is an extremely difficult question to answer simply because there are nine planets in the solar system, not to mention all the moons, asteroids, and comets, each having its own little say in how things turn out. So although the solar system is held up as a prime example of the clockwork universe and of the principle of determinism, it is not always possible to give an exact prediction of its future.

In the early part of the 20th century, these complications in the motions of the planets played a major role in the observational verification of the general theory of RELATIVITY. The orbit of Mercury, like the orbit of all the other planets, is elliptical (*see* KEPLER'S LAWS). If the solar system consisted only of Mercury and the Sun, Mercury would go around the same ellipse time and time again. Because of the influence of the other planets, however, the ellipse tends to get twisted a little each time the planet goes around the Sun. Mercury's point of closest approach to the Sun (its perihelion) moves over time as the planet traces out its orbit. The perihelion point doesn't move much—each century, it moves around the Sun by about 1000 seconds of arc, about a quarter of a degree. Almost all of this could be explained in terms of the influence of the gravitational attraction of the other planets—all but about 43 seconds of arc per century.

Before Einstein formulated his general theory of relativity, the advance of Mercury's perihelion was simply one of those little unexplained mysteries in the universe—nobody knew what caused it and, to be perfectly honest, not many astronomers paid much attention to it either. Nevertheless, when general relativity was applied to the orbit of Mercury it introduced a very small correction to Newtonian gravitation which was exactly enough to explain the planet's perihelion shift. The orbits of all the planets, including Earth, exhibit the same kind of perihelion shift that Mercury does—it's just that Mercury's is largest and easiest to measure because Mercury orbits closest to the Sun and therefore has the highest orbital speed (as explained by Kepler's laws). Today, using sophisticated radar ranging techniques, the perihelion shifts of all the inner planets have been measured and shown to confirm the predictions of general relativity.

So, if the stakes are high enough, scientists will work their way through the complexities of the gravitational forces in the solar system to work out something like the perihelion shift. But the re-entrance problem remains unsolved; it may indeed be intractable, but there are very few brownie points to be had for solving it. The example of the solar system shows that, even for systems that are completely deterministic in the classical Newtonian sense, the ability to make predictions is a separate issue.

Deterministic Chaos

There are systems in nature in which the outcome of a particular situation is extremely sensitive to measurements of the input and whose future behavior is, for all practical purposes, unpredictable

The principle of DETERMINISM is one of the most important in modern science. It says that if we know the current state of any system in nature, then we can use our knowledge of the laws of nature to predict the future behavior of that system. The classical Newtonian "mechanical" universe, in which the positions of the planets were seen as being analogous to the hands on a many-handed clock, and our knowledge of the laws of nature amounted to an understanding of the clock's mechanism, is perhaps the best visual representation of this concept.

During the 20th century, scientists came to realize that there are systems in nature that, though they are completely deterministic in the Newtonian sense, nevertheless have futures that are not, for all practical purposes, calculable. The appearance of fast digital computers in the 1980s made this phenomenon, which has come to be known as deterministic chaos, or *chaos theory,* an active field of research. The best analogy for deterministic chaos is white water on a rapidly flowing section of a mountain stream. If you set two leaves in motion next to each other on the upstream side of the white water, they will most likely be widely separated by the time they reach the downstream side. In a system like this, a small difference in the initial conditions (the position of the leaves) can result in a large difference in the outcome.

Most systems in nature are not like this. For example, if you drop a ball from a height of 5 meters and measure its velocity when it hits the ground, and then drop the same ball from 5.001 meters, its velocity at impact will not be very much different. In systems such as this, small changes in the initial conditions lead to small changes in the final outcome. Most familiar systems in nature are of this type.

In frothy, turbulent water like this, deterministic chaos reigns.

Actually, even for simple systems like the classical Newtonian billiard balls, it can be very difficult to make predictions about future states. It is a standard problem for graduate students in physics, for example, to show that even the problem of a billiard ball bouncing between cushions on a perfect table eventually dissolves into indeterminacy because of uncertainties in measuring the angle at which the ball approaches the cushion at the beginning.

The system of whitewater rapids is different, however, and the discovery of deterministic chaos is a good illustration of how these systems work. By modern standards, early digital computers were very slow and had very little memory. In the 1960s, Edward Lorenz (1917–) and his colleagues at the Massachusetts Institute of Technology were running computer models of the Earth's climate. Their computers would often go to some intermediate state in a calculation, store those intermediate results on paper tape overnight, and finish the calculation the next day. They began to notice that calculations done continuously from start to finish were giving very different results from calculations that were interrupted. They discovered that the difference lay in the way the computer rounded off numbers in its intermediate results. For example, it would store a number as 0.506 for transfer to tape, but as 0.506127 when it was continuing with a calculation. This difference was enough to give completely different predictions for future states of the climate. We now know of systems that are far more sensitive to initial conditions—a difference in the eighth decimal place has been known to have a significant effect on the outcome. (Technically, a chaotic system is defined to be one in which the outcome changes exponentially with changes in the input.)

The point is that when we talk about "determining" an initial state, we are actually talking about a measurement. Every measurement in the real world contains an error—some ambiguity in its actual value. For example, if you measure the length of a table with a ruler on which the smallest division is a millimeter, then there will of necessity be an error of some fraction of a millimeter in your determination. In the same way, if you wanted to determine the position of a leaf upstream of the rapids in the example above, you might measure the distance between the leaf and a spot on the shore. Again, there will always be some small error in that measurement, depending on the accuracy of the measuring device you use. If the system is chaotic, you can repeatedly put the same leaf in what you think is the same position and still get different outcomes, because you will never be able to put it in *exactly* the same place twice.

So, for chaotic systems it is theoretically possible to predict the future outcome, but only if the initial state can be determined with absolute accuracy. Since such accuracy is impossible to achieve, these systems are, for all practical purposes, unpredictable. It is important to realize, however, that the existence of deterministic chaos does not violate the principle of determinism. It simply tells you that under certain circumstances you will not be able to make the kinds of measurements you need to determine the current state of the system with sufficient accuracy to predict its future states.

In chaotic systems, in other words, there is a kind of divorce between determinism (our understanding of the laws that govern the system) and prediction (our ability to tell what the system is going to do). It's not that this divorce didn't exist in Newtonian physics—we've seen that it did. It's just that until recently people didn't pay a lot of attention to it, probably because they felt that it was only a matter of time before the problem of prediction would be resolved. What chaos theory has taught us is that the divorce is not only real, it's permanent. We now understand that it is possible for a system to be deterministic and predictable in principle while at the same time being unpredictable in practice.

Recently, some scientists have tried to apply chaos theory to other areas, including the orbits of planets in the solar system over extremely long time periods, and the stock market. Although a group of physicists left their research labs a few years ago to exploit chaos theory in selling advice about stocks, I haven't seen any of them driving around in a Mercedes yet, so perhaps more work is needed to bring the theory into the practical world.

Differential Resource Utilization

It is possible for species using different amounts of the same resources to maintain a population equilibrium

There are many instances in nature of many species inhabiting the same area. Sometimes the principle of COMPETITIVE EXCLUSION applies, and one species crowds another out. Sometimes—a lawn is a good example—species manage to find a way to coexist and share the resources. Neighboring species may simply use different resources, or they may have very similar needs. A model known as differential resource utilization explains how species can share the same resource base.

To see how the model works, we can start with a simple example and work our way up. Suppose there is a single species of plant that requires two resources—call them A and B—to survive. These resources might be specific chemicals—potassium and phosphorus, for example, or water and carbon dioxide. In the absence of plants, the ecosystem will supply these resources at a constant rate, and there will be some limit below which the supply of each is too low to sustain the plants.

In order to have a stable equilibrium, both components of the ecosystem—the plants and the resources—have to be stable. The plants will then consume exactly as much of each of the two resources as are being replaced. Too little consumption and the resource base grows; too much and it drops. In each case, the consumption will change so as to move the system back to equilibrium (by increasing or lowering the number of plants, for example).

When there are two species present, each of which uses resources A and B, there are several possibilities:

— the supply of A and B can be too low for either species to survive;
— the supply of A or B can be high enough to allow one species to exist but not the other;
— the supply of A or B can be such that competitive exclusion can operate, so that one species drives out the other; or
— the supply of A and B can be such that both species can survive. This is the domain of differential resource utilization.

If both species are to survive, some special conditions have to be met. For example, the first species might be in a location where it has all of resource B it needs, but is limited by resource A. At the same time, the second species has to be in a domain where it has all of resource A it needs but is limited by resource B. In this case, one species is able to consume enough of its limiting resource to survive while leaving enough for the second species. The two can thus coexist in equilibrium within the same ecological niche.

Obviously, this can be made to work for as many species and as many resources as you like.

Diffraction

Diffraction patterns result from the bending of light around obstacles or the passage of light through many apertures

JOSEPH VON FRAUNHOFER
(1787–1826) German physicist and optician. He was born in Straubing, now in Germany, the eleventh child of a master glazier. Orphaned at an early age, he was apprenticed to a glasscutter. When he was 14, the building in which he was working collapsed, trapping him for several days and making him something of a public figure. With a cash gift from the Elector of Bavaria, he was able to set himself up in business as a glassmaker, and in 1806 he joined the Bavarian firm of Utzscheider, then Europe's leading makers of optical instruments. There Fraunhofer began his lifelong quest for the perfect achromatic lens: one that would not produce colored fringes around the image.

The idea that light is a wave (*see* ELECTROMAGNETIC SPECTRUM) was greatly strengthened in the early 19th century by studies of diffraction. The conventional wisdom, associated with Isaac Newton, that light was a stream of particles—the so-called *corpuscular theory of light*—was being seriously questioned, and work on diffraction phenomena established the wave nature of light as the new orthodoxy. With the advent of quantum mechanics and the explanation of the PHOTOELECTRIC EFFECT, we now know that light, like other quantum objects, can behave as both a particle and a wave (*see* COMPLEMENTARITY).

The basics of diffraction can be understood in terms of HUYGENS' PRINCIPLE, which says that each point on a light wave can be regarded as the source of secondary wavelets, and that the "downstream" pattern of diffracted light is produced by the INTERFERENCE of those wavelets. When an advancing light wave encounters an obstacle, some of the Huygens wavelets are blocked. If light from a lamp encounters a razor blade, for example, then the parts of the advancing wavefront above the blade will send out Huygens wavelets, whereas those below the blade will not. As a result, there will be points below the blade where constructive interference occurs and light can be seen. In essence, the wave "bends" around the obstacle. This phenomenon can often be seen in harbors, where incoming waves bend around breakwaters, causing boats anchored in the harbor to bob up and down.

If the source of the light and the point of observation are fairly close to the obstacle, so that the light entering and leaving the system is not made up of parallel rays, we say that we see *Fresnel diffraction*. (Augustin Fresnel (1788–1827) was a French civil engineer in the Department of Bridges and Roads who, during his frequent leaves of absence, worked to establish the wave nature of light.) If the source of light and point of observation are far from the obstacle, we say that we are seeing *Fraunhofer diffraction*. Fraunhofer pioneered the production of high-quality glass and optical instruments, including a device known as a *diffraction grating*. This is a sheet (originally of glass) with a large number of very closely spaced and carefully ruled parallel lines, so that light striking the surface will be reflected from each line. Think of each line as a source of Huygens wavelets which reinforce one another only at certain angles.

The diffraction grating proved to be the most important tool used by scientists from the mid-19th century onwards to sort out the chemical composition of objects which emitted light (*see* SPECTROSCOPY). Fraunhofer's most famous discovery was that when sunlight is passed through a prism, dark lines are seen to be superimposed on the spectrum. Today we understand that they are created by the absorption of light by atoms in the Sun's atmosphere.

Dispersion, atomic theory of

The properties of the atoms of a substance determine the speeds at which different frequencies of light travel through the substance

We know that light travels more slowly in a material medium than it does in a vacuum. This fact is usually expressed in terms of a quantity known as the refractive index, defined by the equation

$$n = c/v$$

where v is the velocity of light in the material and c is the speed of light in a vacuum.

Light moves more slowly in materials because it is continually interacting with the electrons within the atoms. Just as it takes longer to reach your destination if there are periodic traffic jams than if you have the freedom of an open road, so light, by accelerating and decelerating the electrons, travels more slowly than it would in the emptiness of a vacuum. The refractive index of glass, for example, is around 1.5, which means (from the above equation) that the speed of light is a third slower in glass than it is in a vacuum.

Different materials have different refractive indexes. In addition, different wavelengths of light—different colors—may travel at different speeds in the same material. This phenomenon is known as dispersion. According to SNELL'S LAW, the angle through which a beam of light is bent when it enters a medium depends on the refractive index, so dispersion has the effect of making different colors of light bend through different angles when they enter a material like glass. In general, shorter wavelengths of light are bent more when they enter a material than are longer wavelengths.

This is the principle behind the working of a *prism*. When ordinary "white" sunlight (which consists of many colors) strikes a prism and encounters the first interface between air and glass, the red component of the light is bent less than the blue. When the light reemerges at the second, glass/air interface, the white light will have been split into its constituent parts. This is what gives the familiar rainbow effect when the light from a prism strikes a screen or wall.

Speaking of rainbows, dispersion explains them as well. A rainbow is created when sunlight hits raindrops in the air. The light is refracted when it enters the raindrop, is reflected off the back surface, and is then refracted again when it reemerges from the front. This is why we see rainbows when there is rain falling in front of us and the Sun is at our back. Because of dispersion, each color of light is concentrated around a different angle, which is why we see the colors arrayed in an arc. (The colors are not pure since the raindrop does send some light off at angles different from that of the main beam.) Each part of the rainbow, of course, brings light to your eye from a different raindrop, but since the raindrops are falling toward the ground, their angle to our line of sight changes with time, so each raindrop will send each color of the rainbow toward the eye at some instant as it passes through the rainbow-producing region.

The less common double rainbow forms when light bounces twice off the back surface of the raindrop, and the very rare triple rainbow when there are three reflections.

Dispersion is a phenomenon that results from the interaction of light with atoms. Like all forms of radiation in the ELECTROMAGNETIC SPEC-TRUM, light waves consist of electric and magnetic fields. The electric field in a light wave exerts a force on electrons in the atom, causing them to accelerate. According to MAXWELL'S EQUATIONS, the accelerated electron must then radiate a light wave of its own. When light moves through a medium, it is constantly being absorbed and reemitted, as described above.

Electrons in atoms are captives of their nuclei. It can help to understand the outcomes of some subatomic interactions to think of the electrons as being attached to the nucleus by a stiff spring. When a light wave starts to shake the electron, its response depends on how the frequency of the wave measures up against the frequency at which the imaginary spring would bounce back and forth if left to itself. Calculations indicate that the acceleration of the electron (and hence the intensity of the wave it radiates) will grow rapidly as the wavelength of the light becomes shorter. In fact, the intensity of the light given off by the atom usually varies as the fourth power of the wavelength (i.e., the wavelength squared, then squared again).

Light scattered in a medium must come from precisely this interaction between atoms and light, since light that does not interact with atoms will go straight through the material. This means that when you look at white light that has been scattered, you will see the shortest wavelengths—the blue light.

That is why the sky is blue. When you look at the sky away from the Sun, you are seeing sunlight that has been scattered to your eye from a beam of light going somewhere else. The Sun appears yellow because the blue light has been removed from the beam by its travel through the atmosphere.* At sunset, when the light is traveling through more of the atmosphere, the Sun appears red because not only blue, but green and yellow have been scattered from the beam.

Sunset over the Adriatic Sea. The sky is reddened by the dispersion of sunlight as it passes through more of the atmosphere than when the Sun is higher in the sky. Blue and green wavelengths are scattered out of the beam, leaving only red.

---

* Whatever you do, don't try to verify this for yourself by looking directly at the Sun. The intensity of sunlight is such that even a brief glance can cause temporary blindness and permanent damage.

DNA

The DNA molecule has the shape of a double helix, and replicates by a process in which each side of the double helix acts as a template for the assembly of new molecules

Today, we know that the molecule we call DNA carries the code that runs the chemistry of living things (*see* MOLECULAR BIOLOGY), and DNA's double-helix structure has become a familiar scientific icon. Like nearly all great discoveries, this was not the work of a lone genius, but followed from a long chain of experimental results. The HERSHEY–CHASE EXPERIMENT, for example, had established that it was DNA, and not PROTEINS, that carry genetic information in cells. In the 1920s, the Russian-American biochemist Phoebus Levene (1869–1940) had established that the basic building blocks of DNA are the five-carbon sugar known as deoxyribose (this molecule is the "D" in DNA); a phosphate group; and the four bases named thymine, guanine, cytosine, and adenine, usually denoted by the letters T, G, C, and A. In the late 1940s, the Austrian-American biochemist Erwin Chargaff (1905–) established that in all DNA the amounts of T and A are the same, and, similarly, the amounts of G and C are the same. The relative proportions of T/A and G/C in the DNA molecule, however, varies from one species to the next.

In the early 1950s, two further pieces of evidence about the nature of DNA became available: The American chemist Linus Pauling (1901–94) showed that long molecules such as proteins can form links that twist them into a helical shape, and the laboratory of Maurice Wilkins and Rosalind Franklin in London produced X-ray data (based on advanced use of BRAGG'S LAW) that suggested that DNA might have a helical structure.

It was at this time that a young American biochemist, James Watson, went to spend a year at Cambridge University, hooking up with a young English theoretical physicist, Francis Crick. ("I was almost totally unknown at the time," Crick would reminisce later, "and Watson was regarded … as too bright to be really sound.") Working with metal models, they tried to find how the various molecular components would fit into a three-dimensional DNA molecule.

The easiest way to visualize their results is to imagine a tall ladder. The uprights of the ladder are made of molecules of sugar, oxygen, and phosphorus. The important working information in the molecule is carried on the rungs of the ladder. The rungs are made of two molecules, one attached to each upright. These molecules are the four bases. The bases are basically single or double rings containing carbon, nitrogen, and oxygen atoms, so constructed that they have either two or three positions where they can form hydrogen bonds (*see* CHEMICAL BONDS) with other bases. Because of the shape of these molecules, only certain kinds of links—certain completed rungs—are allowed. These are the links between A and T, and between G and C. No other links are allowed. Each rung, then, consists of either A–T or G–C. Once this ladder is assembled, imagine grabbing the two ends and twisting, producing the familiar double helix of DNA.

The Meselson–Stahl Experiment

Once Crick and Watson had suggested the double-helix structure of DNA, it had to be verified experimentally, as does any scientific hypothesis. Two molecular biologists at the California Institute of Technology, Matthew Meselson (1930–) and Franklin Stahl (1910–), carried out the relevant series of experiments in 1957. Their technique depended on being able to distinguish between the masses of very similar molecules. They began by growing bacteria in a medium in which the only nitrogen available was the isotope ^{15}N (normal nitrogen, ^{14}N, is slightly lighter). After several generations, all of the bacterial DNA was constructed from the heavier nitrogen. The bacteria were then transferred to an environment in which all of the nitrogen was in the form of ^{14}N. (Nitrogen appears in the base pairs of DNA, and will therefore be taken up by any organism creating new strands of the molecule.) After one cell division cycle, the DNA of the bacteria had a weight midway between that associated with ^{15}N and ^{14}N. After two cell divisions, one DNA strand in four contained heavy DNA, and so on. By this ingenious experiment, then, Meselson and Stahl verified that with each cell division, the complementary strands of DNA contained half of the old (heavy) DNA and half of the new (light) DNA, just as the Watson–Crick hypothesis predicted.

A section of DNA replicates by unzipping itself and assembling new strands.

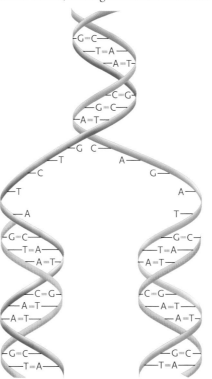

If you read down the rungs on one side of the DNA molecule, you will find a sequence of bases. Think of this as a message written in an alphabet that has only four letters. It is this message which determines how the cell runs its chemistry, and which therefore determines the characteristics of the living thing of which that cell is part. There is no new information contained in the other side of the helix, because if you know what base is on one side, you know what the other half of the rung has to be. In a sense, the two strands of the double helix bear the same relationship to each other as a photograph does to its negative.

When Watson and Crick discovered the double-helix structure of DNA, they also realized that there was a straightforward way in which a molecule of DNA could replicate itself—as it must when a cell divides. In their words, "It has not escaped our notice that the specific pairing we have postulated suggests a possible copying mechanism for genetic material."

This "possible copying mechanism" depends on the structure of DNA. When a cell starts to divide, so that extra DNA is needed to supply daughter cells, enzymes (*see* CATALYSTS AND ENZYMES) start to "unzip" the DNA ladder, leaving the individual bases exposed. Other enzymes cause appropriate bases in the surrounding fluid to attach to the appropriate exposed bases—A to T, C to G, and so on. As a result, each of the two split strands of DNA assembles a match for itself from the surrounding medium, creating two double helices from the original molecule.

Just as every great discovery rests on the work that preceded it, so too does it give rise to fruitful research as scientists use the new information to move on. The discovery of the double helix can be said to have generated a half century of progress in molecular biology, culminating in the success of the HUMAN GENOME PROJECT.

FRANCIS HARRY COMPTON CRICK (1916–)
English molecular biologist (right). He
was born in Northampton, where his
father was a shoe manufacturer. He
graduated in physics from University
College, London, in 1938, and spent the
war designing acoustic and magnetic
mines. He then decided to study "the
mystery of life." In 1951 he was
examining the structure of proteins at a
new unit set up by the Medical Research
Council at Cambridge's Cavendish Laboratory, when a visiting student, James
Watson, suggested that if the function of
the DNA molecule was to be understood,
its structure needed to be found.
Their success in this quest earned
them a share of the 1962 Nobel
Prize for Physiology or Medicine.
Crick's later work included the
formulation of the central dogma
of molecular biology. In 1977 he
moved to the Salk Institute at San
Diego to continue his pursuit of
"the mystery of life," this time
investigating consciousness.

JAMES DEWEY WATSON (1928–)
American biochemist. He was
born in Chicago, Illinois, and
entered the University of Chicago

at the age of 15, graduating four years
later. He received his Ph.D. in 1950 from
the University of Indiana for studies of
viruses. His visit to the Cavendish Laboratory the next year led to the collaboration with Francis Crick that led to the
discovery of the structure of DNA. Their
1962 Nobel Prize for Physiology or Medicine was shared with Maurice Wilkins
(1916–), whose X-ray diffraction results
had been instrumental in elucidating the
double-helix structure. Such recognition
eluded Rosalind Franklin (1920–58),
whose contribution was felt by many to
be also significant.

Doppler Effect

The perceived frequency of a wave depends on the velocity of the source of the wave

You've probably had the experience of being at a roadside when an emergency vehicle has passed by at speed with its siren wailing. When the vehicle is approaching, the pitch is high, but the pitch drops as it passes, producing the familiar e e e e e e E E E A A A O O O w w w w w sound. Although you may not have realized it at the time, you were actually hearing one of the most fundamental (and in many ways one of the most useful) properties of waves.

Waves are strange things. Think of a bottle bobbing up and down near the shore. The water near the bottle is moving in the same way as the bottle, and this means that the water is moving up and down even as the wave moves in toward the beach. In other words, the motion of the medium carrying the wave is not the same as the motion of the wave itself. You can see the same things when fans at a football stadium do "the Wave" (known in the UK as "the Mexican Wave"). Each fan stands up and sits down, but the wave moves around the stadium.

Waves are usually characterized by their frequency (the number of crests passing a given point each second) and their wavelength (the distance between two neighboring crests). These two quantities are related to each other, so if you know one of them and the velocity of the wave, you can calculate the other.

Once a wave starts out from its source, its velocity is determined only by the medium through which it moves—it doesn't depend on the source any more. A wave on water, for example, moves because of the interplay between water pressure and gravity, while sound waves depend on the creation of pressure differentials in air (or whatever other medium the sound moves through). None of these mechanisms depends on how the wave started or what its source does after the wave is emitted. This is what is behind the Doppler effect.

Think about our example of the wailing siren, and let's assume to start with that the emergency vehicle is standing still. The siren produces a sound by periodically compressing the air around it. These compressions, which are the "crests" of the sound wave, then move away from the source to your ear, where they strike your eardrum and trigger the process we call hearing. The frequency with which the crests arrive at our ear are what we call pitch—440 crests per second, for example, corresponds to the note known to musicians as middle A. So long as the vehicle is standing still, the waves from the siren will move out in concentric circles, the crests will continue to arrive regularly, and we will hear a particular pitch of sound.

When the vehicle is moving toward you, however, a new effect is added. Between the time the siren emits one wave crest and the next, it will have moved. Each outgoing sound wave is now centered on the place where the car was when that particular crest was emitted. The result is that the crests will arrive at your ear more frequently than they did when the vehicle was standing still, and you will perceive the pitch of the sound to be higher. Similarly, if the vehicle is moving away from you, the crests will arrive less frequently and you will hear a lower pitch. This explains the change in pitch you hear as the vehicle passes by.

We have discussed this effect for sound, but it is applies to any other wave. If a source of visible light is moving toward you, the light will appear to have a shorter wavelength and we say that it is blueshifted (blue light has a shorter wavelength than most other visible light). If the source is moving away from us, the light will appear to have a longer wavelength, and we say that it is redshifted.

These apparent shifts in frequency are named for Johann Doppler, who first predicted them. I have always been particularly interested in the Doppler effect because it was first verified by what I consider to be one of the most elegant experiments ever conducted. The Dutch scientist Christian Buys Ballot (1817–90) placed a group of trumpet players on an open railroad car and, alongside the tracks, assembled a group of musicians blessed with perfect pitch. (Perfect pitch is the ability to hear a note and say exactly what the note is.) As the train passed the observers, the trumpeters played a specific note and the listeners wrote down the notes they heard. The results showed that, as expected, the apparent wavelengths depended on the speed of the train just as Doppler had predicted.

The Doppler effect has many uses in science and in everyday life. One mundane use is in police traffic radar. The radar gun emits an electromagnetic wave (normally in the microwave or radio range) which is

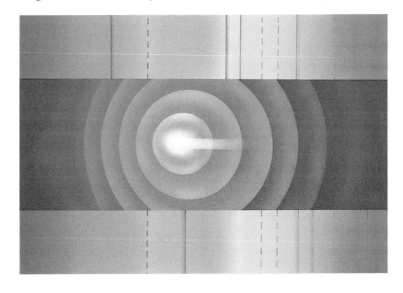

An illustration of the Doppler effect for galaxies. The dashed lines show where a given spectral line would be for a stationary source. The upper spectrum shows a blueshift (source moving toward the observer), while the lower one shows a redshift (source moving away from the observer).

reflected by the metal in your car. The returned beam will be Doppler shifted by the car's motion. By comparing the frequencies of the outgoing and returning waves, the device can measure the speed of your car.

On a more esoteric note, when Edwin Hubble first measured the distances to other galaxies (*see* HUBBLE'S LAW) he found that the light from those galaxies was redshifted compared with light emitted from the same atoms on Earth. He concluded that the redshift indicated that the galaxies were moving away from us, much as you would conclude that a car had passed you after you heard the pitch of the engine noise drop. When Hubble added the observation that the farther away a galaxy was, the greater its redshift (and hence the greater the speed of recession), he realized that his discovery meant the entire universe was expanding. This was the first step on the road to the big bang theory, and a long way from those trumpeters on the railroad car.

CHRISTIAN JOHANN DOPPLER (1803–53) Austrian physicist. He was born in Salzburg, the son of a stonemason. After studying mathematics at the Polytechnic in Vienna he held various tutoring posts until in 1835 he was offered a chair in mathematics in Prague, just as he was about to emigrate to America, convinced that he would never find an academic position. He finished his career as professor of experimental physics at the Royal Imperial University of Vienna, dying of a lung disease just two years after his appointment.

Drake Equation

The number of extra-terrestrial civilizations trying to contact us right now can be predicted by a simple equation

There are very few instances of a major change in scientific thinking that can be dated precisely, but a conference held in Green Bank, West Virginia, on November 1 and 2, 1961, is one of them. That was when a small group of scientists debated a point that had been raised in a paper written by physicists Philip Morrison (1915–) and Giuseppe Cocconi (1914–). The argument: Human scientists, who had only recently completed the first big radio telescopes, were, for the first time, able to detect radio signals from space. If an extraterrestrial civilization were out there trying to contact us, we would, for the first time, be aware of it. The task facing the conference: Estimate the number of extraterrestrial civilizations trying to contact us right now.

The main result they came up with has come to be known as the Drake equation, after the American astronomer Frank Drake. If N is the number of civilizations, then the equation says that

$$N = RPN_eLCT$$

where R is the number of stars formed in the Galaxy each year; P is the probability that a star will have planets; N_e is the probability that an Earth-type planet, capable of supporting life, is among them; L is the probability that life will actually develop on that planet; C is the probability that life will lead to a technological civilization capable of sending radio signals into space; and T is the length of time for which those signals will continue to be sent. The idea of the Drake equation is, if you like, to compartmentalize our ignorance—to link the estimate of N to estimates of other, in some cases less mysterious, quantities.

At the time of the Green Bank conference, we didn't know much about any of the terms on the right side of the Drake equation (except perhaps the first, R). It was thought, for example, that N_e for the solar system could be as high as 5 (in addition to Earth, it was thought that life could exist on Mars, Venus, and some of the moons of Jupiter and Saturn). With such optimistic estimates for this and other terms, values of N in the millions were obtained, leading to the idea that the Galaxy is simply teeming with advanced civilizations and that we are, at best, very junior members of the "galactic club." This concept immediately entered the public consciousness, and in some sections of the media it has become an unquestioned truth about the universe we inhabit.

Since 1961, however, the general trend of new data has tended to temper the original optimism of the Green Bank group. We now know, for example, that it is extremely unlikely that life exists elsewhere in the solar system (except, perhaps, in a frigid ocean beneath the thick icy surface of Jupiter's fourth-largest satellite, Europa). And, though we know of many planets that orbit other stars (knowledge that was not available in 1961), it seems that solar systems like ours, with planets in roughly circular orbits roughly equidistant from their star, are rare. In fact, the conditions that must prevail upon a planet for it to keep liquid water on its surface over billions of year are quite stringent. It turns out that this is possible only on planets not too different in size from the Earth, located within a few percent of Earth's distance from a star like the Sun.

Science-fiction writers and artists have always had fun imagining monsters from outer space, even before the Drake equation.

In 1981, I revisited the Drake equation with my colleague, astronomer Robert Rood (1942–), in the light of this new knowledge. We came up with values of N of roughly 0.003. This implies that extraterrestrial signals have been out there for only about 1/300th of the time for which the Galaxy has been around, so the odds against there being such signals around right now are 300:1 against. Nothing has happened since we did this calculation to change these odds very much. The search for extraterrestrial intelligence (SETI) has gone on for decades, with varying degrees of government and private funding. To date, no contact with extraterrestrials has been made. Nevertheless, the data we have accumulated over the years are enough for us to start to make clear statements about what *isn't* out there.

FRANK DONALD DRAKE (1930–) American astronomer. Drake was born in Chicago and studied electronics at Cornell University. There he heard a lecture by astronomer Otto Struve (1897–1963) on the formation of planets around other stars that ignited a lifelong interest in extraterrestrial life. After serving in the Navy, he held positions at the National Radio Astronomy Observatory (NRAO), Cornell University, and the University of California at Santa Cruz. With Struve's backing, Drake launched Project Ozma, using an 85-ft (28-m) radio telescope at the NRAO in the first major search for extraterrestrial life (*see* FERMI PARADOX).

Early Universe

In the very early history of the universe there were relatively long periods of expansion and cooling punctuated by short periods of fundamental change

Since the discovery of HUBBLE'S LAW, the prevailing view of the universe's origin is that it began in a hot, dense state and has been expanding and cooling ever since. Only since the early 1980s, however, have cosmologists paid serious attention to how the earliest stages of that expansion unfolded. They now have a broad outline of the universe's early history that goes back to an unimaginably tiny fraction of a second after the BIG BANG, and explains the origin of the ELEMENTARY PARTICLES and the chemical elements. Let's start 1 billion years after the big bang (all times are approximate) and work our way back in time.

1 billion years

Galaxies began to form. For the first time in its history, the universe looked something like it does today. The next generation of giant telescopes will be able to look back in time to the very first galaxies.

300,000 years

When the universe was about 300,000 years old its temperature had fallen sufficiently for an electron that had hooked onto a nucleus, forming an atom, to stay attached, so that the atom would survive the next collision with a nucleus. From then on, atoms formed from the sea of electrons and nuclei that had filled the universe up to that point.

Until the first atoms were formed, the universe was composed of a dense, opaque *plasma* of nuclei and electrons. If any of them started to clump together under the influence of gravity, the clumps would absorb radiation and be blown apart. After atoms formed, the matter in the universe became transparent and gravity could start to pull things together. Unfortunately, by that time the expansion would have carried the material in the universe too far apart for galaxies to form. This used to be known as the *galaxy problem*, because it seemed to undermine the big bang theory. It ceases to be a problem, however, in the DARK MATTER scenario, where the clumping occurs in dark matter (not in the ordinary matter) before the atoms form, and ordinary matter accretes onto dark matter after atoms form.

3 minutes

Before three minutes, if two elementary particles—a proton and a neutron, for example—came together to form a nucleus, they would have been torn apart at the next collision. After three minutes they would stay together. Thus, before three minutes the universe was a sea of elementary particles, while after three minutes it contained nuclei.

Nuclei would be built up through successive collisions, a proton or a neutron being added each time. But only the lightest nuclei had time to

---

EARLY UNIVERSE

form in this way before the expansion thinned things out so much that hardly any collisions could happen. The fact that the big bang theory predicts the relative amounts of these light nuclei created during this "window" provides a crucial (and remarkably unforgiving) test of the theory—a test that it passes beautifully.

10^{-5} seconds

At this time—a fraction of a second after the initial event—quarks came together to form the elementary particles (*see* QUARKS AND THE EIGHT-FOLD WAY). Before this time, the universe was a sea of quarks and leptons; afterward it was a cooling sea of elementary particles.

10^{-10} seconds

This time marked the beginning of a new set of transitions—those associated with the unifications of the fundamental forces (*see* THEORIES OF EVERYTHING). It is the point at which the electromagnetic force unified with the weak force. Before this time, there were three forces acting in the universe; afterward there were four. This time also represents the highest energies that can be reached by accelerators on Earth today, so up to this point every statement I've made can, in principle at least, be verified by experiment.

10^{-35} seconds

At the temperatures that prevailed at this time, the strong force unified with the electroweak force. Before this time, there were two forces acting in the universe; after this time there were three. This is also the moment when the universe began a huge increase in size called *inflation* (*see* INFLATIONARY UNIVERSE) that continued until 10^{-32} seconds. It is also the point at which ANTIPARTICLES disappeared from the universe.

The theory of QUANTUM CHROMODYNAMICS and the STANDARD MODEL describe the behavior of matter and energy at the unbelievably high energies of the universe at 10^{-35} seconds. These theories have been tested, but at lower energies. This time represents the current frontier for theories of the early universe.

10^{-43} seconds

Theorists believe that gravity unified with the other forces at this time. Before the universe was this old, there was a single unified force. The transition from one to two fundamental forces is something that theories of everything attempt to describe. As to what happened before this time, we have only speculation to guide us. Like medieval map-makers, we can do no more than write, "Here be monsters!"

Ecological Succession

After disturbances, ecosystems recover in well-defined stages

The marram grass represents the first wave of ecological succession, stabilizing the dune for other plants to take hold.

An ecosystem can be disturbed in many ways—fire, flood, and drought are all common examples. After such a disturbance a new ecosystem establishes itself, and the process by which this happens is regular and repeatable across a wide variety of situations. What happens is that a series of species and ecosystems develops on the spot where the disturbance occurred, and the order in which these species appear will be the same for similar disturbances to similar areas. This is the essence of the idea of ecological succession.

For example, in much of the northeastern United States, lands that were once forests were cleared and turned into farms in the 18th century, farmed in the 19th, then abandoned and allowed to return to forests in the 20th. The plants that moved into the fields over time followed a regular, recognizable, and repeating sequence. In the first year short-lived weeds along with a few tree seedlings appear. In the next few years certain species (called "pioneer species" or, more formally, *early successional species*) start to dominate—white pine is a common pioneer species, for example. These trees grow fast and spread their seeds widely. In a few decades these pioneer trees create a dense forest.

The next step is the appearance of other trees that grow well in the shade of the pioneer forest—maples, for example. After a half century, the pioneer trees will have matured and died. Their seeds cannot grow in the shade of the forest, so the population of trees shifts to the slower-growing newcomers, called *late successional species*. Eventually, the entire forest will be made up of these kinds of trees, a fact that is brought home to New England residents each fall, when the leaves turn and the entire forest becomes a blaze of color.

This pattern of fast-growing pioneers followed by slower-growing species can be seen in many ecosystems. In coastal sand dunes, for example, dune grasses are the first to appear on newly formed dunes. These grasses help to stabilize the dune so that successor species (first shrubs, then trees) can be established.

By studying succession in ecosystems, ecologists have identified three mechanisms at work. They are:

Facilitation When pioneer species move into a new ecosystem, they can make it easier for other species to come in later. For example, when a glacier retreats, the pioneers are mosses and a few plants with very shallow roots—the sorts of plants that can survive on the thin, nutrient-poor soil. As these plants die, they build up the soil so that late successional plants can take over. Similarly, early trees provide shade and shelter for the seedlings of late successional trees.

Inhibition It can also happen that the pioneer species create conditions that make it difficult or impossible for late successional plants to take over. When new surfaces become available near the ocean (from the construction of concrete piers or breakwaters, for example), it can happen that the pioneer species of alga grows quickly and excludes other species. The exclusion comes about simply because this species reproduces rapidly and covers all the available surface, leaving no space for later arrivals. A more active kind of inhibition can be seen in Russian napweed, an Asiatic plant that has invaded the American West. Napweed actually turns the soil in its vicinity highly alkaline, making it unsuitable for many native grasses.

Tolerance Finally, the pioneer species may have no effect, either helpful or harmful, on species that follow them. This can happen, for example, if the various species utilize different resources and, in effect, grow in parallel (*see* DIFFERENTIAL RESOURCE UTILIZATION).

It is important to appreciate that the final state of a forest or dune is not ecologically stable (*see* BALANCE OF NATURE). A final-stage forest is usually characterized by a zero net production of organic material. This means that over time, as processes like erosion remove material, the forest will decline. In fact, most forests are at their most productive about halfway through the succession cycle.

Electrical Charge, conservation of

The total electrical charge of an isolated system remains constant

Humans have known about electrical charge ever since ancient Greek philosophers observed that objects touched with a piece of amber that had been rubbed with cat's fur would move away from one another. Today, we recognize that electrical charge, like mass, is one of the fundamental properties of matter. Every one of the elementary particles that compose the tangible universe has an electrical charge, either positive (e.g., the proton, found in the nucleus of atoms), negative (e.g., the electrons that surround the atomic nucleus), or neutral (e.g., the neutron, also found in the nucleus). (It helps to regard the neutron as having a charge of zero, rather than no charge at all.)

Physicists often search for properties of systems that remain constant even as the systems undergo change. Such properties, in the language of physicists, are said to be "conserved," and the discovery that a certain property is conserved leads to a *conservation law*. A conservation law is simply a restatement of the fact that the property in question remains unchanged.

Electrical charge is a conserved quantity. The net electrical charge of any system has to remain the same, whatever changes that system undergoes. In chemical reactions, for example, the total number of negative charges carried by the electrons stays the same while the electrons are shifting allegiance between the atoms (*see* CHEMICAL BONDS) and, similarly, the total number of positive charges on the nuclei, carried by the protons, stays the same as well. This is the simplest way for charge to be conserved—for the number of both negatively and positively charged particles to stay the same.

At higher energies, electrically charged particles can take on new identities as they react with one another, and the conservation law gets a little more subtle. An isolated neutron, for example, will spontaneously decay via the reaction

$$n \rightarrow p + e + \nu$$

where p is the (positively charged) proton, n is the (neutral, i.e. lacking any charge) neutron, e is the (negatively charged) electron, and ν is a neutral particle known as the neutrino. In this reaction the total electrical charge on both sides is zero, as demanded by charge conservation (the charges on the right are +1, –1, and 0), but the total number of charged particles changes. This reaction, which is a form of RADIOACTIVE DECAY, obeys the law of charge conservation. It is typical of interactions among elementary particles, however, in that it creates new kinds of electrically charged particles.

Electrical Properties of Matter

Substances can be categorized as conductors, insulators, or semiconductors, depending on how the electrons in the material are arranged

When atoms come together to form substances, the electrons in any one atom are influenced by the presence of the electrons in the other atoms (*see* BAND THEORY OF SOLIDS; MOLECULAR ORBITAL THEORY). The ability of electrons to move around in a material therefore depends on the details of how a substance is formed. We can identify three different kinds of substances, each with a different kind of electrical behavior.

Conductors
In some substances, electrons are free to move around. In metals, for example, individual atoms share their electrons with all the other atoms in the material (*see* CHEMICAL BONDS; FREE ELECTRON THEORY OF CONDUCTION). If a substance of this type is subjected to a voltage (e.g., by being connected across the poles of a battery), then these electrons will be free to move. It is the movement of electrons in a substance that constitutes an electrical current. A substance in which an electrical current flows in response to the application of a voltage is called a conductor. Metals are the commonest conductors—the electrical cables in buildings are usually made of copper, one of the better cheap conductors. Other sorts of materials can carry electricity besides metals. Fluids with ions dissolved in them (e.g., salts dissolved in water), plasmas, and certain long organic molecules can carry electrical current as well.

It is not only electrons, however, that can constitute an electrical current. For example, normal tapwater has many different kinds of salts dissolved in it, and many ions (both positive and negative) floating around. Water is therefore a conductor as well, because these ions can move in response to an applied voltage and constitute an electrical current. That is why it is extremely dangerous to be in contact with water when you are working with any electrical equipment.

Insulators
In other materials (e.g., glass and plastics), electrons are bound very tightly to individual atoms or molecules. They are not free to move if the material is subjected to a voltage. These materials are called insulators.

The commonest insulators in use today are plastics. You can think of plastics as being a mix of long-chained hydrocarbon molecules (a plate of spaghetti is a good mental image of the arrangement of molecules in a plastic). In these substances the electrons are firmly tied to the individual molecules (the strands of the spaghetti) and therefore will not move in response to an applied voltage. Other substances, such as rubber and glass, are also occasionally used as insulators.

Both conductors and insulators have important roles to play in our modern electrically driven society. Conductors carry electrical energy from generators to homes and other end-use points, and insulators are used to shield human beings from that electrical current.

Semiconductors

In a few substances—silicon is probably the most familiar—all the electrons are taken up in the chemical bonds linking atoms. You would expect these materials to be insulators, since there are no free electrons left over. But that is not quite what happens, because some of the electrons are bound very loosely to their atoms. This means that thermal motion (the constant vibration of atoms in solids at any temperature above absolute zero) can shake these electrons loose and allow them to move around. Such a substance will conduct electricity when a voltage is applied across it. It will not conduct electricity very well—copper conducts electricity a million times better than silicon, for example—but it will conduct a bit. A substance like this is called a semiconductor.

Actually, there are two ways in which charge can move around in a semiconductor. The first, already mentioned, is the motion of electrons that have been shaken loose from the structure and are roaming around. The second is rather unusual. When an electron is shaken loose from the structure, it leaves behind a vacancy in that structure—something that scientists call a *hole*. An electron from somewhere else in the structure can jump over and fill this hole, but in vacating its original position it leaves behind another hole. This process can go on—one electron jumping and filling a hole left by another—and its net effect is that the hole, or absence of charge, can move around the semiconductor. This, effectively, is also a movement of electrical charge, and can be regarded as an electrical current.

The best analogy for conduction by holes is to think of a traffic jam in which one car moves up to fill an empty space, then another car moves up to fill the space occupied by that car, a third car moves up to fill that space, and so on. There are two ways of describing this: in terms of the actual forward motion of the cars, or in terms of the absence of a car (the empty space) moving back along the line. In the same way, physicists talk about the absence of a negatively charged electron moving through the semiconductor in terms of the motion of a positively charged hole in the opposite direction. Thus, in semiconductors, both electrons and holes can conduct electricity.

Because the current in semiconductors can be controlled very closely, semiconductors are widely used in computers and in other modern electronic devices.

Electro-magnetic Spectrum

There are many types of electromagnetic waves, from radio waves to gamma rays. All travel at the speed of light and differ from each other only in their wavelength

When MAXWELL'S EQUATIONS were first written down, it was clear that they predicted the existence of a wave phenomenon previously unknown in nature. It appeared that these *electromagnetic waves* were electromagnetic fields moving through space. When James Clerk Maxwell first showed that his equations led to this prediction, he also discovered that the speed of these waves, which is related to the force between electrical charges and the force between magnets, is exactly the same as the speed of light. This speed is so important that it is given a special letter, c, to distinguish it from other speeds.

When Maxwell made this discovery, he was able to identify visible light as a type of electromagnetic wave. Now, it was known that visible light has wavelengths spanning a range roughly from the size of about four thousand atoms, for violet light, to about eight thousand atoms, for red light. (In the units used by physicists, this is about 400 to 800 nanometers. One nanometer is a billionth of a meter, or 10^{-9} m.) All the colors of the rainbow were known to correspond to wavelengths within this rather narrow band of values. But Maxwell's equations said nothing about restricting the range of possible wavelengths of electromagnetic waves. In fact, the discovery that these waves ought to exist, coupled with the fact that visible light has such a small range of wavelengths, has been likened to listening to a symphony orchestra, with all its myriad instruments, and being able to hear only the piccolo.

Very soon after Maxwell's prediction of the existence of electromagnetic waves, other kinds of waves began to be discovered. The first of these was the discovery of radio waves in 1888 by the German physicist Heinrich Hertz (1857–94). The only difference between radio waves and light waves is that the wavelength of a radio wave can be anything from many miles to several feet. According to Maxwell's theory, electromagnetic waves are emitted whenever an electrical charge is accelerated. Electrons made to move back and forth in a tall transmission tower create radio waves which are broadcast into the atmosphere. Charges accelerated in other ways produce other kinds of radiation.

Like light waves, radio waves can travel a long distance through the atmosphere without being absorbed, and that makes them very useful for communication. It was as early as 1894—only a few years after the discovery of radio waves—that the first successful transmission of a signal by radio was made by the Italian physicist Gugliemo Marconi (1874–1937).

Once it was found that Maxwell's electromagnetic waves really existed, the rest of the picture filled in very quickly. Electromagnetic waves of all wavelengths are now known, and the great majority have had uses found for them. The frequencies of the waves and the energies of their associ-

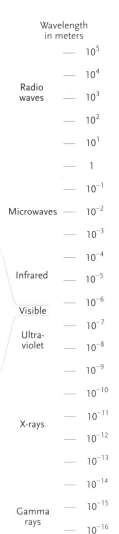

— 10^5

— 10^4

Radio
waves — 10^3

— 10^2

— 10^1

— 1

— 10^{-1}

Microwaves — 10^{-2}

— 10^{-3}

— 10^{-4}

Infrared — 10^{-5}

— 10^{-6}

Visible

— 10^{-7}

Ultra-
violet — 10^{-8}

— 10^{-9}

— 10^{-10}

— 10^{-11}
X-rays

— 10^{-12}

— 10^{-13}

— 10^{-14}

Gamma — 10^{-15}
rays
— 10^{-16}

Electromagnetic waves form a continuous range of wavelengths and energies, from radio waves to gamma rays.

ated photons (*see* PLANCK CONSTANT) increase as their wavelength decreases. The collection of all these waves is what is called the electromagnetic spectrum. From the longest to the shortest wavelengths, the spectrum is divided into the following regions:

Radio waves

As indicated above, the wavelengths of radio waves can be long—even longer than the radius of the Earth, which is nearly 4000 miles (around 6400 km). They are used extensively for communication. Both AM and FM radio broadcasts are at wavelengths in the radio region (FM radio uses shorter wavelengths than AM). Because long wavelengths can actually bend around the curvature of the Earth as they go, they were once used extensively for communication over long distances—a function now largely taken over by satellite communications. Short wavelengths cannot bend in this way, and because they travel in straight lines they have a more limited range. If you've ever driven out of a city while listening to your favorite FM station, you know that the signal will break up after you've traveled fifty miles or so (80 km). This is because at this distance most of the energy in the broadcast signal is aimed into space and can't be picked up by your car's antenna. For wavelengths of a few hundred feet (about 100 m), it is possible to send radio signals a long distance by bouncing the waves off layers of charged particles in the upper atmosphere—this is how short-wave radio signals can be picked up far from the transmitter.

Microwaves

Microwaves have wavelengths from a few feet to a fraction of an inch (from about 300 mm down to 1 mm). Longer-wavelength microwaves, like radio waves and light, can travel through the atmosphere without being absorbed and are therefore very useful in telecommunications. Virtually all signals sent to and from satellites are in this range. The size of a typical satellite TV antenna corresponds roughly to the middle of the wavelength range of microwaves.

Shorter-wavelength microwaves also have many industrial uses, but you are most likely to have met them in the shape of the microwave oven in your kitchen. In a microwave oven, electrons are spun around in a device called a klystron. These electrons emit microwaves at a frequency which is easily absorbed by water molecules. When food is placed in a microwave oven, the water molecules in the food absorb the microwaves, move faster, and thus heat the food. A microwave oven cooks food from the inside out, while an ordinary oven cooks food from the outside in.

Infrared

A division of the electromagnetic spectrum contains radiation whose wavelengths run from a fraction of an inch to about eight thousand atomic diameters (1 mm to 800 nm). This is the domain of infrared radiation, which can be detected directly by the human body. When you stretch out your hand out toward and feel the heat, you are detecting infrared radiation. Some animals (e.g., pit vipers) have special sensors that allow them to detect the infrared radiation given off by their prey.

Since most objects at the surface of the Earth are at a temperature at which they give off infrared radiation, infrared detection plays many important roles in technology. Infrared glasses allow people to "see in the dark," and it is often possible to detect things like motors and metal that have been heated by the Sun during the day when they radiate this heat away at night. Infrared detector are also used by rescue workers, for example to search for survivors in the rubble of buildings that have collapsed during an earthquake.

Visible light

As indicated above, the wavelengths of visible light run from about eight thousand to about four thousand atomic diameters (800–400 nm). The human eye is a superb instrument for detecting electromagnetic radiation in this wavelength range. There are two reasons for this: First, as mentioned above, the atmosphere is transparent to visible light, and second, the surface temperature of the Sun (about 5000°C) is such that much of the Sun's radiation is in the visible range. Thus, our primary source of light produces a lot of electromagnetic radiation in this range, and the environment in which we live transmits it easily. Small wonder, then, that the eye evolved as it did (*see* EVOLUTION).

I have to stress that there is nothing that makes this particular range of wavelengths special from the point of view of the physics of electromagnetic radiation. It is simply a small band in a very wide spectrum. It is important to us only because we are equipped with sensors that can detect it.

Ultraviolet

Wavelengths from a few thousand atoms to a few atoms long (from about 400 nm down to 100 nm) comprise ultraviolet radiation. As we move into this band, we are beginning to encounter radiation whose energy is high enough to cause harm to living things. Ultraviolet radiation is often used in hospitals, for example, to sterilize surgical equipment—in other words, this radiation will kill any living bacteria on the equipment.

Organisms dwelling on the Earth's surface are protected from much of the Sun's ultraviolet radiation by the presence of a layer of ozone molecules in the atmosphere (*see* OZONE HOLE). If it were not for this protection, life on our planet would probably never have come out of the sea to colonize the land. Despite the protection afforded by the ozone layer, sufficient ultraviolet radiation can reach the surface to cause skin cancer in pale-skinned people.

X-rays

Radiation with wavelengths from a few atoms across down to sizes typical of hundreds of nuclei are called X-rays. This radiation can penetrate living tissue, and is therefore enormously useful in medical diagnosis. In fact, as was the case with radio waves, it was not long after these rays were discovered in 1895 that they were put to use, to create the first X-ray photograph in a hospital in Paris. (The Parisian newspapers of the time seemed too interested in the fact that X-rays could penetrate clothing to recognize their potential in medicine.)

Gamma rays

The lowest-wavelength, highest-energy part of the electromagnetic spectrum is home to gamma rays—extremely energetic photons. Gamma rays are routinely used in cancer therapy to obliterate tumors, but extreme care has to be taken not to harm the surrounding healthy tissue.

It is important to emphasize that even though all of these types of radiation appear to be very different, they are brothers under the skin. They all consist of electric and magnetic fields oscillating in space, they all move at *c*, the speed of light, and they differ from one another only in their wavelengths, and consequently in the energies they carry. I ought to add too that the wavelength ranges I have given do not indicate fixed boundaries (and you may find slightly different values in other books). For example, the longest-wavelength microwaves can equally well be regarded as short-wavelength radio waves.

Electron, discovery of

The electron is a subatomic particle that can be affected by both electric and magnetic fields

In the second half of the 19th century, scientists were studying a phenomenon which they called cathode rays. The experimental apparatus they used was very simple. It consisted of a sealed glass tube that had a very low-density gas of some type in it. At one end of the tube was an electrical wire through which a current could be passed. This wire was called a *cathode*. At the other end of the tube was a plate attached to another electrical wire. This plate, known as the *anode*, could be brought to a high voltage. When an electrical current was passed through the cathode and the voltage on the anode was stepped up, a glowing discharge would form in the gas, first around the cathode and then extending toward the anode. The cause of the discharge was attributed to *cathode rays*.

By the late 1880s, the debate among scientists about the nature of cathode rays was hotting up. In Germany, the general feeling was that they were some kind of radiation, similar to light. In England, the favored explanation was that they were a stream of some kind of particle—perhaps of atoms that had lost some of their electrical charge. Each side was able to cite observational support for its point of view. For example, those who believed cathode rays to be particles pointed out that the lines of glowing gas could be deflected by bringing a magnet near the tube. Light isn't affected by a magnetic field, but atoms conceivably could be. The cathode-rays-as-radiation camp countered with the argument that no known particle could produce other observed effects. For example, these rays cast shadows, so they had to be some sort of radiation.

In 1897 a young English researcher, J.J. Thomson, resolved this issue once and for all, in the process gaining credit for discovering the particle we now know as the electron. Thomson's apparatus was an extension of the cathode ray tube described above. He put a large coil of wires around the tube and passed a current through them, thereby immersing the tube in a magnetic field (*see* AMPÈRE'S LAW). He then installed a set of electrical plates parallel to the lines of the tube. When he turned on the cathode rays, he could now watch how they reacted under both electric and magnetic influences.

He first confirmed that, in the presence of a magnetic field alone, the cathode rays followed curved paths. He also confirmed that if he turned on the electric field between the two plates in the absence of a magnetic field, the cathode rays would also be deflected. Finally, he showed that it was possible to arrange things so that the cathode rays would move in a straight line when both the magnetic field and the electric field were turned on. In other words, he showed that the deflection that would be caused by a magnetic field could be canceled exactly by a countervailing deflection caused by the electric field.

It turns out that when you do the calculation for this cancellation effect, the relation between the electric and magnetic fields needed to produce exact cancellation depends on the speed of the particle. From his measurements, Thomson was therefore able to determine how fast the cathode rays were traveling. His result, that they traveled at only a fraction of the speed of light, proved conclusively that cathode rays could only be particles, since light and all other electromagnetic radiation moves at the speed of light (*see* ELECTROMAGNETIC SPECTRUM). Cathode rays had to be caused by an as yet unknown particle. Thomson called them "corpuscles," but they were soon renamed electrons.

It was immediately clear that electrons had to exist inside the atom—where else could they come from but from inside atoms? The atom, until then thought to have been the fundamental constituent of matter (*see* ATOMIC THEORY), therefore had to be divisible. This, together with the subsequent discovery of the nucleus (*see* RUTHERFORD EXPERIMENT), is at the center of our modern concept of the atom.

Cathode rays and cathode ray tubes of the type described here were the primitive forerunners of television sets. In a TV, precisely controlled amounts of electrons are boiled off cathodes and steered by magnetic fields so that they strike certain spots on the inner surface of the screen, creating a pattern. The arrival of the electrons triggers the PHOTO-ELECTRIC EFFECT, which results in the emission of visible light, forming the picture you see when you look at the screen.

JOHN JOSEPH THOMSON (1856–1940) English physicist, known always as "J.J. Thomson." He was born in Cheetham Hill, near Manchester, the son of an antiquarian bookseller. He won a scholarship to Cambridge in 1876, becoming Cavendish Professor of Experimental Physics in 1884, and turning the Cavendish Laboratory into a renowned research institute. Thomson won the 1906 Nobel Prize for Physics for his discovery of the electron and subsequent work on the conduction of electricity in gases. His son, George Paget Thomson (1892–1975), won the same prize in 1937 for discovering the diffraction of electrons by crystals.

Elementary Particles

There are many different kinds of subatomic particles inside the nucleus

There has been a steady progression in our understanding of the basic structure of matter. The ATOMIC THEORY showed that things are not always what they seem, and that complexity at one level can be explained by the existence of simplicity at the next level down. During the 20th century, the discovery of the structure of atoms (which the model known as the BOHR ATOM sought to capture) focused the attention of scientists on the nucleus of the atom.

At first, it was thought that there were only two kinds of particles in the nucleus—protons and neutrons. Starting in the 1930s, however, a series of unexpected experimental results showed that this simple notion wouldn't do, and that the nucleus was a dynamic system inhabited by all manner of particles whose fleeting existence played a crucial role in nuclear processes. By the 1950s, the study of these so-called elementary particles had become the frontier field of physics.

The general technique for studying elementary particles is to direct high-velocity streams of protons or electrons toward target nuclei, and then to observe the fragments thrown out by the resulting collisions. According to the theory of RELATIVITY, the kinetic energy of the fast-moving particles can be converted into mass via the famous equation $E = mc^2$, so that new kinds of particles can be (and are) created copiously.

From the 1930s, scientists began to observe the collisions of cosmic rays with nuclear targets. *Cosmic rays* are high-speed particles, mostly protons, generated by a variety of processes in the universe, which are constantly raining down on the Earth's atmosphere. Here was a beam of high-energy projectiles that nature was providing free of charge, and experimenters were not slow to make use of cosmic rays in their research. In the 1950s, machines called particle accelerators came into their own, and precisely controlled beams of high-energy particles became

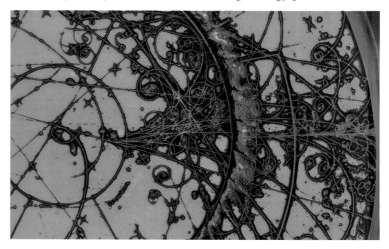

A photograph of a collision between elementary particles in the Big European Bubble Chamber at the European Centre for Nuclear Research (CERN) in Geneva, Switzerland. The colors have been added to make the paths of the particles clearer. The blue lines trace a chain of bubbles which form around atoms that have been disturbed by the passage of a charged particle.

available on tap. In the decades that followed, physicists discovered over two hundred different kinds of elementary particles.

With the exception of the proton and electron, all of these particles are unstable—that is, they last for only a short while before they decay into other particles (away from the nucleus, even the neutron soon decays). But all that matters for a particle's participation in nuclear phenomena is that it lasts long enough to move from one side of the nucleus to the other.

The elementary particles fall into two classes:

Leptons

These are particles, like the electron, that do not participate in the nuclear maelstrom. There are six such particles known. There is an electron family consisting of the electron itself and two other particles, the *mu* and the *tau,* which are like the electron except that they are more massive. These heavier particles decay, eventually, into a set of products that includes the electron. There are also three massless (or nearly massless—the particle physics jury is still out on this), electrically neutral particles known as *neutrinos,* each one paired with a member of the electron family. The word lepton comes from the Greek *leptos,* meaning "small."

Hadrons

These are the particles that exist inside the nucleus of the atom. The best known are the proton and neutron, but there are (literally) hundreds of short-lived relatives. With the exception of the neutron, they are all unstable, and they can be classified by the particles into which they decay. If the decay products of a particle ultimately produce a proton (along with other stuff), the particle is called a *baryon* (from *barys,* Greek for "heavy"), while if no proton is produced in its decay, the particle is called a *meson* (from *mesos,* Greek for "middle"). The word hadron itself comes from the Greek *hadros* ("bulky").

The confused picture of the subatomic world, added to by the discovery of each new hadron, gave way to a new simplicity with the arrival of the concept of the quark (*see* QUARKS AND THE EIGHTFOLD WAY). In the quark model the hadrons (but not the leptons) are made from particles more elementary still. The baryons are made from three quarks, the mesons from a quark–antiquark pair (*see* ANTIPARTICLES).

The particles described above are those that make up the nucleus— think of them as the bricks of the universe. Another group of particles, called *gauge bosons* (the photon is one of them), are associated with the forces that hold particles together (*see* THEORIES OF EVERYTHING), and can be thought of as the mortar of the universe.

Equilibrium

Equilibrium is the state of a system when the forces acting on it are in balance. Equilibrium can be stable, unstable, or neutral

The concept of equilibrium is ubiquitous in the sciences. It can be applied to any system, from a planet circling its star to the population of fish in a tropical lake. But the easiest way to think about equilibrium is in simple mechanical terms. In mechanics, a system is in equilibrium if all the forces on it balance, canceling one another out. If you are sitting in a chair and reading this book, for example, the downward force of gravity on your body is being canceled by the upward force exerted on you by the chair. You neither fall nor rise—you are in equilibrium.

There are several different types of equilibrium, each corresponding to a different physical situation.

Stable equilibrium

This is what most people mean when they use the word "equilibrium." Imagine a ball rolling around on the bottom of a hemispherical bowl. When the ball is dead center in the bowl, it, like you in your chair, is in equilibrium because the downward force of gravity is balanced by the upward push of the bowl. If the ball is displaced slightly from this equilibrium position by being pushed up the side of the bowl, the force of gravity will pull it back down, toward its equilibrium position. Any system in which forces will act to restore equilibrium if it is disturbed is said to be in stable equilibrium.

You, sitting in your chair, are also in stable equilibrium. If you suddenly add to your weight (e.g., if a child sits on your lap), then the atoms in the chair will readjust their positions to exert a greater upward force. You may sink a little deeper into the cushions, but you will remain in equilibrium.

There are many examples of stable equilibria in nature. Consider, for example, the relation between predator and prey in an ecosystem (*see* PREDATOR–PREY RELATIONSHIPS). The numbers of predator and prey will come to an equilibrium—so many foxes for so many rabbits, for example. If for some reason the equilibrium number of prey is disturbed (e.g., by the rabbits having a successful breeding season) then the number of predators will increase and drive the rabbit population down again. Although these aren't physical forces, they are still forces, in a broader sense of the term, operating in the ecosystem to drive it back toward equilibrium following a disturbance.

You can see the same sort of effect at work in economic systems. If the price of a stock dips suddenly, then bargain hunters may enter the market and drive the price back up. In this case, the stock price would be in stable equilibrium. (In a real stock market, as in a real ecosystem, it is also possible to have external factors whose effect is to drive the system away from equilibrium—in effect, to change the shape of the bowl.)

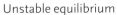

Unstable equilibrium

Not every equilibrium is stable. Imagine, for example, a ball balancing on a knife blade. The downward force of gravity on the ball is clearly balanced by the upward force exerted by the blade, so the ball is in equilibrium. But if you displace the system ever so slightly by pushing on the ball with the tiniest of forces, the force of gravity acts to pull the ball farther away from the blade, moving it farther from equilibrium. Such a system is said to be in unstable equilibrium.

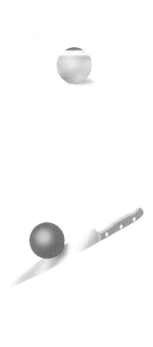

An example of unstable equilibrium can be seen in the Earth's ice ages (*see* MILANKOVIĆ CYCLES). The surface temperature of our planet is determined by the balance between the incoming sunlight and the energy radiated away into space. Under certain circumstances, there can be an unstable equilibrium. It works this way: In a given winter, a little more snow than usual falls. The following summer is then a little cooler than usual, so more snow stays on the ground, reflecting more sunlight back into space. This makes the next year cooler, so even more snow falls and stays on the ground, reflecting more sunlight back and making the next year even cooler. It's not hard to see that the further this system departs from the original equilibrium, the more the dynamics of the system push it further still. The end result is that the snow builds up into layers of ice several miles high and moves to lower latitudes, pushing the planet into an ice age. It's hard to think of an equilibrium more "unstable" than that!

One type of unstable equilibrium that deserves special attention is called *metastable equilibrium*. Picture this as a ball rolling in a small valley on the side of a steep slope. A small deviation from equilibrium will generate forces that move the ball back toward dead center, but a large push will move the ball out of the little valley and plunging down the hillside. Systems in metastable equilibrium tend to remain in equilibrium for a while, then leave it as they "climb the lip" and move away.

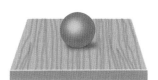

A common example of metastable equilibrium is found in atoms in certain types of lasers. Electrons move to metastable orbits in atoms, then stay there until a photon of light goes by. This has the effect of pushing the electron out of its "valley," so the electron "falls down" and the atom emits another photon that, ultimately, may become part of the laser beam.

Three types of equilibrium illustrated by a ball: in a bowl, it will always return to its original position if disturbed; balanced on a knife edge, the slightest force will disturb the equilibrium; on a tabletop, the ball will assume equilibrium wherever it is moved.

Neutral equilibrium

The transition between stable and unstable equilibrium occurs when deviations produce no force to move the system at all. Think of a ball on a level tabletop—no matter where it is moved, no force will arise to return it to its original position.

Equivalence Principle

It is impossible to tell by experiment whether you are in an accelerated frame of reference or a gravitational field

You have probably had the experience of standing in an elevator and feeling momentarily heavier, pushed down into the floor, as it started upward (or perhaps feeling momentarily lighter as is started down). Although you may not have realized it at the time, you were experiencing the equivalence principle. As the elevator started upward, it pushed the floor into your feet, and you felt that you were being pushed down. Had you been standing on a scale, your registered weight would have increased during the acceleration. When the elevator reached "cruising speed" the acceleration stopped and your weight fell back to normal. Thus, acceleration produces the same effect as gravity.

Imagine that you are in an elevator in outer space, well away from any gravitational field, accelerating upward at 32 feet (9.8 meters) per second each second. If you were to stand on a scale, you would find that you weighed the same as would be indicated by your bathroom scale on Earth. If you held a ball out and let it go, the floor would accelerate upward and hit the ball. As far as you were concerned, the ball would be accelerating downward at 32 feet per second per second, just as it would if you dropped it on Earth. The equivalence principle simply states that there is no experiment you could do in that elevator that would tell you whether what you saw was due to acceleration or to the force of gravity.

The equivalence principle makes some interesting predictions about the behavior of light in a gravitational field. Imagine riding in your elevator and shining a pulse of light toward a mark on the far wall. That mark will accelerate upward during the time it takes the pulse to get from your flashlight to the wall. Thus, it will hit the wall below the mark,* and you will interpret this effect as the light falling downward during transit. If we now invoke the equivalence of gravity and acceleration, we will have to predict that light should be deflected when it encounters a gravitational field, just as a projectile would be. For a beam of light from a distant star that just grazes the edge of the Sun, Einstein predicted that the deflection would be 1.75 seconds of arc (about one two-thousandth of a degree), whereas Newtonian physics predicted just half of that. Thus the measurement by Sir Arthur Eddington (1882–1944) during the 1919 total solar eclipse of a deflection of 1.6 arc seconds was a triumphant experimental confirmation of the theory of general RELATIVITY.

The equivalence principle also predicts, by similar reasoning, that light moving upward in a gravitational field should have its frequency shifted toward the red, another effect that has been experimentally verified.

The equivalence principle is built into the theory of general relativity, but the two are not the same thing. The equivalence principle is restricted to discussing the effects of gravity and acceleration, but every verification of equivalence also verifies relativity.

---

* Only very slightly below the mark of course, but you have to imagine here that you can perceive such minuscule differences in position.

Evolution, theory of

Life on Earth arose through physical and chemical reactions and developed through the process of natural selection

Before we start our discussion of evolution, arguably the most important concept in the life sciences, I would like to remind you of a point that was made in the Introduction. The word "theory" as used by scientists does not necessarily imply a lack of confidence in the ideas being treated. It comes instead from custom and historical usage, and in fact many theories (such as general relativity) rank among the most widely accepted components of the scientific world view.

The fact that evolution has taken place is no longer questioned by serious scientists, although there are several theories competing to describe just how it happened. In this respect, evolution is similar to gravity. There are several theories of gravity—NEWTON'S LAW OF GRAVITATION, general relativity, and, someday perhaps, a THEORY OF EVERYTHING. There is also, however, the *fact* of gravity—the fact that if you drop something, it will fall. In the same way, there is the *fact* of evolution, even as there are arguments among scientists on the details of the theory.

When it comes to the history of life on Earth, there are two stages to consider, each governed by different principles. In the first, inorganic materials on the early Earth formed the first living cell in a process governed by chemical evolution. In the second stage, the descendants of that first cell diversified and produced all of the living things on the planet today. The governing principle here was natural selection.

Chemical evolution

The notion that we can understand the process by which non-living materials organize themselves into simple living systems is a fairly recent one in human thought. An important milestone in this realization was the MILLER–UREY EXPERIMENT of 1953, in which it was shown for the first time that ordinary chemical reactions could give rise to the basic MOLECULES OF LIFE. Since that time, scientists have suggested many other ways in which chemical evolution could have taken place. Some of these ideas are listed below, but it is important to remember that there is no consensus on which of them might be right. We do know, however, that one of these processes, or one that no one has thought of yet, produced the first living cell on our planet (unless life originated elsewhere—the idea of *panspermia,* discussed under ACIDS AND BASES).

Primordial soup Processes such as those replicated in the Miller–Urey experiment produced molecules in the atmosphere, which rained down into the ocean. There (or perhaps in a tidal pool) they organized themselves into the first cell by some as yet unknown process.

Primordial oil slick Miller–Urey processes can produce *lipids,* molecules that form spontaneously into little spheres (you can often see such spheroids floating in a bowl of soup). Each sphere contains a random collection of molecules. In the billions of bubbles on the ocean surface, eventually one could have the right collection of molecules to take in energy and materials, and divide. This would be the first cell.

RNA world One of the problems in evolutionary theory involves the development of the coding system using the molecule DNA (*see also* MOLECULAR BIOLOGY). The problem is that DNA codes for PROTEINS, but the action of proteins is required in order for the DNA code to be read. Recently, scientists have discovered that RNA, which today plays a role in turning the DNA code into proteins, can also play the kind of role that proteins play in living systems. The formation of RNA molecules seems to have been crucial to the development of life.

Ocean vents At the tremendous pressure prevailing on the ocean floor, chemistry is different from what it is at the surface. Scientists are investigating this chemistry to see whether it would make the development of life easier. If this turns out to be so, then perhaps life developed first on the ocean floor, then migrated to land.

Autocatalytic sets This concept comes from the theory of COMPLEX ADAPTIVE SYSTEMS. The suggestion is that the development of the chemistry of life was not a step-by-step process, but an emergent property of the chemistry of the primordial soup.

Clay world The first template for life may not have been chemical reactions, but static electrical charges on the surface of clays on the ocean floor. In this scheme, the complex molecules of life are assembled not by random combinations, but by electrons on the surface of clays holding small molecules together while they assembled themselves into larger ones.

As you can see, there is no shortage of ideas about how life might have developed from inorganic materials. Until the late 1990s, though, the origin of life was not a high-priority field, and progress in sorting through these theories was slow. In 1997, NASA decided that the investigation of the origin of life was one of its primary missions. I suspect that before long scientists will be able to create in their laboratories simple organisms like those that might have existed on our planet 4 billion years ago.

Natural selection

Once the first living, reproducing organism appeared on the planet, life "shifted gear," as it were, and change began to be governed by the process of natural selection. It is natural selection that most people have in mind when they use the term "evolution." The concept was introduced in 1859 when the English naturalist Charles Darwin published his monumental book *On the Origin of Species by Means of Natural Selection, or the Preservation of Favoured Races in the Struggle for Life*. The idea of natural selection, which had been conceived of independently by Alfred Russel

Darwin's Finches

The diversity of finches in the Galapagos islands is one of the best examples of natural selection in action. Darwin's theory of evolution was based strongly on the observation of nature. During his tenure as naturalist on the *Beagle,* Darwin visited the Galapagos Islands, home to some of the most remote habitats on Earth. About 40% of the bird species on the island are finches. Presumably, they are the descendants of a single species of finch that arrived on the islands long ago.

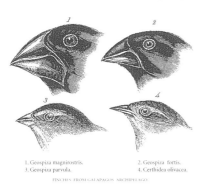

1. Geospiza magnirostris.   2. Geospiza fortis.
3. Geospiza parvula.   4. Certhidea olivacea.

FINCHES FROM GALAPAGOS ARCHIPELAGO.

Darwin noticed that the finches had evolved to fill different ecological niches. The original ancestor of the Galapagos finches was a ground-dwelling, seed-eating bird. Its descendants today include birds that live on the ground, ones that live in trees, birds that feed on seeds, on cacti, on insects. It was this diversity in closely related birds that is supposed to have suggested to Darwin the process of natural selection. Because of this, Darwin's finches have become something of an icon in the history of science.

The Peppered Moth

According to the theory of evolution, the characteristics of populations change in response to their environment, and those characteristics which give organisms a better chance of reproducing will be

favored. One of the best studies of natural selection in action was carried out on an insect called the peppered moth (*Biston betularia*). Living in central England, these moths were most often found on lichen-covered trees. The lichen in this area is light-colored, and the moths that matched the lichen were less likely to be seen by predators.

During the 19th century central England became heavily industrialized, and much of the moth's home territory became severely polluted by smoke and soot. The tree trunks turned black, a significant change in the moth's environment. The moth population started to change, with darker colors being favored in polluted areas. Eventually, entire populations turned black. The change took place just as evolutionary theory predicted —the small number of dark moths in the normal population gained an enormous competitive advantage because of the change in the environment, and gradually their genes came to dominate.

The explanation of changes in the peppered moth population, like all scientific hypotheses, had to be verified by experiment. The experimenter was an amateur entomologist, Henry Bernard David Kettlewell (1907–79), who carried out his studies in the 1950s. He marked peppered moths on the underside of the wings, where the marks would be invisible to a predator. He then released a sample of marked light- and dark-colored moths near Birmingham, in the heart of the Black Country, and another in rural Dorset, in the relatively unpolluted southwest of England. After release, he would come back to the area at night, using a light to attract and recollect his moths. He found that near Birmingham he recovered 40% of the dark moths, as against 20% of the light, while in Dorset the recovery rates were 6% dark and 12% light. It was clearly advantageous for a moth to be dark in the polluted region near Birmingham, but light in unpolluted Dorset.

The story of the peppered moth isn't over yet, though. Starting in the 1960s, air pollution controls began to be instituted in England, and the soot accumulation in the Black Country began to decrease. In response, the moth population has started to shift back to light from dark—a result that would, once more, be predicted from Darwinian arguments.

Wallace (1823–1913), depends on two things: Members of a species will differ in some respects from one another, and there will always be competition for resources. The first of these postulates is obvious to anyone who has observed any population (humans included). Some members are bigger, some can run faster, some have coloration that blends in better with the surroundings. The second postulate reflects a sad fact about the natural world—there are many more organisms born than can survive, and so there is constant competition for resources.

Taken together, these two postulates lead to an interesting conclusion. If some individuals have a trait that allows them to compete better in a given environment—predators with musculature that allows them to hunt more efficiently, for example—then they are more likely to survive into adulthood and produce offspring. And the offspring are likely to carry the same trait. In modern terms, we would say that the individuals are likely to pass on the genes that are responsible for the fast running. Slow runners, on the other hand, are less likely to survive and have offspring, so that their genes are less likely to be passed on. Because of this, the next generation will have more individuals with the "fast" genes than the current one, and the generation after that even more. In this way a trait that increases the probability of survival eventually spreads throughout an entire population.

This process is what Darwin and Wallace called natural selection. Darwin likened it to artificial selection, the process by which humans breed plants and animals with desirable traits by selecting the adults which are allowed to mate. If humans can do it, Darwin argued, why can't nature? The increased survival rate of adaptive traits for successive generations over long periods of time would be more than enough to produce the diversity of species we see on our planet today.

Darwin, following the doctrine of UNIFORMITARIANISM, felt that the production of new species would be a gradual process—two populations would grow further and further apart until they could no longer interbreed. More recently, scientists have noted that this pattern is not always followed. Instead, a species will stay the same for a long time, then change suddenly, in a process called *punctuated equilibrium*. In fact, we see both kinds of *speciation* in the fossil record, which should not surprise us, given our current understanding of genetics. We now understand that the basis for the first of the two postulates above is the fact that different individuals have different versions of their genes written into their DNA. Changing the DNA can produce anything from no effect (if the change is in DNA that isn't used by the organism) to a huge effect (if the change is in a gene that codes for a crucial protein). Once the gene has changed, producing either a gradual or a sudden effect, natural selection acts either to spread it throughout the population (if it is advantageous) or wipe it out (if it is deleterious). The rate of change, in other words, depends on the genes, but once the change is there it is natural selection that operates to produce widespread changes in populations.

Like all scientific theories, the idea of evolution has to be tested against observation of the real world. There are, in fact, three large classes of observations that support the theory.

The fossil record

When an animal or plant dies, its remains will most likely end up being scattered across the environment. Occasionally, though, one will be buried—in sediment, perhaps, as the result of a flood—and protected from the forces of decay. Over time, as the sediment turns to rock (*see* ROCK CYCLE), slow chemical processes replace the calcium in the skeleton or other hard body parts with minerals from the surrounding rock. (On rare occasions, conditions are such that softer body structures, such as skin or feathers, may be preserved as well.) Eventually this process produces a perfect replica in stone of the original part—a fossil. The sum of all the fossils that have been found is called the fossil record.

The fossil record goes back approximately 3.5 billion years, to impressions of what was once green pond scum on ancient rocks in Australia. It tells a dramatic story of the gradual growth in complexity and diversity that has produced the huge variety of life forms that now inhabit the Earth. For most of the ecological past, life was relatively simple, consisting of single-celled organisms. About 800 million years ago, multicelled life forms began to appear. Because they were soft bodied (think of jellyfish), they didn't leave much in the way of fossils, and it is only in the last few decades that scientists have established their presence from the imprints they left in sediments. About 550 million years ago, hard coverings and skeletons appeared, and it is at this point that the fossil record begins in earnest. The first animals with backbones, fish, appeared about 300 million years ago, dinosaurs became extinct about 65 million years ago (*see* MASS EXTINCTIONS), and the first hominids appeared, in Africa, about 4 million years ago. All of these events are clearly written in the known fossil record.

The biochemical record

All living things on our planet share the same genetic code—we are all just different messages written in the universal language of DNA. It is reasonable to expect, therefore, that if life developed as outlined above, then living things today should have differing amounts of overlap in their

DNA, depending on how long ago they shared a common ancestor. Human beings, for example, should share more of our DNA with chimpanzees (with whom we shared a common ancestor about 8 million years ago) than we do with fish (with whom we shared a common ancestor hundreds of millions of years ago). In fact, when we analyze the DNA of living things, we find exactly that pattern—the farther apart two organisms are on the evolutionary tree, the less similar their DNA is. This makes sense, since the latter organisms have had more time to accumulate differences than the former.

The ability of DNA analysis to reveal our evolutionary past is sometimes called the MOLECULAR CLOCK. This is the strongest evidence for the theory of evolution. Human DNA is closer to chimpanzee DNA than to fish DNA. It could have been the other way round, but it isn't. In the language of philosophers of science, this makes the theory of evolution *falsifiable*—it is possible to imagine an outcome that would prove it false. Evolution thus differs from so-called creation science, supposedly based on the biblical Book of Genesis, because there is no observation or experiment that could conceivably convince a supporter of creation science that it is wrong.

Imperfect design

Although it isn't an argument in favor of evolution *per se,* the "argument from imperfect design" is profoundly consistent with the picture of life propounded by Darwin and inconsistent with the notion that living things were designed to carry out specific roles in life. The point is that to get its genes into the next generation, an organism doesn't have to be perfect, just good enough to beat the opposition. Every step on the evolutionary ladder therefore has to build on the one before it, and characteristics that might have been advantageous at one stage will get "frozen in," even when there are better options available.

Engineers refer to this as the QWERTY *effect,* after the letters in the top row of almost all keyboards now in use. When keyboards were originally designed, the main goal was to slow down the typist so that keys wouldn't jam up on mechanical typewriters. This design was frozen in, and it's still with us today, even though more efficient keyboards are available.

In the same way, design features get frozen in at the early stages of evolution and then are retained, even though any new engineering student could do better. Some examples:

— The human eye has some of the circuitry that turns incoming light into nerve impulses in front of the retina, even though those circuits block some of the incoming light.

—The leaves of plants are green, which means that they relect away some of the light that falls on them. Any engineer knows that the proper color for a solar collector is black.

—There are snakes that live in deep underground caverns that have eye sockets located *underneath* their skin. This makes sense if their ancestors once lived on the surface and needed eyes, but no sense if they were created to live in caves.

—Whales have small leg bones imbedded in their bodies. These bones serve no useful function today, but can be understood if the whale's ancestors once lived on land.

—The human appendix has no known function, although the appendices of some herbivores play a part in digesting grass.

The enormous weight of these three converging arrows of evidence long ago convinced serious scientists that not only has Darwinian evolution occurred, but forms the linchpin of any explanation of the working of living systems of our planet.

CHARLES ROBERT DARWIN (1809–82) English naturalist whose theory of evolution by natural selection. Darwin completely revised the human view of nature. He was born in Shrewsbury into a prominent family. His father was a successful physician, and his mother was a member of the Wedgwood family, famous for its pottery. Darwin was an unimpressive student, finding his school education dull and boring. His headmaster criticized him for wasting his time on chemistry experiments, and his father, in one of those devastatingly inaccurate parental broadsides, told him, "You care for nothing but shooting, dogs, and rat catching, and you will be a disgrace to you and all your family."

Darwin was sent to Edinburgh to study medicine, but could not abide attending operations (which were then conducted without anesthetic). He then went to Cambridge to prepare to become a clergyman. There he met people who interested him in geology and natural history, and who later arranged for him to sail (as an unpaid naturalist) on the ship *Beagle* on its five-year voyage of exploration around South America and Australia. It was on this voyage that he made observations of finches that were to lead him to the theory of evolution.

On his return to England, Darwin married his cousin but soon fell ill with what modern scholars have tentatively identified as Chagas' disease, caused by insect bites sustained in Argentina. Retiring to the country, Darwin found himself with ample time and leisure to reflect on the observations and specimens amassed by himself and others. He began to question the accepted notion that species of plants and animals were immutable, and gradually came to see that the natural world could be much better explained by a system in which species evolved over time in response to their environment. His *On the Origin of Species* was published in 1859 and caused an immediate uproar. His thesis was seen by some as an attack on Christian teachings (a view that still survives today), and debates on the Darwinian view occupied much of the latter part of the 19th century.

Today the view that life developed by the process of evolution by natural selection is the unifying theme that ties together all of the life sciences, from ecology to molecular biology.

Exponential Growth

When the net increase in a population is proportional to the number of individuals, the population will grow exponentially

The term "exponential growth" has entered the popular lexicon to designate a rapid, usually unbridled, increase. It is often used, for example, to describe urban sprawl or increases in human populations. Mathematically, however, it has a precise meaning and refers to a specific kind of growth.

Exponential growth occurs in populations in which the net birth rate (births minus deaths) is proportional to the number of individuals in the population. In a human population, for example, the birth rate is approximately proportional to the number of reproducing couples, and the death rate is approximately proportional to the number of people in the population, which we can call N. To a reasonable approximation, then,

$$\text{net birth rate} = \text{number of births} - \text{number of deaths}$$
$$\propto N$$
$$= rN$$

(Here r is what is called a *constant of proportionality*, which enables us to write an expression of proportionality as an equation.)

Now, if dN is the net number of individuals added to the population in a time dt, and if there are N individuals all told in the population, then the conditions for exponential growth will be met if

$$dN = rN\,dt$$

Ever since Isaac Newton invented the calculus in the 17th century, we have known how to solve this equation for N, the number in the population at any given time. (For the record, it is what is known as a *differential equation*.) The solution is

$$N = N_0\,e^{rt}$$

where N_0 is the number of individuals in the population when you start counting, and t is the time that has elapsed since then. The symbol e is a number called the *base of natural logarithms* (it has a value of about 2.7) and the right-hand side on the equation is called an *exponential function*.

One way to imagine exponential growth is to think of a population which consists initially of a single bacterium. In a given time (typically hours or minutes) the bacterium will split, doubling the size of the population. In the next time interval, each of the two bacteria will split and the population will double again, this time to four. After 10 such doublings there will be over a thousand bacteria, after 20 over a million, and so on. The growth simply goes on for ever, the population doubling with each splitting.

There is a story (most likely apocryphal) that the man who invented chess so pleased his sultan that the sultan offered him any reward he named. The man asked that the sultan give him one grain of wheat for the first square on the chessboard, two for the second, four for the third, and so on. The sultan, considering the request too small for the services rendered, begged his servant to reconsider, but he would not. In the end, of course, by the time that original grain of wheat had gone through 64 doublings, there wasn't enough wheat in the world to satisfy the request. In a version of the story I know, at this point the sultan had the inventor beheaded. The moral, as I tell my students, is that sometimes it doesn't pay to be too smart!

What the chessboard problem (and the hypothetical bacteria) shows is that no population can grow forever. Sooner or later it simply runs out of resources—space, energy, water, whatever. Thus, populations may grow at an exponential rate for a while, but sooner or later they have to slow down. This is usually dealt with in the equation by allowing for the rate of increase to slow down when the population approaches the maximum number that can be sustained by the environment. If we call this maximum population K, then the equation can be modified to read

$$dN = rN(1 - (N/K)) \, dt$$

When N is much smaller than K, the term N/K can be neglected and we get back to the original equation for ordinary exponential growth. But when N approaches its maximum value, K, the quantity $1 - (N/K)$ approaches zero, and so does the net increase in population. The total population now stabilizes and remains at the level K. The type of curve associated with this equation is called by various names—*S curve, logistic equation, Volterra equation, Lotka–Volterra equation.* (Vito Volterra (1860–1940) was a prominent Italian mathematician and teacher; Alfred Lotka (1880–1949) was an American mathematician and insurance analyst.) By any name, it is a simple representation of a population that takes off exponentially, then slows down as it approaches some limit. Thus it is a better representation of the growth of real populations than a simple exponential.

With exponential growth, the rate of growth increases indefinitely.

Faraday's Laws of Electrolysis

The amount of chemical change that occurs during electrolysis is proportional to the amount of charge passing through the electrolysis cell

The amount of chemical change produced in a substance by a fixed amount of electrical current depends on the mass and charge of the substance's ions

The laws of electrolysis are just two of Michael Faraday's many contributions to science. *Electrolysis* is what takes place when an electrical current passes through an *electrolyte*—a molten ionic substance (such as a fused salt) or a solution containing ions. The current enters and leaves the electrolyte through two electrodes. Positive ions in the electrolyte move toward the negative electrode, the *cathode,* and negative ions migrate toward the positive electrode, the *anode.* Chemical reactions take place at the electrodes. Faraday carried out the basic research on electrolysis, and his laws are simply stating that the chemical change depends on the flow of electrons (which is what an electrical current is): the more electrons there are, the more chemical change there will be.

Electrolysis is an important industrial process, used both in the extraction of certain metals, and in giving products a "finish," or coating, by a technique called electroplating. An example of its application is in the refining of copper after it has been removed from its ore. A thin sheet of pure copper acting as the cathode is lowered into an electrolyte consisting of a solution of copper sulfate and sulfuric acid, and a bar of impure copper to be refined acts as the anode. When the current starts to flow, copper ions form at the impure bar, along with some iron and zinc ions, and enter the electrolyte. Everything else in the bar (including significant quantities of silver, gold, and platinum) simply flakes out and falls to the bottom of the electrolysis cell. The copper ions move through the electrolyte to the cathode, which becomes coated with pure metallic copper. The zinc and iron remain in solution.

In an industrial setting, electrolytic cells like this may take a month to refine several tons of copper, but the product is then 99.96% pure. Even better, the precious metals recovered from the sludge at the bottom of the tank pay for the cost of the entire operation. In addition to copper, metals such as magnesium, sodium, and aluminum are refined electrolytically on a commercial scale.

In the refining of copper described above, the copper atom loses two electrons when it enters the electrolyte as an ion. At the anode, therefore, it takes two electrons to convert the ion back into a neutral copper atom (you can think of these two electrons as running through an external wire, as an electrical current). Faraday's first law says that if you want to refine twice as much copper, you have to supply twice as many electrons.

Faraday's Laws of Induction

A changing magnetic flux through an area produces an electric field on the boundary of the area

The size of the electric field is proportional to the change in magnetic flux

Since the discovery in the early 19th century that electrical currents could give rise to magnetic fields (*see* OERSTED DISCOVERY, BIOT–SAVART LAW), scientists had suspected that there had to be a reverse connection—that somehow magnetic fields should be able to produce electrical effects. One day in 1822, Michael Faraday wrote in his notebook that he had to "convert magnetism to electricity." It would be nearly ten years before he achieved his goal.

At various times during those years, Faraday returned to the problem. The experiment he finally hit upon was simple by modern standards. A length of conducting wire was wrapped around one side of a doughnut-shaped piece of iron and connected to a battery. A second length of wire was wrapped around the other side of the doughnut and connected to a galvanometer, an instrument that would detect the flow of electrical current. (The purpose of the iron is just to trap the magnetic field generated by the first wire and make sure it all goes through the second loop.)

At first, Faraday found the results disappointing. When current was flowing in the wire connected to the battery, no current at all was found in the other loop of wire. It seemed as if the "conversion" he had long sought for just wasn't going to happen. Then, to his surprise, he noticed the galvanometer registered a current when he disconnected the battery. This led to his first great insight—it takes a *change* in magnetic field to produce an electrical current. The mere presence of a magnetic field is not enough by itself. Physicists refer to a change generated by a field as *induction*.

Eventually, Faraday found that the crucial quantity which determines that change is what is called *magnetic flux*. Imagine a closed curve drawn on a piece of paper, and a magnetic field going through the paper. The magnetic flux through the curve is then defined to be the area of the curve multiplied by the component of the magnetic field perpendicular to the paper. Faraday's first law of induction says that whenever the flux in an area changes, there will be an electric field around the curve enclosing the area. If the curve is actually a copper wire, then this field will induce an electric current to flow in the wire, even though there is no battery or other power source connected to it. The second law is a simple statement of how the magnetic field affects the strength of the induced current.

There are three ways to change the magnetic flux through a curve:

An engraving of Michael Faraday giving a lecture at the Royal Institution, London, in 1830.

— change the area enclosed in the curve,

— change the strength of the magnetic field, or

— change the relative orientation of this magnetic field and the plane in which the curve lies.

This last method works because it changes the component of the magnetic field perpendicular to the plane of the curve, even though neither the total strength of the field nor the area of the curve changes. It has the most important implications, because it is the basis for the operation of the *generator*. In a simple version of the generator, a loop of wire is rotated between the poles of a large magnet. Because the rotation is constantly changing the relative orientation of the magnetic field and the loop of wire, there is always a current flowing in the loop. Because of LENZ'S LAW, that current will flow one way around the loop for half of each turn, then reverse and flow the other way for the second half of the turn. This, of course, is the familiar alternating current (AC) that flows in common household circuits around the world. It is worth noting that the choice of AC frequency is completely arbitrary—in North America, for example, the loops of wire in generators spin 60 times a second, in Europe 50 times a second.

The electrical generator has played an extremely important role in the development of our technological civilization because it gives us a way of generating energy in one place and using it in another. A steam engine, for example, can convert the energy stored in coal into useful work, but that work has to be done at the place where the coal is burned. An electrical

Believe it or not, this plant in Suffolk, England, generates its power from a mixture of wood chips, straw, and chicken droppings. It is an example of the production of energy from renewable resources. No matter how electricity is generated, however, Faraday's first law has to be obeyed.

generating plant, on the other hand, can be located far from where the electricity is needed: in factories, homes, and so on.

There is a story (most likely apocryphal) that Faraday was showing a prototype of the electrical generator to John Peel, the chancellor of the exchequer. "Well, Mr. Faraday," said Peel, "this is all very interesting, but what good is it?"

"What good is it?" Faraday replied. "Why, one day you will be able to tax it!"

MICHAEL FARADAY (1791–1867) English physicist and chemist. Faraday was born in London, the son of a blacksmith who belonged to a small Protestant sect. The opportunities for education available in early 19th-century England, such as they were, were denied to him. Apprenticed to a bookbinder at 14, he found himself reading the books as he worked on them. The *Encyclopaedia Britannica* stirred his interest in science, and he began attending a series of public lectures and demonstrations given by the leading chemist of the day, Humphry Davy (1778–1829).

He took notes on Davy's lectures, then bound them in leather and used the book as a calling card. In 1813 he was taken on as Davy's assistant at the Royal Institution, after which he rose to become one of the Victorian era's most famous scientists, and indeed a regular confidant of Queen Victoria herself.

Faraday's greatest accomplishments came from his explorations of the connection between electricity and magnetism. His work established the principles of electrolysis, and showed how electricity could be generated. However, his early research was in chemistry: he found how to liquefy chlorine and a number of other gases, and he discovered the compound benzene.

Faraday's outstanding attribute was as an experimentalist not afraid to challenge conventional views. He was also a great popularizer, instituting in 1826 a series of annual scientific lectures for children at the Royal Institution that continue to this day.

Fermat's Last Theorem

If the integer n is greater than 2, then there are no integers x, y, and z, none of which is zero, for which the equation $x^n + y^n = z^n$ has a solution

You probably remember the *Pythagorean theorem* from school—the one which says that the square of the hypotenuse of a right triangle equals the sum of the squares of the other two sides. You may even remember the 3–4–5 triangle, the one in which the two short sides are 3 and 4 units long and the hypotenuse is 5. In this case, the Pythagorean theorem gives you

$$3^2 + 4^2 = 5^2$$

This is an example of a solution to the generalized Pythagorean equation stated at the beginning: *x*, *y*, and *z* are all nonzero integers, and *n* = 2. Fermat's Last Theorem asks whether there are any values of *n* greater than 2 for which this same situation could arise—and says that the answer is no.

The history of Fermat's Last Theorem is a fascinating one, and not just for mathematics. Pierre de Fermat's contributions to mathematics were spread over many fields, but very few were published during his lifetime. For him, mathematics was a recreation rather than a profession. He corresponded with the leading mathematicians of his time, but had no incentive to publish himself. His writings consisted of privately circulated manuscripts and jottings, sometimes in the margins of books. One of these marginal notes, found shortly after his death, concerned the above problem. It read:

> *I have a truly marvelous demonstration of this proposition, which this margin is too narrow to contain.*

Unfortunately, it seems he never found anywhere else to write it, and for centuries mathematicians tried unsuccessfully to find Fermat's "marvelous" proof. Of all the puzzling notes that Fermat scribbled in his books, this was the only one that stubbornly resisted solution—which is why it became known as his "last" theorem.

All sorts of people joined the search for a proof to Fermat's Last Theorem, only to be frustrated. René Descartes (1596–1650) called Fermat a "braggart," and the English mathematician and cryptographer John Wallis (1616–1703) once referred to him as "that damned Frenchman." Fermat himself showed that the theorem was true for *n* = 4. The great 18th-century Swiss mathematician Leonhard Euler (1707–83) disposed of *n* = 3 but could get no further—he even asked a friend to search for a clue in the house where Fermat had lived. In the 19th century, new mathematical techniques, in particular ones involving prime numbers, helped to extend the value of *n* to nearly 200, though there were a few gaps on the way that could not be proved.

In 1908 a prize of 100,000 marks was offered for the first solution of the problem. The money had been left for this purpose in the will of the

German industrialist Paul Wolfskehl, who had once apparently been contemplating suicide, but became so immersed in Fermat's Last Theorem that he forgot to carry out his plans. The prize was his way of thanking Fermat and his theorem for saving his life. As the century progressed, mechanical calculators and then electronic computers were used to push n still higher: 617 in the 1930s, 4001 in 1954, 125,000 in 1976. In the late 20th century, computers at Los Alamos Laboratories in New Mexico were programmed to use their spare time (the equivalent of screen saver time on a personal computer) to search for solutions to the generalized equations. They established that the theorem was true up to impossibly large values of x, y, z, and n, but that, of course, doesn't constitute a mathematical proof, because the next number could always be the one that proves the theorem wrong.

Finally, in 1994, the English mathematician Andrew John Wiles (1953–), working at Princeton, published a paper on Fermat's Last Theorem that, with a few subsequent corrections, amounted to a complete proof of the theorem. The proof was over 100 pages long and relied on modern concepts of abstract mathematics that couldn't possibly have been known to Fermat. So what could Fermat have been thinking of when he wrote his note? Most mathematicians I talk to point out that there have been many incorrect proofs of the Last Theorem over the centuries, and assume that Fermat had struck upon one of these but had failed to see its subtle flaws. Then again, it's always possible that there's an extremely elegant, short proof out there that no one has been able to find. The one thing we can say for certain is that we now know the theorem is true. Most mathematicians undoubtedly agree with Andrew Wiles on this subject when he said, "My mind is at rest."

Mathematician Andrew Wiles presenting his proof of Fermat's Last Theorem at Cambridge, England, in 1993.

PIERRE DE FERMAT (1601–65) French mathematician and magistrate. He was born in Beaumont-de-Lomagne. Fermat studied law and spent his working life as a lawyer and judge. Although mathematics was a spare-time pursuit, he nevertheless made outstanding contributions in many fields, becoming known as "the Prince of Amateurs." Apart from his work on number theory (the name given to the branch of mathematics into which the Last Theorem falls), he helped to pave the way for Newton's calculus and Descartes's geometry, and with Blaise Pascal (1623–62) he founded the theory of probability. In optics, he discovered that light travels faster in denser materials.

Fermat's Principle

*The path of a light beam
between two points
minimizes the travel time
for the light*

Fermat's principle, named after the French scientist and mathematician Pierre de Fermat (whose name is immortalized in FERMAT'S LAST THEOREM) is an example of what scientists call an *extremum principle*. An extremum principle says that a particular system will behave in a way that makes a particular quantity reach either a maximum or a minimum value (i.e., it takes on an extreme value). Typically, these principles turn out to be compact ways of summarizing the laws that govern phenomena of light. Fermat's principle is a simple mathematical statement from which all of these basic laws can be derived. Think of it, therefore, as a neat summary of a large number of experimental findings about the behavior of light.

For example, when a ray of light encounters a glass block, we can use Fermat's principle to tell us how much the ray will be bent at the surface of the glass. The question, of course, is which path will allow the light to get from a point in the air to a point in the glass in a minimum amount of time, given that the light moves more slowly in the glass than it does in the air. A direct line between the two points may represent the shortest distance, but because the light will spend a relatively long time in the glass, that line will not represent the shortest time. On the other hand, any other path will increase the total distance to be travelled by the light. To find the path that will get the light between the two points in the shortest time therefore requires a tradeoff between increasing the total distance to be traveled and decreasing the distance traveled at the slower speed through the glass.

When the geometry of this problem is worked out (it's a straightforward if somewhat cumbersome calculation), we find that it leads us to SNELL'S LAW, the law which governs the refraction of light. In the same way, we can use Fermat's principle to show that when light is reflected from a surface, the angle at which it is reflected is the same as the angle of the incident light beam, the angle at which it struck the surface. This is the law of REFLECTION.

Thus, the collection of laws that govern the behavior of light beams can be derived from a simple extremum principle which says that light will travel between two points in the minimum possible time. It is important to realize, however, that, like all laws of nature, Fermat's principle depends ultimately on experimental verification for its validity.

Fermi Paradox

If intelligent extraterrestrial life exists, why have we received no signals from it or detected other evidence of it?

The search for extraterrestrial intelligence (SETI) first came to the attention of mainstream scientists in 1961, at a meeting at the radio observatory in Green Bank, West Virginia. It was pointed out that, armed with large radio telescopes, human beings could now detect signals beamed our way by extraterrestrials (assuming that they were out there and sending). In those heady early days, SETI enthusiasts talked of thousands of civilizations belonging to a vast "galactic club" that we would soon join (*see* DRAKE EQUATION).

They had to contend, though, with a comment made years before by Enrico Fermi, the Italian-American Nobel laureate. At a luncheon at Los Alamos in 1950, he was told of the arguments about why there had to be many other advanced technological civilizations in the Galaxy. He paused for a while, then asked, "Where are they?"

In various forms, this argument has been a thorn in the side of the SETI community ever since. In a more complete form, it goes like this: The laws of nature are the same everywhere in the universe, so every advanced civilization should be able to do what we can do. There are already credible designs for ships that could carry humans to the stars (moving, perhaps, at 10% of the speed of light). Any race equipped with such ships could colonize the entire Galaxy in a matter of a few million years—a long time by human standards, but a mere blink of the eye in cosmic terms. If thousands of such civilizations exist now, then they must have been there a few million years ago. Michael H. Hart (1932–) argued in 1975 that the absence of ETs on Earth right now is proof enough against the existence of advanced civilizations elsewhere (the paradox is sometimes called the *Fermi–Hart paradox*). So, where are they?

You can't duck out by saying that the extraterrestrials don't want to travel—the *couch potato hypothesis*—or that they are leaving us alone— the *zoo hypothesis* (we are the prize exhibits). Both of these hypotheses, and others like them, suffer from a common fault: It is necessary to assume that every ET civilization shares the trait in question—that they are all sedentary or all share the same high ethical standards (and actually live up to them!). If there really are thousands of them out there, this seems pretty unlikely. After all, we humans establish game reserves, but that doesn't always deter poachers.

Let me give you an example of why I think the zoo hypothesis must fail. During my second marriage, my father-in-law was the game warden of Carbon County, Montana, one of the most isolated and beautiful spots in North America. Yet despite this isolation, my father-in-law had to deal with poachers and illegal fishermen on a daily basis. What are the odds that *every* extraterrestrial civilization would be able to enforce its moral standards on all of its individual members? Not very good, I would think.

Since 1961 there has been an on-again-off-again search for radio signals from ET civilizations. The results are uniformly negative—we have no evidence for ET civilizations at all. We can use the history of the search to set limits on where as yet undetected civilizations might be. We know, for example, that there is no civilization within a thousand light years of Earth capable of generating the kinds of signals that we know how to create.

SETI scientists grade civilizations by their ability to generate energy. A Type I civilization generates as much energy as is in the sunlight striking an Earth-type planet, while a Type II generates as much energy as its parent star. (On this scale, Earth is a "Type 0.7"—70% of the way to having the energy consumption of a Type I civilization.) We know that there are no Type I civilizations within ten thousand light years of Earth, and no Type II civilizations in the cluster of galaxies to which the Milky Way belongs. Presumably, as the search continues, these limits will be pushed out farther.

As you can probably tell, I am pretty skeptical about finding ETs out there. Nevertheless, I believe strongly that the search has to be made. It is the only scientific study I can think of where the results would be fantastic no matter how things turn out.

ENRICO FERMI (1901–54) Italian-American physicist. He was born in Rome and educated at the University of Pisa, and became a professor of physics at the University of Rome while still in his mid-twenties. He moved to America in 1938. Fermi is most famous as the first person to initiate a controlled nuclear chain reaction, in the first "atomic pile"—a pile of graphite blocks, into which uranium rods were inserted, constructed in a squash court at the University of Chicago. Element number 100 is named fermium in his honor.

Fibonacci Numbers

A sequence in which each term is the sum of the two preceding terms has many interesting properties

Leonardo of Pisa, or Fibonacci as he is more usually known, was the first great mathematician of late medieval Europe. Born to a wealthy merchant family in Pisa, Italy, he came to mathematics because of his very practical need to maintain business accounts. The young Leonardo traveled widely, accompanying his father on business trips—we know of extended visits he paid to Byzantium and Sicily, for example. During these travels he spent a great deal of time with local scholars.

The numerical sequence that now bears his name arose from a problem on rabbits that Fibonacci included in his book *Liber abbaci* of 1202:

A certain man put a pair of rabbits in a place surrounded on all sides by a wall. How many pairs of rabbits can be produced from that pair in a year if it is supposed that every month each pair begets a new pair which from the second month on becomes productive?

You can see that the number of pairs of rabbits in each of twelve successive months will be

$$1, 1, 2, 3, 5, 8, 13, 21, 34, 55, 89, 144, \ldots$$

In other words, the number of pairs of rabbits will create a series in which each term is the sum of the previous two. This is known as a *Fibonacci sequence*; the numbers themselves are the Fibonacci numbers. The sequence turns out to have interesting mathematical properties. An example: You can divide a line into two segments so that the ratio of the larger to the smaller segment is the same as the ratio of the line itself to the larger segment. This ratio, approximately 1.618, is known as the *golden ratio*, and was regarded during the Renaissance as a pleasing proportion for use in architecture. If you take successive pairs of numbers from the Fibonacci sequence and divide the greater in each pair by the smaller, you get progressively better approximations to the golden ratio.

Since Fibonacci's time, phenomena in the natural world have been found in which the Fibonacci sequence seems to play a role. One is *phyllotaxis*—how leaves are arranged on a plant stem, for example, or florets on a cauliflower. The seed heads on a sunflower are arranged in two sets of spirals, one set going clockwise and one counterclockwise. The numbers in each case? 34 and 55.

The florets in this sunflower head are arranged in two spirals. The numbers of florets are related to an intriguing mathematical sequence.

FIBONACCI (LEONARDO OF PISA) (*c.*1175– *c.*1250) Italian mathematician. Born in Pisa, he became the first great mathematician of late medieval Europe. It was the need to maintain business accounts that led him to mathematics. He published books on arithmetic, algebra, and other mathematical topics. From Muslim mathematicians he learnt of the system of numerals evolved in India and taken up in the Arab world, and became convinced of their superiority (they were the precursors of what we now call Arabic numerals).

Flame Test

It is possible to identify the presence of metals by the color of the flames they produce

Light is emitted by an atom when an electron makes a quantum jump from one allowed orbital to another (*see* BOHR ATOM). This means that the light emitted by an atom of one element will be different from the light emitted by an atom of another element, simply because the energy levels of the two atoms are different. This is the basis of the science of spectroscopy (*see* KIRCHHOFF–BUNSEN DISCOVERY).

This fact about emission is also the basis for the flame test in chemistry. If a solution containing ions of one of the alkali metals (i.e., elements in the first column of the PERIODIC TABLE) is heated in a flame, the flame will assume a color that depends on the element(s) present. Sodium, for example, betrays its presence with a bright yellow color, while potassium produces violet, and lithium a deep red. These colors come from light emitted when collisions with hot gases in the flame push electrons in the atoms to excited states, from which they drop back down and emit light of a characteristic color.

This property of atoms explains why ocean driftwood is so highly prized for use in fireplaces. During its time at sea the wood absorbs many materials, and these materials impart a variety of colors to the flames when the wood is burned.

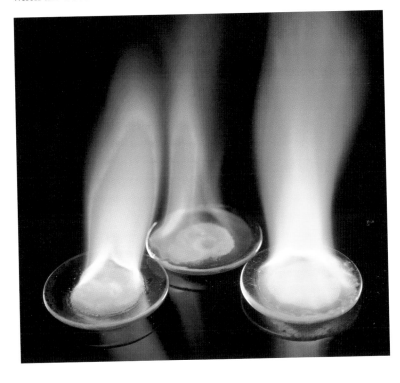

The flame test in action. The three metals (reading from left to right) are sodium, strontium, and boron (contained in boric acid). The flame produced by each substance is a different color.

Fourier Analysis

Any complex wave shape can be analyzed as a sum of simple waves

Joseph Fourier's great desire was to understand in a mathematical way how heat is conducted through solid objects (*see* HEAT TRANSFER). Perhaps his interest in heat was sparked by his time in North Africa—he accompanied Napoleon on the French expedition to Egypt, where he spent some time. Fourier had to develop new mathematical techniques to achieve his aim. The results were published as *Théorie analytique de la chaleur* ("Analytical Theory of Heat") in 1822, in which he described how to analyze complex physical problems in terms of the solutions to a series of simpler problems.

The method of analysis is based on what is called a *Fourier series*. Based on the principle of INTERFERENCE, the series begins by analyzing a complex shape—the disturbance of the Earth's surface by an earthquake, for example, or the shape of a comet's orbit under the gravitational influence of several planets, or the flow of heat through an irregularly shaped barrier of insulating material. Fourier showed that this complex shape could be represented as a sum of simple waves. It usually happens that the equations that describe classical systems can be solved easily for each of these simple waves. Fourier then showed how these simple solutions could be added together to give a solution to the complete complex problem. (Mathematically, a Fourier series is a way of expressing a function as the sum of sine and cosine "harmonics"—in fact, Fourier analysis was formerly known as "harmonic analysis.")

Until the advent of computers in the mid-20th century, techniques such as those pioneered by Fourier were the best weapons in the scientific arsenal for attacking the complexities of nature. From then on, scientists were able to use Fourier's complex techniques to deal with more than the simple kinds of problems that could be solved by a straightforward application of NEWTON'S LAWS OF MOTION and other fundamental equations. Most of the great advances of Newtonian science in the 19th century, in fact, depended on techniques first pioneered by Fourier. The methods have since been extended to problems from astronomy to mechanical engineering.

JEAN-BAPTISTE JOSEPH FOURIER (1768–1830) French mathematician. Fourier, born in Auxerre, was orphaned at the age of nine, and developed a fascination with mathematics at a young age. He was educated in military and church schools, later becoming a lecturer in mathematics. A lifelong political activist, he was arrested in 1794 for his defense of victims of the Terror. After the death of Robespierre, he was released from prison and helped to set up the prestigious École Polytechnique in Paris, a position that served him as a springboard to prominence in the regime of Napoleon. He accompanied Napoleon to Egypt, where he was made Governor of Lower Egypt; on his return in 1801 he was appointed as a provincial governor in France. In 1822, he became permanent secretary of the French Academy of Sciences, a position of some power in French intellectual life.

Free Electron Theory of Conduction

Electrical conduction in a solid is caused by the bulk motion of electrons

Electricity flows through a conductor, carried by electrons that detach themselves from the atoms.

By the end of the 19th century, scientists knew that OHM'S LAW describes the relation between electrical resistance and electrical current and voltage. From the HALL EFFECT, they also knew that electrical currents consist of moving electrons. What was missing was a description of resistance at the atomic level. The first attempt at such a description was supplied by the German physicist Paul Drude (1863–1906) in 1900.

The essence of the free electron model for electrical conduction is that each atom in a metal contributes an electron that is free to roam around in the metal. The atoms in a metal are arranged in a three-dimensional lattice, through which the electrons are free to wander (*see* CHEMICAL BONDS). When a voltage is applied to the conductor (e.g., by placing it across the poles of a battery), the electrons start to accelerate. This acceleration is short-lived, though, because the electrons soon collide with atoms and start to vibrate faster about their average positions in the lattice, an effect we perceive as the conductor heating up.

The effect of the collisions on the electrons is to retard their motion, much as you would be slowed down when crossing a crowded room by collisions with other people. As a result, the average speed of the electron—what is called the *drift velocity*—is actually fairly small. The drift velocity for electrons in the wires leading to a lamp in your house, for example, is just a fraction of an inch per second, comparable to the speed of a snail. The reason that the light goes on immediately when you throw the switch is that all the electrons in the wire start moving at once, including those that light the bulb. This effect is similar to turning on a garden hose that's already full of water and getting a stream out the other end immediately.

Drude took the notion of free electrons seriously. He treated the free electrons in a metal according to the IDEAL GAS LAW, drawing an analogy between the collisions of an electron trying to shoulder its way through a crowd of atoms in a metal and the collisions between molecules in a gas. This allowed him to predict the amount of resistance associated with a particular conductor in terms of the average collision time for electrons. As with many simple theories, this model works well in explaining some of the main features of electrical conductivity, but fails to account for many of the details. In particular, it cannot explain the behavior of electrical conductivity at different temperatures, nor the onset of superconductivity (*see* BCS THEORY OF SUPERCONDUCTIVITY). Today, electrical conductivity is interpreted according to quantum mechanics (*see* SCHRÖDINGER'S EQUATION).

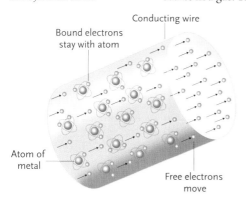

Conducting wire

Bound electrons stay with atom

Atom of metal

Free electrons move

Gaia Hypothesis

The Earth is a living system

Above right: the now-familiar image of the Earth from space—a small, living planet in an ocean of emptiness.

Gaia was the Greek goddess who drew forth the living world from chaos. The hypothesis is the brainchild of English scientist James Lovelock, who worked for NASA in the early 1960s when the search for life in the solar system was just beginning. Starting from the fact that the Earth's atmosphere differs significantly from the atmospheres of planets without life, Lovelock argued that our planet and its biosphere are a living system. In his words, "The Earth is more than just a home, it's a living system and we are part of it."

It is hard to know quite what to make of the Gaia hypothesis, because it does not make unique predictions that can be tested against experiment

—a requisite of every scientific theory. There are those (myself included) who think that it is best regarded as something like a literary metaphor—perhaps useful in talking or thinking about the planet, but lacking precision. Nevertheless, some mainstream scientists, such as the American biologist Lynn Margulis, have been keen supporters of Gaia.

The proponents of the hypothesis point out that it predicts the existence of feedback mechanisms created by living systems that keep the planet hospitable to life. An example is that increasing levels of carbon dioxide in the atmosphere lead to increased plant growth, which in turn lowers the level of the gas. These feedback mechanisms, however, are well known and do not require the Gaia hypothesis for their explanation.

What the hypothesis has done is to encourage a systems approach to the study of the Earth, one in which the planet is seen as a whole, rather than as a series of disconnected parts. It is certainly true that progress in the earth sciences in the last decades of the 20th century was spurred on by a realization that components of the planet—rocks, for example, or oceans—cannot be studied in isolation from the others. This is why university departments now have names like Department of Earth Systems Science rather than Department of Geology. To a large extent, this change was driven by advances in computing power, but it can be argued that the Gaia metaphor helped things along.

JAMES EPHRAIM LOVELOCK (1919–)
English scientist, born in London. After graduating he joined the National Institute for Medical Research. In 1964, after a brief period working for NASA, he set himself up as an independent scientist, determined to be free of any constraints imposed by what he perceived as the influence of multinational companies on the direction of science. Two years later he discovered chlorofluorocarbons (CFCs) in the atmosphere (*see* GREENHOUSE EFFECT). Lovelock is widely known for the hypothesis first set forth in his book *Gaia* (1979).

Gauss's Law

The electrical flux through a closed surface is proportional to the charge contained within the surface

It often happens in the sciences that laws can be stated in several different ways. Restating a law, of course, does not add to its validity, but it can help theorists to find new ways of interpreting the law and tying it in to other things we know about the world. This is the case with Gauss's law, which is really just a restatement of COULOMB'S LAW, which in turn summarizes what we know about the forces between electrical charges.

There are few areas of mathematics, physics, and astronomy that failed to benefit from the remarkable genius of Carl Friedrich Gauss. From 1831, he and Wilhelm Weber (1804–91) began to study electricity and magnetism, and Gauss's law was an outcome of this research. To understand the law, imagine a single electrical charge of magnitude q sitting isolated in space. Now imagine enclosing that charge within an imaginary surface. The surface can be of any shape—you can think of a crumpled balloon surrounding the charge. At each point on that surface there is an electric field associated with the charge, and the product of the strength of this electric field times the area of the little bit of the surface around it is called the *electrical flux*. You can calculate the flux for each of these little *surface elements*. Gauss's law says simply that when you add up all the fluxes, the result gives you the value of the charge.

A simple example will show the connection between Gauss's law and Coulomb's law. Suppose the surface around the charge, of magnitude q, is a sphere of radius r. The electrical field at a distance r from the charge, which is defined to be the force that would be felt by a unit charge at that distance, is given by Coulomb's law as

$$E = kq/r^2$$

This result is the same for any point on the sphere, so the sum of all the electrical fluxes will be the value of the electrical field times the surface area of the sphere (which is $4\pi r^2$). In other words, the total flux is

$$4\pi r^2 \times kq/r^2 = 4\pi kq$$

which is just Gauss's law.

One interesting outcome of Gauss's law comes from applying it to a piece of metal. Picture an imaginary surface which is just below the actual surface of the metal. The total electrical charge enclosed by that imaginary surface is zero, since there are as many positive charges on atomic nuclei as there are negative charges in the electrons. The total electrical flux through the surface must therefore be zero. Since this is true of any surface you can imagine inside the metal, it follows that there can be no electric field inside a metal surface.

This property is often utilized by researchers and communications engineers who have to shield their delicate instruments from outside electrical influences. Typically, they build large cages made of copper screening. From Gauss's law, they know that no external electrical field can penetrate such a cage and interfere with their equipment.

Gauss's law also tells you that if you encounter a lightning storm while driving, the safest place to be is inside the metal skin of your car. Even if there is a direct strike on the car, the charges will flow around the sides and into the ground. The current may incinerate your tires, but you'll be safe.

CARL FRIEDRICH GAUSS (1777–1855) German mathematician, one the greatest, ranking alongside Newton and Archimedes. He was born in Braunschweig to a peasant family. His mathematical genius was apparent from a very early age, and his astonished teacher persuaded his family that he should be trained for a profession instead of learning a trade. At the age of 14 he demonstrated his prowess to the Duke of Braunschweig, who was so impressed that he provided Gauss with financial support. Gauss had made most of his important mathematical discoveries before he was awarded his doctorate, from the University of Göttingen, in 1799, and just two years later he published his most influential book, *Disquisitiones mathematicae,* dedicated to his patron.

This work dealt with a branch of mathematics called number theory, concerned, for example, with prime numbers and relations such as FERMAT'S LAST THEOREM. Gauss would go on to study the mathematics of probability and statistics, and was the first to write down the expected distribution of random points around an average value. The so-called bell curve or normal distribution that resulted from his

calculations is also known among scientists as the *Gaussian distribution.*

Gauss turned to astronomy in 1801, following the discovery of the first asteroid, Ceres. He worked out the method of *least squares,* which he used to calculate the asteroid's orbit from just three observations of its position. Five years later he was appointed director of the observatory at Göttingen, and remained in this post for the rest of his life. Gauss was one of the pioneers of the detailed study of the Earth's magnetism (and, indeed, the basic unit of the magnetic fields is called the gauss).

Genetic Code

Three base pairs in DNA code for one amino acid in a protein

It has become a commonplace that the molecule called DNA contains the blueprint for all living systems. The easiest way to think of this molecule is to imagine a tall ladder. The uprights of the ladder are made of molecules of sugar, oxygen, and phosphorus. The important working information in the molecule is carried on the ladder's rungs. Each rung consists of two molecules, one attached to each upright of the ladder. These molecules are called bases. Their full names are adenine, guanine, cytosine, and thymine, but they are usually represented by the letters A, G, C, and T. Because of the shape of these molecules, only certain kinds of links—certain completed rungs—are allowed in a DNA molecule. These are the links between A and T, and between G and C (these pairings are called *base pairs*). No other links are allowed.

Down each side of the DNA molecule is a sequence of bases. It is the messages in these bases which determine how the cell runs its chemistry, and which therefore determine the characteristics of the living thing of which the DNA is part. The central dogma of MOLECULAR BIOLOGY is that the DNA codes for proteins which, in turn, act as the enzymes (*see* CATALYSTS AND ENZYMES) that run the organism's chemical reactions.

The exact relationship between the base pairs in the DNA and the sequence of amino acids that make up the protein enzymes is what is known as the genetic code. The code was worked out soon after the discovery of the double-helix structure of DNA. It was known that the newly discovered molecule *messenger RNA* (mRNA) contained the information coded in DNA. The biochemists Marshall W. Nirenberg and J. Heinrich Matthaei, then at the National Institutes of Health at Bethesda, near Washington, DC, did the first experiments that led to the unraveling of the code.

They began by constructing artificial molecules of mRNA that contained only a constant repetition of the base uracil (which is analogous to thymine, "T", and binds to the "A" base in DNA). This mRNA was added to test tubes containing a mixture of amino acids, with a different amino acid tagged with a radioactive tracer in each tube. They found that their artificial mRNA triggered the formation of a protein in only one test tube—the one in which the amino acid phenylalanine had been tagged. Thus, they discovered that the mRNA message "–U–U–U–," and therefore its DNA equivalent "–A–A–A–," translated into a protein in which the amino acids were all phenylalanine. This was the first part of the genetic code to be found.

Today, we know that three base pairs on DNA (a group called a *codon*) code for one amino acid in a protein. By carrying out experiments like the one described above, geneticists eventually worked out the code by which the 64 possible codons are translated into amino acids.

Genetic Drift

*The frequency of genes
in a population can vary
because of random events*

The HARDY–WEINBERG LAW tells us that for a theoretically perfect population, the distribution of genes will remain constant from one generation to the next. In a population of plants, for example, there will be just as many "grandchildren" with tall genes as there are "parents" with the same gene. But real populations aren't like this. Random events make the frequency of gene distribution change slightly from one generation to the next—a phenomenon known as genetic drift.

Here's a simple example: Imagine a group of plants isolated in a mountain valley. There are 100 adult plants in the population, and there is a particular version of a gene (e.g., for flower color) that is found only in 2% of the population, so that only two plants carry it in the generation we're looking at. It is quite possible that a minor calamity (e.g., a flood or a tree falling) could destroy both of those plants. That particular version of the gene (or, to use the technical term, that *allele*) would simply disappear from the population. Future generations would not be the same as the current one.

There are other examples of genetic drift. Consider a large breeding population with a well-defined distribution of alleles. Suppose that for some reason part of that population becomes separated and begins to form its own community. The gene distribution in the sub-population may not be typical of that in the wider group, but that atypical distribution would be found in the sub-population from that time on. This is known as the *founder effect.*

A similar type of genetic drift can be seen in the dramatically named *population bottleneck effect.* If for some reason a population is drastically reduced by forces not related to natural selection (e.g., by an unusual drought or by a brief increase in levels of predation) which appear suddenly and then disappear, the effect is a random elimination of many individuals. As in the founder effect, when the population again flourishes it will be found to have the genes characteristic of those random survivors, rather than of the original population.

Californian elephant seals, for example, were hunted almost to extinction in the late 19th century. Today the population (which has rebounded) exhibits an unusually small amount of genetic variation. Anthropologists believe that early modern humans went through a population bottleneck about 100,000 years ago. This would explain why humans are so genetically similar to one another. There is, for example, more genetic variation among members of gorilla clans in a single African forest than there is among all of the human beings on this planet.

Germ Theory of Disease

Infectious diseases are caused by the actions of microorganisms that originate outside the human body

In the mid-19th century a major debate took place in the medical sciences about the origin of infectious diseases. In one camp were the defenders of the ancient notion that diseases arise through some sort of imbalance within the body, perhaps exacerbated by outside influences. Squaring up to them was a group of scientists preaching the revolutionary notion that infectious diseases were the result of the invasion of the body by microorganisms.

A leader in the new wave of thinking was the French scientist Louis Pasteur. He began his investigations in an unlikely way: In 1854 he was professor of chemistry at Lille, where the university had a strong interest in aiding local industry. He took up the study of the process of fermentation, which is, of course, essential in the production of wine. He concluded that fermentation is the action of microbes feeding on the sugar in the grape juice and generating alcohol as a waste product. It was therefore clear to him that this was a biochemical process—not simply a chemical process, as many had believed—that could not take place without microorganisms, specifically yeast.

Pasteur also discovered that wine could be preserved by heating it. Heating killed the microbes that would otherwise initiate further reactions which would spoil the wine. This is the basis of the *pasteurization* process, still used to make the milk supply safe in most of the world.

Like many scientists of his time, Pasteur felt that there was a connection between the process of fermentation and the disease process in humans. The notion that diseases, like fermentation, were caused by microorganisms had a number of advocates in the late 19th century, and

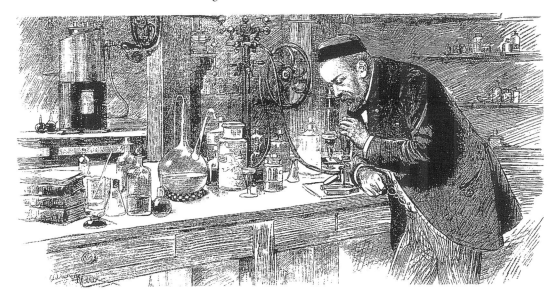

there was accumulating evidence for it. Pasteur himself had established that a disease which had caused great damage to the silkworms in France had a bacterial origin, albeit a rather complex one. In the 1860s, the English surgeon Joseph Lister (1827–1912) had taken Pasteur's ideas and used them to demonstrate the advantages of antiseptic surgery, while the German bacteriologist Robert Koch (1843–1910) was making headway in showing that anthrax, a disease of large animals (and occasionally humans), had a bacteriological basis. Pasteur showed that anthrax could be transmitted even by highly diluted blood, but not by blood that had been passed through a filter (a process that would remove bacteria). He soon found that microbes were the cause of a number of other diseases, including childbed (puerperal) fever, then a major cause of mortality among women. He incurred the wrath of the medical community by establishing that the disease was spread by physicians themselves as they moved from one birth to the next.

Pasteur then turned to the study of cholera in fowl, and discovered (almost by accident) that leaving the microorganisms alone for long periods of time reduced their virulence, so that they could serve as a vaccine. This led to the development of vaccines for anthrax and, more famously, for rabies. By the time of Pasteur's death in 1895, the germ theory of disease was accepted throughout the scientific and medical communities.

LOUIS PASTEUR (1822–95) French chemist and microbiologist. He was born in Dôle, the son of a tanner. He studied chemistry at the École Normale Supérieur in Paris and received his doctorate in 1847. His early work had to do with the optical properties of materials. After short stays at the universities in Dijon and Strasbourg, in 1854 he was appointed professor of chemistry at Lille University, where he carried out his studies of fermentation. He moved to the Sorbonne as professor of chemistry in 1867, and from 1888 until his death he was head of the Pasteur Institute in Paris.

His most significant achievement in chemistry was to discover optical isomers—pairs of chemical compounds with the same formula but which rotated the plane of polarized light in opposite directions. In microbiology, as well as his work on fermentation and putrefaction, he made significant advances in the fight against disease: he was the first to vaccinate sheep against anthrax, and humans against rabies.

Louis Pasteur at work in his laboratory. Although others had come up with the same idea, Pasteur's work was instrumental in establishing that diseases are caused by microorganisms, and his name is preserved in the term "pasteurization."

Giant Impact Hypothesis

The Moon was formed as a result of the impact of a Mars-sized object with the early Earth

1755 ● NEBULAR HYPOTHESIS

1900 ● RADIOACTIVE DECAY

1940s ● **GIANT IMPACT HYPOTHESIS**

1960s ● PLATE TECTONICS

Alone among the inner planets of the solar system, the Earth has a large satellite. How the Moon was formed is an old puzzle in astronomy, but one that many planetary scientists think is now resolved. The basic problem is this: The density of the Moon is about 3.6 times that of water—about the density of the rocks in the Earth's outer layers. But the density of the Earth is about 5.5 times that of water (the Earth's core is made of heavy iron and nickel). In essence, the Moon is like the Earth without its core. The question is how two objects could be so similar yet so different. According to the NEBULAR HYPOTHESIS, we would expect objects forming in the same region of the *solar nebula*—the disk of material from which the planets formed—to be similar in chemical composition.

Historically, two kinds of theories were advanced to answer this question: *capture theories* and *tidal theories*. The first held that the Moon formed elsewhere in the early solar system and was captured into orbit when it came close to the Earth. The problem with this is that it's hard to see how such a capture process could work. The competing theory held that the rotation of the Earth (believed to have been more rapid in the past), or possibly tides raised by a passing body, tore a piece out of our planet, a piece that eventually went into orbit and became the Moon. People used to point to the Pacific Basin, for example, as the "birth scar" of the Moon. The Moon's composition rules out both theories: It's too similar to the Earth's for the Moon not to have formed nearby, but not similar enough for it once to have been part of the Earth.

Over the last few decades, another theory has come to gain wide acceptance. In the early solar system, there was a period of intense bombardment when the Earth and other newly formed planetary bodies suffered collisions with the many asteroid-sized objects that had formed. The kinetic energy of the impacting objects was converted to heat, and the Earth melted, allowing it to *differentiate*—the heavy iron and nickel sank to form the core, and the lighter rocks rose to form the mantle and crust (*see* PLATE TECTONICS). Later, a body about the size of Mars crashed into the Earth. Material from the Earth's mantle and crust was blasted into orbit, where it came together to form the Moon.

This theory, known as the giant impact hypothesis (or more informally as the *big splash*), explains the Moon's low density, since most of its material came from the Earth's mantle and crust. We also now know that the minerals present in the Moon's surface rocks are very similar to those found on Earth. The clincher came with analysis of lunar rock samples brought back by the crews of the Apollo missions. The ratio of certain isotopes of oxygen (*see* RADIOACTIVE DECAY) are identical to those found in terrestrial rocks—conclusive proof of a common origin.

Glycolysis and Respiration

The metabolism of animals and other organisms depends on the chemical processes that extract the energy which is stored in carbohydrates

The process of PHOTOSYNTHESIS traps energy from the Sun and stores it in CHEMICAL BONDS in carbohydrate molecules, most notably in the six-carbon sugar called glucose. When these molecules are ingested by other organisms, the processes of glycolysis and respiration extract that stored energy and use it to run the organism's metabolism. The overall chemical process can be summarized as

$$\text{glucose} + \text{oxygen} \rightarrow \text{carbon dioxide} + \text{water} + \text{energy}$$

A simple way to visualize these processes is to imagine the organism "burning" the carbohydrates to get its energy.

The term "glycolysis" comes from putting together *-lysis*, which means "splitting," with *glucose*. As the name implies, this process begins the chemical extraction of energy by splitting the glucose molecule into two parts, each of which contains three carbon atoms. The three-carbon molecule is known as pyruvic acid, and glycolysis produces two of them from each glucose molecule. In addition, energy from the glucose is stored in "energy currency" molecules (*see* MOLECULES OF LIFE)—two molecules of ATP and two of NADH. Even in the first stage of glycolysis, then, energy is released in a form that can be used in an organism's cells.

What happens next depends on whether oxygen is available. If it is not, so-called *anaerobic* processes convert the pyruvic acid to other organic molecules. In yeast, for example, the pyruvic acid is converted into ethanol. In animals such as humans, when oxygen in muscles is in short supply, pyruvic acid is converted into lactic acid—the stuff that produces the all-too-familiar feeling of muscle stiffness after heavy exercise.

In the presence of oxygen, however, energy is extracted by the process of *aerobic respiration*, which breaks down the pyruvic acid into carbon dioxide and water, at the same time extracting the rest of the energy stored in the carbohydrate molecule. Respiration takes place in special structures in the cell known as *mitochondria*. The first step is a process that splits one of the carbon atoms off of the pyruvic acid, producing carbon dioxide, energy (stored in a single molecule of NADH), and a two-carbon molecule known as an acetyl group. The reaction chain then enters the metabolic clearing house of the cell—the *Krebs cycle*.

The Krebs cycle (or the *citric acid cycle* or the *tricarboxylic acid cycle*) is an example of a fairly common phenomenon in biology—a chemical reaction that starts when a particular input molecule combines with another molecule that can be assigned the role of "helper." This combination initiates a series of other chemical reactions which create further output molecules and then, at the end, recreate a helper molecule so that the entire process can start again. In the Krebs cycle the input molecule is the acetyl group that comes from the breakdown of the pyruvic acid. The helper is a four-carbon molecule of oxaloacetic acid. In the first chemical reaction in the cycle, these two combine to form six-carbon molecules of citric acid (and give the cycle one of its alternate names). From this point on, a series of eight chemical reactions take place, first producing energy-carrying molecules and carbon dioxide, then producing another molecule of oxaloacetic acid. It requires two trips around the cycle to process the energy in a single glucose molecule. The net yield: two molecules of ATP, four carbon dioxide molecules, and ten other energy-carrying molecules (about which more later). The carbon dioxide eventually diffuses out of the mitochondrion and is exhaled.

The Krebs cycle is crucial to life not only for energy generation. For one thing, many molecules other than glucose can feed into it, creating their own pyruvic acids. When you go on a diet, for example, your body doesn't have enough glucose to run its metabolism, so lipids (fats) are broken down and used to run the Krebs cycle. This is why you lose weight. Also, molecules can be shunted out of the cycle to provide the building blocks of new proteins, carbohydrates, and lipids. Thus, the Krebs cycle can accept energy from molecules in many forms, and create many kinds of molecules as output.

In terms of energy, the net effect of the Krebs cycle is to complete the removal of energy stored in the chemical bonds of the glucose, put a little of it into ATP molecules, and store the rest in other energy-carrying molecules. (When you think of energy being stored in chemical bonds, remember that work needs to be done to separate the atoms that are linked.) In the final stage of respiration, this remaining energy is removed from the carriers and also stored in ATP. The energy-storing molecules move about inside the mitochondria until they encounter specialized proteins embedded in the inner membranes. These proteins remove the electrons from the energy carriers and start passing them down a series of molecules—a molecular bucket brigade—as they remove the energy associated with the bond. At each step, energy is removed and stored in ATP. Finally, the electrons are attached to oxygen atoms that go on to combine with hydrogen ions (protons) to make water. The electron bucket brigade produces no fewer than 32 ATP molecules—90% of the energy that was stored in the original glucose.

Glycolysis

Pyruvic acid (3) ⇌ Lactic acid (3)

Acetyl-CoA (2)

Citric acid (6)

Oxaloacetic acid (4)

Isocitric acid (6)

Malic acid (4)

α-Ketoglutaric acid (5)

Fumaric acid (4)

Succinyl-CoA (4)

Succinic acid (4)

The Krebs cycle is a circular sequence of biochemical reactions responsible for the process of respiration in animals, plants, and many microorganisms. This is a simplified depiction of the process. The numbers in circles show how many carbon atoms are in each organic molecule.

The process of the energy transfer in the Krebs cycle involves a rather complex biochemical process known as *chemiostomic coupling*, a term that implies that the energy release proceeds by both chemical reactions and osmosis. Basically, the electrons from the energy carriers that are the product of the Krebs cycle and are stripped away and passed down via the "electron bucket brigade," are passed through proteins embedded in the membrane that separates the inner and outer compartments of the mitochondrion. The energy released by the electrons is used to move hydrogen ions (protons) to the outer compartment, where it acts as an energy reservoir, much like the water stored behind a dam. When these protons flow back through the membrane, the energy is used to create ATP molecules, just as water stored behind a dam is used to generate electricity when it falls through a generator. Finally, in the inner compartment of the mitochondrion the hydrogen ions join up with oxygen molecules to make water, one of the end products of metabolism.

This discussion of glycolysis and respiration illustrates an important point about the modern understanding of living systems. A simple statement about a particular process—that metabolism requires the "burning" of carbohydrates, for example—quickly leads into incredibly detailed molecular descriptions of complex processes, with a huge cast of molecular characters. In a sense, understanding modern molecular biology is a little like reading a classic Russian novel. Each interaction between characters is easy to understand, but by the time you get to

The energy that allows these runners to perform comes from the biochemical reactions in the Krebs cycle.

page 1423, you may well have forgotten how Pyotr Pyotrovich relates to Alexei Alexeivich. In the same way, each chemical reaction in the chain we've just described is quite straightforward, but by the time you've gone through the whole story, the sheer complexity will have overwhelmed you. If it's any consolation, I feel the same way.

HANS ADOLF KREBS (1900–81) German-British biochemist. He was born in Hildesheim, Germany, the son of a Jewish ear, nose, and throat specialist. He received his MD from the University of Hamburg in 1925, and began his research at the University of Freiburg. But Krebs decided to leave Germany when the Nazis came to power, and moved to England in 1933, working at the University of Sheffield (1935–54) and subsequently at Oxford. At Sheffield he measured the abundances of various molecules in pigeon tissues after respiration, and in 1937 worked out the chemical cycle that now bears his name, for which he shared the 1953 Nobel Prize for Physiology or Medicine.

Gödel's Incompleteness Theorems

Every mathematical system beyond a certain level of complexity must either be incomplete or contradictory

In 1900, at a mathematical conference in Paris, the German mathematician David Hilbert (1862–1943) laid out a list of 23 unsolved problems that he felt would be solved during the new century. The second problem on his list was one of those simple questions whose answer seems obvious until you start to think more deeply about it. In modern language, he asked whether mathematics is itself consistent. In mathematical terms, the challenge of Hilbert's second problem was to show that *axioms,* the basic assumptions that form the foundations of mathematical systems, are exact and complete, and that all of mathematics can be derived from them. The main thing to show was that the axioms are not contradictory, and that logical arguments based upon them could never lead to contradictory statements.

Take an example from school geometry. You may recall that in standard geometry (what mathematicians call *Euclidean geometry*) you can prove that the statement "The sum of the angles in a triangle is 180°" is true, and that the statement "The sum of the angles in a triangle is 137°" is false. In fact, every statement in Euclidean geometry can be proved true or false, and it probably seemed a reasonable expectation to those mathematicians back in 1900 that the same had to be true of every logical system.

In 1931, however, a slight, bespectacled Viennese mathematician by the name of Kurt Gödel published a short paper that turned the world of mathematical logic upside down. After some complex mathematical preliminaries, Gödel examined statements of the form "Proposition number 247 in this logical sequence cannot be proved." Call this sentence "Statement A." He then proved an amazing proposition:

> If Statement A can be proved, then so can Statement not-A

In other words, if the statement "Proposition 247 *cannot* be proved" can be proved, then so can the statement "Proposition 247 *can* be proved." In other words, if every statement in the system can be proved, as Hilbert suggested, then the system must be contradictory.

The only way out of this dilemma is to say that Statement A, though it is true, cannot be proved within the context of the logical system. Every logically consistent system, in other words, must contain at least one statement that cannot be proved within that system. If there is no such statement, then the system will contain logical contradictions. Every logical system containing arithmetic must therefore be either incomplete (in the sense that not every statement can be proved) or contradictory.

The statement that "every formal axiomatic system contains undecidable propositions" is known as the *first incompleteness theorem.* Gödel took his arguments a stage further with the *second incompleteness theorem.*

This, a corollary of the first theorem, states that the logical consistency (or otherwise) of a mathematical system cannot be demonstrated within that system, but only by recourse to a higher, stronger system.

You would be forgiven for thinking that Gödel's incompleteness theorems are relevant only on the farthest fringes of mathematical logic, but in fact they have turned up in the debate on the nature of the human brain. The English mathematical physicist Roger Penrose (1931–) has pointed out that Gödel's theorems can be used to show that the brain is not a computer. The argument is simple: A computer can operate only by the laws of logic, so it can never decide whether Statement A is true or false, since the statement cannot be proved. The human brain, however, can easily recognize that Statement A is true, even if it can't be proved logically. Thus, there is (at least) one thing the brain can do that a computer cannot. The brain can recognize the truth of the Gödel statement, while a computer, bound to logic as it is, is unable to. Therefore the brain, whatever it is, cannot be just a computer. This issue is known as *decidability*, and it lies behind the TURING TEST.

I wonder what Hilbert would have thought if he could have known where his question would take us?

KURT GÖDEL (1906–78) Austrian-American mathematician. He was born in Brünn (now Brno, in the Czech Republic) and educated in Vienna. He took his doctorate in mathematics from the University of Vienna in 1929 and was appointed to the faculty in 1930. A year later he published the theorem that now bears his name. A non-political man, he was deeply disturbed when a fellow faculty member was murdered by a Nazi student, falling into the first of many depressions in his life. During the 1930s, he visited America, but returned to Austria and married. In 1940, with war in Europe a reality, he escaped to America via Russia and Japan, and joined the Institute for Advanced Studies at Princeton. Unfortunately his mental problems deepened, until, convinced that people were trying to poison him, he starved himself to death.

Goldbach Conjecture

Every even number greater than 2 can be written as a sum of two primes

CHRISTIAN GOLDBACH (1690–1764) German mathematician. He was born in Königsberg, Prussia (now Kaliningrad, Russia). In 1725 he became professor of mathematics at St Petersburg, and three years later went to Moscow as tutor to the future Tsar Peter II. On Goldbach's European travels he met many of the leading mathematicians of the day, including Gottfried Leibniz, Abraham de Moivre, and the Bernoulli family. Much of his work in number theory was developed in correspondence with the great Swiss mathematician Leonhard Euler (1707–83). The first mention of what is now called Goldbach's conjecture is in a letter he wrote to Euler in 1742.

Sometimes the simplest ideas in mathematics turn out to be the hardest to prove. For example, it took hundreds of years before FERMAT'S LAST THEOREM was finally proved, at the end of the 20th century. There is another idea out there, somewhat similar to Fermat's, that mathematicians have not been able to prove. It is called the Goldbach conjecture, and it is as simple an idea as you can imagine. It simply says that any even number greater than 2 can be written as a sum of two prime numbers. (A word of explanation: A *prime number* is a number which is divisible only by 1 and by itself, so 2, 3, 5, 7, and 11 are prime numbers, whereas 4 (2 × 2), 6 (3 × 2), and 9 (3 × 3) are not.) The conjecture was first put forward in 1742 by Christian Goldbach. It says, to take a simple example, that 10, an even number, can be expressed as 7 + 3, 7 and 3 both being prime numbers. Another conjecture, rather less well known, is that all odd numbers greater than or equal to 9 are the sum of three primes (e.g. 13 = 7 + 3 + 3 = 5 + 5 + 3).

Ever since it was first put forward, mathematicians have believed that the Goldbach conjecture, like Fermat's Last Theorem, is true. Unlike the Last Theorem, however, no one has ever claimed to have proved it. There is a brute-force approach to this conjecture, of course, which is simply to run computer programs for long periods of time, checking the truth of the conjecture for successively larger and larger even numbers. If the theorem were not true, it might be possible to *disprove* it by this means. But you cannot *prove* the theorem in this way, for the simple reason that you can never know that the next number you would have checked if you had let your program run one step further wasn't the first exception to the rule. In point of fact, we know that the conjecture is true for even numbers at least up to 100,000.

In the 1930s, a group of Russian mathematicians established that there is an upper limit to the number of primes that had to be added together to make an even number, and that the conjecture is true for a large class of even numbers. Nevertheless, a proof of the conjecture has still not been found.

Why do mathematicians spend so much time working on things like Fermat's Last Theorem and the Goldbach conjecture? Solving them has no practical importance, and there is no way to turn the solution into profits. This sort of work, is, I think, an example of a very old and very human quest—the search for self-evident, unquestionable truth. Philosophers have sought this for millennia. By looking at systems that deal with pure logic, mathematicians hope to uncover such truths. That the proofs are so elusive probably tells us more about the nature of logic, about the difficulty of finding truth in this uncertain world, than it does about mathematics itself.

Graham's Law

The rate at which an ideal gas leaks out through small openings in a container is lower for denser gases

THOMAS GRAHAM (1805–69) Scottish chemist. He was born in Glasgow, the son of a prosperous manufacturer, and chose to study chemistry against his father's wishes (his father wanted him to become a clergyman). After finishing his studies at the University of Glasgow, he worked at various scientific institutions in that city, following a varied career—for a time, he held Isaac Newton's former position as Master of the Mint. He became well known in the scientific world, and founded the chemistry of colloids (something like solutions, but where the particles are much larger than molecules).

Effusion is the process by which gases leak out of containers through small (often microscopic) holes. If you've ever thrown a birthday party complete with helium-filled balloons, only to find the balloons collapsed the next morning, then you have encountered effusion. While you were sleeping, the helium in the balloon leaked out through microscopic pores in the material of the balloons.

In 1829, Thomas Graham carried out a series of effusion experiments that showed that, for a given temperature and pressure, the rate of gas loss, r, is inversely proportional to the square root of the gas density, d. In the language of equations, this result is expressed as

$$r \times \sqrt{d} = k$$

where k is a constant. In other words, the higher the density at a given temperature and pressure, the lower the effusion rate. Perhaps the most surprising thing about Graham's law is that the constant k on the right-hand side of the above equation is approximately the same for all gases at the same temperature and pressure.

From the IDEAL GAS LAW, it follows that the density of a gas at a given temperature and pressure is proportional to the relative molecular mass, M, of the gas. Knowing this, we can rewrite Graham's law as

$$r \times \sqrt{M} = k$$

In this form, the law tells us something interesting about the relative rates at which different gases will escape from identical containers. It tells us that the smaller the relative molecular mass of the gas, the higher will be the effusion rate. This is why a rubber balloon filled with helium (which has a relative molecular mass of 4) will lose that gas overnight, whereas if it is filled instead with air, which is a mixture mainly of nitrogen molecules (relative molecular mass 28) and oxygen molecules (relative molecular mass 32), it will stay inflated for several days. (Balloons made of metal foil, which has much smaller pores than does the rubber used for ordinary balloons, can retain helium for several weeks.)

A rather surprising application of Graham's law is to the construction of spacecraft on which humans will be spending long periods of time. A spacecraft may be very different from a balloon, but given enough time, air will leak through the materials of which the spacecraft's hull is made, just as it leaks out of a birthday balloon. This may not be the first concern of those who think about the future of the human race in space, but it's one that will have to be dealt with eventually, most likely by providing a means of generating gases on board to replace those lost to the vacuum of space.

Green Revolution

Population growth proceeds exponentially, the production of food arithmetically. Therefore, populations are always in danger of outstripping their food supply. Between 1950 and the present, world food production increased dramatically because of the introduction of new grain plants in agriculture

The history of human population growth has always been a battle between the ability to grow food and the appearance of new individuals to feed. In the words of British cleric and economist Thomas Malthus (1766–1834), "I think we may fairly make two posulata. First, that food is necessary to the existence of Man. Secondly, that the passion between the sexes is necessary and will remain nearly in its present state." In other words, populations will always show EXPONENTIAL GROWTH, and so must always outstrip the food supply. This dour forecast for humanity's future, known as the *Malthusian dilemma,* has now been lowering over two centuries of unprecedented economic growth.

This dilemma has divided thinkers into different intellectual camps, which I call the Malthusians and the techno-optimists. The Malthusians argue that, sooner or later, the human population will pass the limits of food production set by the environment, worldwide famine will result, and each new human being will be an additional drain on the biosphere. The techno-optimists (a group in which I include myself) argue that technology constantly raises our ability to produce what they need to survive, including food. Since technology is a product of the human mind, each new human being has the potential to push the Malthusian limit further into the future, and is therefore to be regarded as an asset.

The past two centuries have seen unrelenting triumphs for the techno-optimists, and the Green Revolution has been arguably the most striking of them. Since around 1950, improved strains of agricultural grains have drastically increased the world's food production. Amazing statistics abound: between 1950 and 1990, for example, the crop yield of India grew by 2.8% per year, at a time when the population was experiencing an annual growth rate of 2.1%. The grains on which the green revolution were built came not from modern genetic engineering but from plain, old-fashioned plant breeding. Genetic engineering promises more green revolutions in our future, and the yield of grain plants (especially rice) is expected to increase dramatically.

Most Malthusians concede the success of the green revolution, but question whether this sort of progress is sustainable into the future. They point out that one reason for the success was the massive use of fertilizers (*see* NITROGEN CYCLE), and that human disruption of the nitrogen cycle cannot go on forever. They also point to the large *monocultures*—field after field planted with the same type of plants—that the revolution has brought about. These monocultures are susceptible to sudden destruction by pests and disease. At the start of the Irish potato famine of 1845–7, for example, almost the entire Irish potato crop was wiped out in a week.

And so it goes. I suppose a hundred years from now some version of this argument will still be going on.

Greenhouse Effect

The atmosphere around the Earth (or other planet) raises its average surface temperature

Gardeners are familiar with this piece of physics. It is always warmer inside a greenhouse than outside, and that helps to nurture plants, especially during the colder months of the year. You can feel the same effect when you are inside a car. The reason is that the Sun, with a surface temperature of around 5000°C, radiates primarily visible light—the part of the ELECTROMAGNETIC SPECTRUM to which our eyes are sensitive. Because the atmosphere is largely transparent to visible light, radiation from the Sun easily penetrates to the surface of the Earth. Glass is also transparent to visible light, so the Sun's rays pass into the greenhouse and their energy is absorbed by the plants and any other items inside. Now, according to the STEFAN–BOLTZMANN LAW, every object radiates energy in some part of the electromagnetic spectrum. Objects at a temperature of about 15°C, the average temperature at the Earth's surface, radiate energy in the infrared range of wavelengths. So the objects in the greenhouse emit infrared radiation. However, infrared cannot pass easily through glass, so the temperature inside the greenhouse rises.

A planet with a substantial atmosphere, such as the Earth, experiences just the same effect, but on a global scale. In the long term, to maintain a stable temperature the Earth has to radiate away as much energy in the infrared as it absorbs from the visible light sent our way by the Sun. The atmosphere behaves just like the glass of the greenhouse—it is not as transparent to infrared radiation as it is to visible light. Various types of molecules in the atmosphere (carbon dioxide and water are the two most important) absorb infrared radiation, acting as *greenhouse gases*. Thus, infrared photons emitted from the Earth's surface do not always travel directly out into space. Some of them are absorbed by the greenhouse-gas molecules in the atmosphere. When these molecules re-radiate the energy they have absorbed, they may radiate it outward into space, but they may also radiate it inward, back toward the surface. The presence of these gases in the atmosphere has the effect of putting a blanket around the Earth. They do not keep heat from leaking out, but they retain heat at the surface for longer, so the surface of the Earth is significantly warmer than it would be otherwise—without the atmosphere, the average surface temperature would be a chilly –20°C, well below the freezing point of water.

It is important to understand that there has always been a greenhouse effect on the Earth. Without a greenhouse effect from the carbon dioxide in the atmosphere, the oceans would have frozen long ago, and higher life forms would not have evolved. The current debate about the greenhouse effect is about *global warming*: whether, by burning fossil fuels and other activities, we humans are adding enough extra carbon dioxide to the atmosphere to significantly change the energy balance of the planet. At

The planet Venus has experienced a runaway greenhouse effect, and the apparently benign white clouds hide a searingly hot surface.

the moment, the scientific consensus seems to be that we are responsible for adding to the natural greenhouse effect by a few degrees.

The greenhouse effect is not something that affects only planet Earth. In fact, the biggest greenhouse effect we know of is on our sister planet, Venus. The Venusian atmosphere is almost all carbon dioxide, and the planet's surface is a steamy 475°C as a result. Climatologists believe that we have escaped this fate because of the presence of oceans on the Earth. Because of the oceans, carbon is taken out of the atmosphere and stored in rocks such as limestone, thereby pulling carbon dioxide out of the atmosphere. Venus had no oceans, and all of the carbon dioxide injected into its atmosphere by volcanoes stayed there. Consequently, on Venus we see an example of what is called a *runaway greenhouse effect*.

Hall Effect

An electrical current flowing through a metal in a magnetic field produces a voltage which is perpendicular to both the current and the magnetic field

When an electrical charge moves in the presence of a magnetic field, a force is exerted on that charge. This simple fact is behind the operation of machines such as the synchrotron, used in research into particle physics, in which charged particles are kept moving in a circular path by magnets as they are accelerated. On a significantly smaller scale, it is also the operating principle of a microwave oven, a device in which electrons circling in a magnetic field emit microwave radiation.

Imagine a piece of metal wire lying on a table, and a magnetic field oriented in a direction perpendicular to the table. If electrons flow in the wire, the force associated with the magnetic field will cause the charges to move to one side of the wire. Once some charges have moved to the edge, they will accumulate there and begin to exert a force on other electrons according to COULOMB'S LAW. Charges will accumulate at the edge until this force cancels the magnetic force and the charges once again flow undeflected through the wire. The force associated with the accumulation of charge is equivalent to a voltage applied across the wire in a direction perpendicular to both the flow of current and the direction of the magnetic field.

Edwin Hall, in an experiment carried out in 1879, measured the voltage across a wire that was part of the arrangement described above. He also realized that the direction of the voltage depended on the electrical charge of whatever was carrying the electrical current. Thus, he was the first to demonstrate that what we think of as electrical current is actually the flow of negatively charged electrons. Before this experiment, people were not sure whether the force was acting on the charge carriers or on the conductor itself.

A century after Hall's work, the German physicist Klaus von Klitzing (1943–) discovered the quantum mechanical version of the Hall effect, for which he was awarded the 1983 Nobel Prize for Physics.

EDWIN HERBERT HALL (1855–1938) American physicist. He was born in Great Falls, Maine, and was one of the first to study physics at Johns Hopkins University in Baltimore, the first American institution to be modeled on the German research university. His discovery of the Hall effect grew out of his doctoral research on electricity and magnetism. After obtaining his Ph.D., he spent most of his career at Harvard University, where he became a major innovator in physics education, particularly at school level.

Hardy–Weinberg Law

It requires more than genetic recombination to change the composition of the gene pool

It's not often in the sciences that two different people stumble upon the same law independently, but there are enough instances to make one believe in the concept of the zeitgeist. The Hardy–Weinberg law (or equilibrium rule, as it is also known), one of the foundations of population genetics, is one of those instances. It has to do with how genes are distributed in a population. Suppose that a particular gene can have two variations—or, to use the technical term, *alleles*. They might be the "tall" and "short" genes in Mendel's peas, for example (*see* MENDEL'S LAWS), or the propensity to have twins or not to have them. Weinberg and Hardy showed that if mating were random, and there were no outflow or inflow of individuals and no mutations, then the proportion of individuals with each of these alleles in the population will remain fixed from one generation to the next. In other words, there will be no GENETIC DRIFT.

We can illustrate the rule with a simple example. We'll call the two alleles X and x. There are then four possible combinations of these alleles in individuals: XX, xx, xX, and Xx. If we denote by p and q the percentage of individuals with alleles X and x, respectively, then the Hardy–Weinberg equilibrium rule states that

$$p^2 + 2pq + q^2 = 100\%$$

where p^2 is the percentage of individuals who are XX, $2pq$ is the percentage who are xX or Xx, and q^2 is the percentage who are xx. These percentages will, under the conditions stated, remain constant from one generation to the next, regardless of how much shifting around there is among individuals or how large (or small) p and q are. The rule, then, provides geneticists with a tool against which they can measure changes in the distribution of genes in a population caused by phenomena like mutation and migration. It is, in other words, a theoretical benchmark against which changes in distribution can be measured.

GODFREY HAROLD HARDY (1877–1947) English mathematician, born in Cranleigh, Surrey, the son of an art teacher. He was educated at Cambridge University, and studied mathematics there and at Oxford. Hardy is probably best remembered for his collaborations, first with John Edensor Littlewood (1885–1977), and later with Srinivasa Aaiyangar Ramanujan (1887–1920), a clerk in Madras who was self-taught in mathematics. In 1913 Ramanujan sent Hardy a list of theorems that he had proved. Recognizing genius, Hardy brought the young clerk to Oxford and, in the few years before Ramanujan's untimely death, the two published a series of brilliant papers together.

WILHELM WEINBERG (1862–1937) German physician who had a large private practice in Stuttgart. He is reported to have delivered over 3500 babies in his career, including no fewer than 120 sets of twins. His experience with twins, together with the new availability of MENDEL'S LAWS of genetics, led him to conclude that the tendency to bear dizygotic (nonidentical) twins is an inherited trait.

Heat, atomic theory of

Heat is a form of energy associated with the random motion of atoms

Heat is a mysterious thing. You can pick up a piece of wood and hold it in your hand without discomfort. But throw it into a fire, and it begins to burn, producing its own heat. Where does this heat come from? People used to think that heat was a kind of fluid (it was given the name PHLO-GISTON) which was contained in wood and other flammable substances, and that burning released it. By the end of the 18th century, however, experimental data were coming in to show that this notion was wrong.

One of the early champions of our modern notion of heat was Benjamin Thompson. He always had a technological turn of mind, and throughout his life was interested in the science associated with artillery. While in Bavaria, he was in charge of a factory where cannon barrels were made. The roughly cast barrels were mounted on a kind of lathe, and the inside of the barrel was smoothed out by a metal-cutting tool. Rumford noticed that the barrels seemed to heat up when the cutting tool was dull, but not when the tool was sharp. By measuring the heat-storing proper-ties of the metal shavings, he was able to show that the heat was not something residing in the cannon barrel, but was produced by friction. He is even supposed to have demonstrated his ideas by immersing the entire cannon-boring apparatus in a vat of water and cutting away until the water boiled.

Today, we understand heat (more precisely, thermal energy) as a form of energy associated with the motion of the atoms or molecules that make up a material. When heat flows into a material, the atoms or mol-ecules from which it is made begin to move faster. If the material is a gas or a liquid, in which the atoms or molecules are not bound tightly together, then they simply move around and rebound from one another more vigorously. If the material is a solid, the atoms vibrate more ener-getically around their average positions. In both cases, though, what we perceive as heat, or thermal energy, is actually associated with the motion of atoms and molecules. Like other forms of energy governed by the first law of THERMODYNAMICS, this heat can be converted into other forms of energy, and this is exploited in, for example, the internal-combustion engine and the electrical generator.

BENJAMIN THOMPSON (LATER COUNT RUMFORD) (1753–1814) American administrator and scientist. He was born in Woburn, Massachusetts. He sided with Britain during the Revolutionary War, and in 1775 he abandoned his family and fled to London to escape charges of espionage. Later he settled in Bavaria, where he became a high government official. In 1790 he was rewarded for his service by being made a count, taking his title from the former name of Concord, New Hampshire, where he had lived with his first wife. Back in London, he founded the Royal Institution in 1800. In 1804 he moved to Paris and married the widow of Antoine Lavoisier (*see* PHLOGISTON). It was not a happy union; Rumford com-mented that Lavoisier was fortunate to have died at the guillotine.

Heat Transfer

Heat can be transferred by conduction, convection, or radiation

The second law of THERMODYNAMICS says that heat always flows from hot objects to cold objects, but it says nothing about the mechanism by which this flow is accomplished. The movement of heat from one place to another is extremely important to scientists and engineers, and has been studied extensively. As stated above, there are three different ways that heat can be transferred, each of them via a different physical process.

Conduction

If you place one end of a metal bar in a fire, the other end will eventually become hot to the touch, even though it is not itself in contact with the flames. The heat is flowing from the fire to your hand through the metal bar. This is an example of heat transfer by conduction.

Here's how it works: The atoms in a metal are arranged in a three-dimensional lattice, and are constantly vibrating about their average positions. The atoms at the end of the bar which is in the fire collide with fast-moving molecules in the coals and in the heated gas around them and so vibrate more. This increased motion manifests itself as a rise in the temperature at that end of the bar.

The atoms at the end of the bar that have been made to move faster will now collide with their neighboring atoms. Again, each of these collisions will transfer energy to the more slowly vibrating neighbors, which will vibrate faster in turn. The atoms at the end of the bar, which lost energy in the initial collisions, regain it in subsequent collisions with atoms in the fire.

Thus, a chain of atomic collisions transfers heat up the bar, the energy lost in the collisions being constantly replenished by the energy of the fire. Eventually, this chain reaches the end of the bar you are holding, causing the atoms there to vibrate more. This you feel as an increase in the temperature of that end of the bar.

Conduction, then, is a process that operates through the medium of individual atomic collisions. Heat moves through a material and individual atoms vibrate more around their normal positions, but there is no bulk motion of material.

The equation for the flow of heat by conduction is

$$Q = A \times \Delta T / R$$

where Q is the amount of heat energy that passes through the object, A is the cross-sectional area of the object, ΔT is the temperature difference across the object, and R is the *heat resistance*, a quantity that characterizes the ease with which heat will flow through the material of which the object is made. In the above example of the metal bar held in the fire, ΔT is the difference in temperature between the fire and the air at the other end of the bar, *A* is the cross-sectional area of the bar, and *R* is characteristic of the metal. In general, this equation tells us that the higher the temperature difference and the larger the area, the greater will be the amount of heat transferred. For a given temperature difference and cross-sectional area, the higher the value of *R*, the lower will be the amount of heat transferred. Thus, materials with high *R* values, such as fiberglass and feathers, make good insulators.

Convection

Consider a pan of water being heated on a stove. At first the water is stationary, and heat is transferred from the burner to the upper surface of the water by conduction. However, as the water gets hotter, heat begins to be transferred by a different process, known as convection.

When water near the bottom of the pan is heated, it expands. Consequently, a given volume of water at the bottom of the pan will weigh less than the same volume of cooler water at the top. This creates an instability, and the system responds by lowering the energy, which it does by having the hot and cold volumes exchange their positions. The warm water rises to the top, and the cold water sinks to the bottom.

Once this has happened, though, the situation begins to reverse itself. The cold water that has descended to the bottom of the pan begins to absorb heat from the burner and expands, while the warm water that has risen to the top loses heat by radiation (*see* below) and begins to cool and contract. Soon the original instability situation has reappeared, with dense water at the top and less dense water at the bottom.

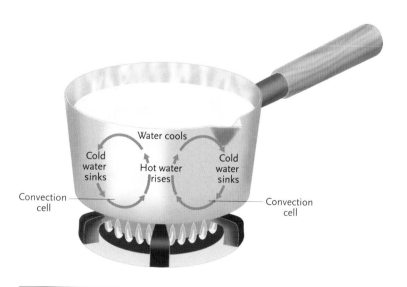

Convection in action in a heated pan of water. Heat is transferred as water circulates in convection cells.

It's not hard to see how this situation gives rise to a continuous movement of the water, with water being warmed, rising to the top, cooling off, and sinking back down to begin the cycle again. Any region in which this up-and-down cycle operates is called a *convection cell*. When you look at the surface of a boiling pan of water, you can see convection cells in operation—the clear areas are places where water is rising from the bottom, while the places where bubbles congregate mark where water is sinking back down, leaving at the surface anything that floats.

Convection cells are one of nature's commonest ways of transferring heat. They operate inside the Sun, in the region between the core and the surface, to bring heat from the fusion reactions in the core to the surface. They operate in the interior of the Earth, driving the motion of the plates that make up the surface of our planet (*see* PLATE TECTONICS). They play an important role in the Earth's weather, moving heat from the equator to the poles through the motion of the atmosphere and the oceans. They are even responsible for small-scale climatic conditions in urban areas, where the cement and bricks of the city heat up and play the role of the stove in the above example, creating small "heat islands." In effect, they allow cities to create their own weather.

Convection, then, is the transfer of heat by the bulk motion of matter. Material absorbs heat in one place and then moves to another place, carrying the heat with it. This is in contrast to conduction, in which the medium remains stationary while heat is transferred through it.

Radiation

Unlike the first two modes of heat transfer, the third does not require the presence of matter, either in atomic or bulk form. Instead, it is related to the fact that every object with a temperature above absolute zero radiates energy into its environment (*see* STEFAN–BOLTZMANN LAW). The type of radiation emitted depends on the temperature of the object. This notion is part of our common experience—a piece of metal in a blacksmith's forge will first glow orange, then red, then white-hot as its temperature changes. This is an indication that as the temperature of the object increases, so too does the frequency of the radiation it emits. Cooler objects will emit infrared radiation (which human beings can feel but not see) or even microwave radiation.

Perhaps the most famous example of the detection of radiation from an object was the discovery of the cosmic microwave background, one of the main pieces of evidence in favor of the BIG BANG. In effect, this radiation is emitted by the universe itself as it expands and cools from its initial colossally high temperature.

Heisenberg's Uncertainty Principle

The position and velocity of a quantum particle cannot both be known with absolute accuracy at the same time

We human beings spend our lives interacting with objects whose sizes aren't too different from our own: cars, buildings, grains of sand, and so on. Our ideas about how the world is the way it is (what we call our intuition) are built up by observing how such objects behave. Because we all have a lifetime of experience telling us that what we can see of the universe behaves in a certain way, we have come to expect that every other corner of the universe should behave in just the same way. We are perplexed, even shocked, when we encounter a corner where the rules aren't the ones we're used to, and where our intuition doesn't work.

In the early part of the 20th century, physicists had just this reaction when they began to investigate the behavior of matter at the atomic and subatomic levels. The rise of QUANTUM MECHANICS presented us (and still presents us) with a system that just doesn't seem to make sense, that just doesn't square with our intuition. We need to keep in mind that our intuition has its basis in the behavior of normal things, while quantum mechanics deals with things of which no human being has ever had any direct experience. If we don't, we shall easily succumb to puzzlement and frustration. The way to approach quantum mechanics, I have found, is this: Whenever a voice in the back of your head says, "But things can't be that way!" ask yourself, "Why not? How do I know what things are supposed to be like inside an atom? Have I even been inside an atom?" If you do this, the pages of this book that deal with quantum mechanics will be a lot easier to handle.

The Heisenberg uncertainty principle plays a central role in quantum mechanics, if for no other reason than that it provides an easy way to see why and how the world of the atom differs from the world of the familiar. To understand the principle, start by thinking about what it means to measure something. If you want to know where this book is when you enter a room, for example, you look around until you see it. In the language of physics, you perform a measurement (by looking at the book) and determine an outcome (its location). What actually happens is that light from somewhere—a lamp, for example, or the Sun—travels through space, interacts with the book and is reflected toward your eye, where it is detected. It is the interaction of the light and the book which is the crux of the process: In any measurement you can imagine there will be an interaction like this between a probe (in this case the light) and the object being measured.

In classical physics—the kind of physics developed by Isaac Newton to apply to our ordinary world—we customarily ignore the fact that the interaction aspect of measurement can change the object being measured. You don't, for example, expect a book to move from table to chair just because you turn the lights on. Our intuition tells us (in this case, quite rightly) that the act of measurement does not affect the object being measured. Now think about how the same process works on the subatomic level. Suppose I want to find out where an electron is. I still have to send a probe to interact with it and carry information to my detectors. This is where the difficulty arises, for the only probes at my disposal are other subatomic particles. While it may be reasonable to assume that light can interact with a book without affecting it, it is certainly not reasonable to assume that one electron can interact with another in the same way.

The analogy I like to use is this: Imagine you have to check whether a car is parked in a tunnel, and that the only method at your disposal (your means of making a measurement), however improbable it sounds, is to send another car into the tunnel and listen for a crash. This measurement would certainly tell you whether the tunnel were clear; if it weren't, by noting the time of the crash you could find out where the car had been parked. The one thing you cannot do, however, is to assume that the car in the tunnel was the same after the interaction as it was before. The same applies in the subatomic world. We cannot make a measurement without an interaction, and we cannot have interactions that do not affect the thing being measured.

Werner Heisenberg was a young German physicist in the early 1920s, when the great burst of creativity that led to quantum mechanics was taking place. Starting from a complex mathematical formulation for the behavior of the subatomic world, he came to a precise statement about the effects of the kind of measurement interactions we have been discussing. His result, which now bears his name, can be stated as

uncertainty in position $\times$ uncertainty in velocity $> h/m$

or, in the language of mathematics,

$$\Delta x \times \Delta v > h/m$$

The symbols on the right sides of these equations have the following meanings: m is the mass of the particle, while h is the PLANCK CONSTANT, named for the German physicist Max Planck, another one of the founders of quantum mechanics. The Planck constant is a very small value (in normal units it's a decimal point followed by 33 zeroes and a 6).

The phrase "uncertainty in position" represents the error made in determining where a particle is. For example, if you used the Global Positioning System* to locate this book, it would tell you the location to within a matter of feet (a meter or so). As far as the GPS measurement is concerned, the book could be anywhere in an area of perhaps a couple of dozen square feet, and we would call that the uncertainty in the book's

* A system of 24 satellites used for navigation. If you have a GPS receiver, you can use the signals beamed out by the satellites to accurately locate your position on the Earth's surface.

position. You could improve on this, for example by measuring the position of the book with a ruler—it might be 6 feet 5¾ inches from one wall and 3 feet 9½ inches from another, for example. But the ruler will have a smallest division—⅛ inch, perhaps—and its position couldn't be determined more accurately than that. As far as the measurement by the ruler is concerned, the center of the book could be anywhere in a ⅛-inch square, and that would be the uncertainty in position. If you used a more accurate measuring device, this uncertainty could be lowered. In principle, in our ordinary world, the uncertainty could be reduced to zero, at which point we would know exactly where the book was. We could go through a similar exercise to determine the uncertainty in the velocity of a moving object, and that too could in principle could be reduced to zero.

And now we get to the essence of the difference between the ordinary world and the world of the atom. In the ordinary world, measuring the position or velocity of an object has no effect on that object. Thus, both the position and velocity can, in principle, be measured with unlimited accuracy (or, equivalently, to zero uncertainty). We can know both of these quantities at the same time as accurately as we want.

In the quantum world, though, every measurement we make changes the system. If we choose to measure the position of an object, the very fact that we have made that measurement will introduce an uncertainty in the velocity (and vice versa). This is what it means to have the number on the right side of Heisenberg's equation not equal to zero. As the uncertainty in one variable (say, Δx) gets smaller and smaller, the uncertainty in the other (Δv) has to get bigger and bigger to keep the product on the left side of the equation from getting too small. In fact, if one uncertainty is reduced to zero, the other must become infinite to balance it. In other words, if we could know exactly where a particle is in the quantum world, we would know nothing whatsoever about how fast it is moving; and if we could know exactly how fast it is moving, we would have no idea about where it is. In practice, of course, physicists are almost always dealing with measurements between these two extremes, where there are uncertainties in both position and velocity.

There are actually many uncertainty principles, not just the one relating position and velocity, though that is the easiest to discuss. The same sort of reasoning tells us, for example, that we cannot know both the energy of a system and the time at which it has that energy. In other words, if we measure a quantum system to determine its energy, that measurement will take a certain amount of time to complete—call this Δt. During this time, the energy of the system could fluctuate without our being able to detect it. Call the amount of fluctuation ΔE. Then the same sort of reasoning that led to the previous equation for position and velocity leads to the result.

$$\Delta E \, \Delta t > h$$

There are a couple of things that need to be said about the uncertainty principle:

—it does not imply that we cannot know either the position or the velocity of a particle, and

—it does not require the presence of a conscious mind.

Sometimes you will see it stated that the uncertainty position implies that quantum particles have no positions or velocities, or that these quantities can never be known. As we have seen, though, the principle tells us that we can know either quantity as accurately as we like. All it says is that we cannot know them both at the same time with high accuracy. As in so many things in life, there is a tradeoff. Similarly, New Age writers sometimes argue that because the making of measurements implies a conscious observer, then somehow at its most fundamental level the universe is bound up with human consciousness. As we have pointed out, the key element in the Heisenberg relation is the interaction of the particle being measured and the probe. The fact that a conscious scientist initiated the measurement is irrelevant—the probe would have exactly the same effect regardless of whether a conscious entity is present.

WERNER KARL HEISENBERG (1901–76) German physicist. Heisenberg was born in Wurzburg. His father was professor of Greek at the University of Munich. The young Heisenberg not only displayed a high level of aptitude in mathematics, but also became an accomplished pianist. As a student, he was part of a civilian militia that maintained order in Munich during the chaotic days following the end of World War I. He entered the University of Munich in 1920 intending to study mathematics, but when the professor of that subject refused to let him enter an advanced seminar, he switched to theoretical physics. By this time, the new insight into atomic structure provided by the BOHR ATOM was permeating the world of physics, and scientists were aware that something strange was going on inside the atom.

Heisenberg received his doctorate in 1923 and spent time at Göttingen work-ing on problems in atomic theory. Following a severe attack of hay fever, he spent several months on the remote island of Helgoland where, like Isaac Newton in the plague year of 1665, he used his enforced isolation to complete several important scientific projects, the most important of which was a new mathematical formulation of what was to become the science of QUANTUM MECHANICS. Called "matrix mechanics," it was mathematically equivalent to the wave mechanics approach embodied in SCHRÖDINGER'S EQUATION. It was, however, more difficult to use, and physicists generally adopted the wave mechanics approach to the new subject.

In 1926, Heisenberg became Neils Bohr's assistant in Copenhagen. It was here, in 1927, that he discovered the Heisenberg uncertainty principle, arguably his most important contribution to science. Also in that year, Heisenberg was appointed to a professorship in Leipzig, becoming the youngest professor in Germany. There he began a lifelong quest, ultimately unsuccessful, to find a way of unifying the theories of quantum mechanics and relativity. (For a modern approach to this problem, *see* THEORIES OF EVERYTHING.) For his major role in the development of quantum mechanics he received the 1932 Nobel Prize for Physics.

The historical figure of Werner Heisenberg has been, and probably always will be, associated with another kind of uncertainty. The coming of the Nazi party to power opened what has become the most difficult part of his life to understand. At first, as a theoretical physicist, he was caught up in an ideological struggle in which his field was portrayed as "Jewish physics," and Heisenberg himself was labeled a "white Jew." Only appeals to high officials in the Nazi party stopped the campaign. More problematic was Heisenberg's role in the German nuclear physics program during World War II. At a time when many of his colleagues had already fled or been forced out of Germany by Hitler's regime, Heisenberg was placed in charge of the German nuclear effort.

Under his direction, the program concentrated on building a nuclear reactor, although in a famous meeting with Bohr in 1941, he may have given the impression that weapons research was part of the program. Did Heisenberg consciously direct the German program away from weapons research as a matter of conscience, as he later claimed, or because he made some mistaken assumptions about the physics of nuclear fission? As it happened, the German program produced no atomic weapons. But the question of Heisenberg's motives, as explored in Michael Frayn's excellent play *Copenhagen,* is an historical riddles that is likely to keep investigative authors in business for a long time to come.

After the war, Heisenberg became a major spokesman for developing West German science and reintegrating it with the international scientific community. He was also instrumental in keeping the West German armed forces non-nuclear.

Helium, discovery of

The element helium was first discovered in the Sun, and only later on the Earth

Most of the elements in the PERIODIC TABLE were discovered in the 19th and 20th centuries. Most of them are too rare (at least in their native form) to have become known to peoples lacking advanced technology. Each element has its own discovery story. The discovery of helium is perhaps the most fascinating, because until the late 1930s it was not certain whether there were chemical elements found elsewhere in the universe but not on Earth. Had this turned out to be so, it would have cast doubt on one of the central tenets of modern science—that the laws of nature as we know them here and now apply throughout the universe and through all time (this is what is behind the COPERNICAN PRINCIPLE).

The chief figure in the story of helium is Norman Lockyer, founder of one of the world's foremost scientific publications, *Nature*. While preparing for the launch of this journal, he became acquainted with the London scientific establishment and developed an interest in astronomy. This was a time, following the KIRCHHOFF–BUNSEN DISCOVERY, when astronomers were first beginning to study the spectra of the light emitted by stars. Lockyer made a number of important discoveries—he was the first, for example, to show that sunspots are cooler than the surrounding solar surface, and the first to identify and name the outer layer of the Sun, the

Lighter-than-air craft, like the Goodyear airship Eagle, *seen here over Las Vegas, are filled with helium.*

Coronium and Nebulium

The question of whether there were elements in the universe not found on Earth persisted into the 20th century. Astronomers investigating the Sun's outer atmosphere, the highly tenuous *corona,* found spectral lines that they could not identify with any known terrestrial element, and they ascribed them to a new element which they named *coronium.* Similarly, when they examined the spectra of certain *nebulae,* distant clouds of gas and dust in the Galaxy, they found other mysterious lines, and attributed them to another "new" element, *nebulium.* In the 1930s, the American astrophysicist Ira Sprague Bowen (1898–1973) discovered that both these sets of lines came from oxygen, but could do so only under the extreme conditions that existed in the Sun and nebulae—conditions that could not then be attained in Earth-bound laboratories. For this reason, they are known as *forbidden lines.*

chromosphere. In 1868, as part of an investigation into the light emitted by atoms in prominences—huge eruptions of plasma from the solar surface—he saw a series of previously unknown spectral lines (*see* SPEC-TROSCOPY). When attempts to duplicate these lines in the laboratory failed, Lockyer concluded that he had found a new chemical element. He named it helium, from *helios,* the Greek word for "Sun."

Scientists were at a loss as to what to make of helium. Some suggested that there had been a mistake in interpreting the spectra of the prominences, but this position became harder to maintain as more people observed Lockyer's lines. Others argued that there were elements in the Sun not found on the Earth—which, as pointed out above, contradicted a major assumption about the laws of nature. Still others (a distinct minority) held that helium would someday be found on Earth.

In the late 1890s, Lord Rayleigh and Sir William Ramsay were carrying out the work that led to the discovery of ARGON. Ramsay adapted their technique to examine the gases given off by minerals containing uranium. Spectra of these gases showed strange lines, and Ramsay sent samples to several colleagues for analysis. When Lockyer got his sample, he immediately recognized the lines he had seen in the light from the Sun over a quarter century earlier. The mystery of helium was solved: It was indeed to be found in the Sun, but it existed here on Earth as well. Today, of course, it is best known in everyday life as the gas used to inflate blimps and party balloons (*see* GRAHAM'S LAW), and in science for its use in *cryogenics*—the technology of achieving very low temperatures.

JOSEPH NORMAN LOCKYER (1836–1920) English scientist. He was born in Rugby, the son of an apothecary-surgeon. He entered the ranks of science through an unusual route, having started work as a clerk in the War Office. To supplement his salary, he took advantage of the public interest in science by publishing a general-interest science magazine. In 1869, the first issue of the journal *Nature* appeared; Lockyer remained its editor for 50 years. He was a member of many expeditions to observe total eclipses of the Sun, and it was one of these that led to his discovery of helium. He also founded the field of archeoastronomy— the study of the astronomical significance of ancient structures, such as Stonehenge—and wrote many popular-science books.

Henry's Law

The mass of gas that will dissolve in a liquid depends on the pressure of the gas just above the liquid

Henry's law applies to the process by which gases dissolve in liquids. We are familiar with gases in liquids in the form of, for example, soft drinks, mineral water, and (if we are lucky) champagne. In all of these drinks, the dissolved gas is carbon dioxide, and the fizz of the drink appears when dissolved carbon dioxide comes out of the liquid to form bubbles when the bottle is opened and the pressure drops.

What the law says is that the higher is the pressure of the gas above the liquid, the harder it will be for the gas in the liquid to escape. This makes sense from the point of view of the KINETIC THEORY, since gas molecules leaving the surface of the liquid will have a harder time escaping if there are more gas molecules to collide with there.

Henry's law also explains another phenomenon associated with carbonated drinks—the characteristic hiss and fizz when you open a soft drink or (depending on your fortunes) uncork a bottle of champagne. In order to get lots of carbon dioxide into the drink, manufacturers usually maintain a high pressure in the gas just above the liquid in the bottle (in champagne the pressure is created by carbon dioxide given off by the fermentation of the wine itself).

When you pull the tab or open the bottle, this high-pressure gas escapes with a familiar hiss or pop. The pressure above the liquid then rapidly falls to the atmospheric pressure, and the liquid can no longer hold all of its carbon dioxide in solution. The gas starts to come out in the form of bubbles in the fluid, giving the drink its fizz. Eventually, the carbon dioxide dissolved in the liquid drops to the amount appropriate for atmospheric pressure, and no more bubbles appear. This is why a drink goes flat if it's left for too long.

Physics is there to be found even in opening a can of drink.

WILLIAM HENRY (1774–1836) English physician and chemist. Henry was born into a chemical manufacturing family in Manchester. He trained as a physician at Edinburgh University and worked at Manchester Infirmary before taking over the family chemical business and beginning his researches. As well as the law that bears his name, Henry discovered the formula for ammonia, and distinguished methane from ethylene. In addition to his research (supported by the family fortune), he wrote *Elements of Experimental Chemistry*, the most successful chemistry textbook of the early 19th century. He had a close friend and collaborator in John Dalton, who originated DALTON'S LAW. Henry's son, William Charles Henry, wrote one of the earliest and most influential biographies of Dalton.

Hershey–Chase Experiment

Hereditary information is coded in DNA

DNA has had a long and interesting history. First identified in 1869 by the Swiss chemist Johann Miescher (1844–95), it languished in relative obscurity for decades. In 1914 a German chemist discovered that DNA would take up a red dye, but he considered this result so unimportant that he didn't publish the fact for 10 years. Later, however, this staining was used to establish the fact that DNA appears in all cells and was characteristically located in the chromosomes. In the 1920s, the Russian-American biochemist Phoebus Levene (1869–1940) analyzed DNA and established the basic building-blocks from which it is made (a phosphate group, a sugar, and one of four types of molecules known as a *base*). He correctly deduced that the DNA molecule was built up from units (called *nucleotides*) made up of a combination of these three components.

Starting in the 1940s, the Italian-American Salvador Luria (1912–91) and the German-American Max Delbrück (1906–81), two microbiologist refugees from a Europe then dominated by Adolf Hitler, added a crucial technique to the study of genetics. They examined the properties of a set of viruses known as *bacteriophages* ("bacterium eaters"). Every known bacterium is preyed upon by at least one of these viruses, which consist of a core of DNA surrounded by a shell made of proteins. Bacteriophages are easy to maintain in a laboratory, and their action on their hosts is striking—within a few minutes of being infected, the host bacterium breaks open and releases a hundred or more identical copies of the original virus. Clearly, something in the virus was transmitting genetic information to those offspring, but was it the proteins or the DNA?

That was the question that the Hershey–Chase experiment resolved. The procedure used by Alfred Hershey and his colleague Martha Cowles Chase (1927–) is simple to describe. Two groups of bacteria were grown, one in a medium stocked with radioactive phosphorus-32, the other in a medium stocked with radioactive sulfur-35. When bacteriophages were allowed to attack these bacteria, the radioactive tracers were taken up into the viruses. To understand what happens next, you have to realize that phosphorus appears in DNA (in the phosphate group in the nuclei), but not in the proteins in the shell of the virus. Similarly, sulfur appears in the protein, but not in DNA. The twin radioactive tracers, then, provided convenient ways to differentiate the roles of the two components of the virus during reproduction.

The two groups of viruses—one with the DNA marked, the other with the protein—were then allowed to attack host bacteria. While the process of infection was still going on, the bacteria were separated from the rest of the material by using a centrifuge, then analyzed for the presence of the radioactive tracers. The results were very clear: The phosphorus-32 had been taken into the bacteria, while the sulfur-35 remained outside. It had to be the DNA that was causing the viruses to reproduce, not the proteins, since reproduction took place inside the bacteria, where the proteins did not penetrate.

Today we know that this process operates as follows: The virus attaches itself to the bacterium, injecting it with viral DNA inside and leaving the protein coat on the outside. The viral DNA incorporates itself into the bacterial DNA, co-opting the genetic machinery of the bacterium to make more copies of the virus. When the resources of the bacterium have been exhausted, the cell splits open and the newly minted viruses are released. The Hershey–Chase experiment established clearly that genes were arranged on molecules of DNA, a central tenet of modern science.

ALFRED DAY HERSHEY (1908–97) American biologist. He was born in Michigan and received his Ph.D. from Michigan State University. After working at the Washington University School of Medicine, he joined the staff of the Carnegie Institution at Cold Spring Harbor, New York, in 1950 and eventually became director of the Genetics Research Unit there. For showing that DNA carries hereditary information, Hershey, Salvador Luria, and Max Delbrück shared the 1969 Nobel Prize for Medicine or Physiology.

Hertzsprung–Russell Diagram

Stars plotted on a diagram according to their physical characteristics fall into distinct groups that correspond to different stages in their evolution

Stars come in a huge variety of types. There are, for example, stars that are 30 times the Sun's diameter, and stars that could fit comfortably inside the city limits of any major metropolis. There are stars that burn blue-hot and stars that barely glow a dull red. In the 19th century, astronomers began a major transition in their field, from classical astronomy ("Where is it and how does it move?") to astrophysics ("What is it and how does it work?"). One of their first tasks was to find some order among all the different stars. This they achieved through the Hertzsprung–Russell diagram.

The HR diagram, as it is usually known, is one of those developments in the sciences that took place around the same time and independently in two different places. Henry Norris Russell, one of the big names in American astronomy in the early 20th century, had a longstanding interest in the life cycle of stars and apparently had the main idea for the diagram in 1909, although the work wasn't announced until 1913. The Dane Ejnar Hertzsprung, had earlier come to the same conclusions as Russell, but he published his work in 1905 and 1907 in the *Zeitschrift für Wissenschäftliche Photographie* ("Journal of Scientific Photography"), not something you would expect many astronomers to have read. Although his work came to light soon after Russell's announcement, the diagram continued to be known as the "Russell diagram" until the 1930s, when the present name came into use.

The HR diagram is a graph on which the vertical axis represents a star's intrinsic brightness, or luminosity, and the horizontal axis is the star's surface temperature. Both of these quantities can be measured for any star whose distance from the Earth has been determined. For historical reasons which I have no interest in defending, it is customary to plot surface temperature on the horizontal axis in reverse, so that the farther to the right you go on the graph, the lower the temperature of the star. What one does is to plot on this diagram the data for a large number of stars, so that every star is represented by a point.

When we do so, we notice straight away that the distribution of points is by no means random: all stars fall into one of three regions. The first is a band that stretches from the upper left to the lower right. This is called the *main sequence.* The Sun is an example of a star on the main sequence. Stars at the top of this line have a high surface temperature and a high luminosity, while stars at the bottom have a low surface temperature and a low luminosity. Stars on the main sequence generate their energy by the nuclear fusion of hydrogen into helium (*see* STELLAR EVOLUTION).

Above the main sequence lie stars that have high luminosity but have very cool outer surfaces, and below the main sequence are stars that have very high surface temperatures but don't give out much light. Each of these regions corresponds to different stages in the life cycle of a star. At the upper right of the HR diagram are the so-called *red giants* and *supergiants,* stars whose outer envelopes have ballooned to a very large size indeed. (When the Sun becomes a red giant about 6½ billion years from now, its outer boundary will be outside the current orbit of Venus.)

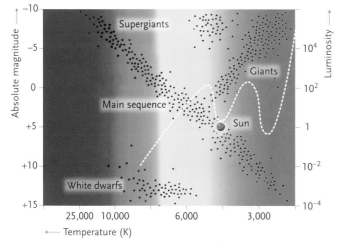

All stars lie somewhere on the Hertzsprung–Russell diagram. "Normal" stars lie on the diagonal main sequence. Above them are giants and supergiants; below them are the white dwarfs. The evolutionary tracks of stars can be plotted on the diagram. The Sun is shown in its present position, and the dashed line shows its history and its destiny.

These stars are giving out similar amounts of energy to main-sequence stars, but are doing so through a very much larger surface area, which accounts for their low surface temperature. At the lower left of the diagram are stars called *white dwarfs* (*see* CHANDRASEKHAR LIMIT). These stars have a very high temperature but are very small—typically, about the size of the Earth. Consequently, even though they are sending out less energy into space, they have a very small surface area through which to do the radiating, so each piece of the surface is at a higher temperature.

The life cycle of a star can be traced as a trajectory on the Hertzsprung–Russell diagram. A Sun-like star condenses (*see* NEBULAR HYPOTHESIS) from a cloud of gas and dust until the temperature and pressure in its core are high enough for fusion reactions to ignite, entering the diagram on the right-hand side and joining the main sequence. While it is burning hydrogen (as the Sun is doing now) it remains on the main sequence and moves relatively little. Once it has burned up all its hydrogen, it moves up the diagram and becomes a red giant, and then down into the realm of the white dwarfs.

EJNAR HERTZSPRUNG (1873–1967)
Danish astronomer, born in Frederiksberg. He qualified as a chemical engineer at Copenhagen Polytechnic in 1898, and then studied the chemistry of photography. He spent most of his astronomical career at the observatory at the University of Leiden in the Netherlands. Hertzsprung's background in photochemistry allowed him to develop techniques of measuring the luminosities of faint stars from their images. It was the correlations of these results with the spectra of stars that led him to classify giant and dwarf stars.

HENRY NORRIS RUSSELL (1877–1957)
American astrophysicist. He was born in Oyster Bay, New York, the son of a Presbyterian minister, and educated at Princeton University, where he became professor and director of the observatory. He had a longstanding interest in understanding the life cycle of stars. Like Ejnar Hertzsprung, Russell arrived at the HR diagram from studies of stellar images on photographic plates. But he got from it the wrong idea about stellar evolution —he thought that all stars began life as giants, and moved down the main sequence to become dwarfs.

Hooke's Law

*The force exerted by
a spring depends on the
amount of stretching
or compression*

Imagine grabbing hold of the end of a spring and pulling or pushing on it. The more you stretch or compress the spring, the harder it will resist being stretched or compressed further. This is the principle of the spring scale, in which a weight is either hung from a spring (in which case the spring stretches) or is put on a spring (in which case the spring is compressed). In either case the force exerted by the spring opposes the action of the weight, either pushing up or pulling down until the force exerted by the spring equals the force of gravity on the weight. Thus, the stretching or compression of the spring provides a way of measuring the weight of the object placed on the scale.

The first to investigate this kind of stretching in a properly scientific way was Robert Hooke. In his original experiment he used a wire rather than a spring, and measured how it gradually lengthened when the force applied to one end of it was increased, the other end being fixed. He found that the extension was proportional to the force, as long as the material of the wire did not reach a point beyond which it underwent permanent deformation (*see* below). In equation form, Hooke's law is written as

$$F = -kx$$

As applied to a spring, F is the force exerted by the spring, x is the amount by which the spring is stretched or compressed, and k is a quantity known as the *spring constant*. The higher the value of k, the stiffer the spring and the harder it is to stretch or compress. The minus sign indicates that the force exerted by the spring opposes the changes induced by an external force—stretch a spring, in other words, and it tries to exert a force to make itself shorter; compress it, and it tries to lengthen.

Hooke's law is the foundation of the branch of science known as the theory of *elasticity*. It turned out to have wider applications because the atoms in solids can be thought of as being held together by springs. Thus, if a solid material is stretched, each little interatomic spring stretches and exerts a force as dictated by Hooke's law. In the language of elasticity theory, we say that the force is the *stress* and the amount of deformation is the *strain*.

Thomas Young (*see* INTERFERENCE) first wrote down the version of Hooke's law that applies to solid materials. In equation form, it is

$$\sigma/\eta = E$$

where σ is the stress, η is the strain, and E is a quantity, analogous to k in the previous equation, known as the *elastic modulus* or *Young's modulus*. It varies from one material to another, and is a measure of the amount by which a solid will compress under pressure or stretch under tension.

*The spring in the scale this
biologist is using to weigh an
Atlantic sturgeon operates
according to Hooke's law.*

Young is actually best known in the sciences for his advocacy of the wave nature of light. He devised the double slit experiment (*see* COMPLEMENTARITY; INTERFERENCE) to prove his thesis, and was instrumental in gaining acceptance for the theory (though he never worked out the full mathematics of his ideas). Young's modulus is actually one of three numbers used to characterize the way a solid responds to a force. The others are the *shear modulus* (which indicates how much a material will bend in response to a sideways force) and the *Poisson ratio* (which indicates how much an object thins down when its stretches). The latter is named for the French mathematician Siméon-Denis Poisson (1781–1840).

Hooke's and Young's laws don't give a full description of everything that can happen to a solid when a force acts upon it. Think about stretching a simple rubber band. If you stretch it by a small amount, it will exert a Hooke's law force, bounce back, and be as good as new. Stretch it too far, however, and the rubber band loses its elasticity. Now you can stretch it all you want, and there will be no restoring force. We say that the rubber band has exceeded its *elastic limit*.

In other words, Hooke's law applies only when the stretching or compression is relatively small. As long as this is the case, the force is proportional to the deformation and the system is linear—if the force is increased in regular steps, the stretching or compression will also increase in regular steps. Once it's stretched too far, however, the interatomic springs are broken and the simple linear form of Hooke's law no longer applies. We say that the system has become nonlinear. The study of nonlinear systems is now a major area of research in the physical sciences.

ROBERT HOOKE (1635–1702) English scientist. He was born in Freshwater on the Isle of Wight, the son of a clergyman, and studied at Oxford University. At Oxford he assisted Robert Boyle, building the air pump that was used in the experiments that led to BOYLE'S LAW. Hooke was a contemporary of Isaac Newton and was very active in the newly formed Royal Society, becoming its secretary in 1677. Like many scientists of that time, his interests and contributions spanned a wide range of fields. He published *Micrographia,* a book containing images of microscopic life and coined the modern term "cell." In geology, he was one of the first to understand the significance of fossils, and was the first scientific catastrophist (*see* UNIFORMITARIANISM). He also was one of the first to suggest that gravity decreased as the square of the distance between objects, a crucial ingredient of NEWTON'S LAW OF GRAVITATION and a source of dispute between the two men as to who had thought of it first. Hooke designed and built a wide variety of scientific instruments—indeed, some consider this to be his most important contribution to science. He was, for example, the first to put cross-hairs into the eyepiece of a telescope, the first to suggest that the freezing point of water should be used for the zero of the temperature scale, and he was the designer of the universal joint.

Hubble's Law

The speed at which a galaxy appears to be receding from the Earth is proportional to its distance from us

Edwin Hubble's first job after his return from World War I was at the Mount Wilson Observatory near Los Angeles, then the premier astronomical observatory in the world. Using the observatory's new 100-inch (2.5-m) reflecting telescope, he carried out a series of measurements that changed our view of the universe forever.

The basic problem he attacked was an old one in astronomy and concerned the nature of objects called nebulae. The word comes from the Latin for "cloud," for these objects appeared as fuzzy, cloudy smears of light in the telescopes of the 18th and 19th centuries. By the beginning of the 20th century, several had been resolved into stars, while others remained truly nebulous. The question then being asked was this: Were they all inside the Milky Way, or were some of them, in the evocative language used at the time, other "island universes"? Before the Mount Wilson telescope came on line in 1917, this question could not be answered because there was no way to measure their distances.

Hubble began his investigations by using the new telescope to study what was then called the Andromeda Nebula. In 1923, he was able to see that its outer regions consist of vast swarms of individual stars, and he found that some were of the type astronomers call *Cepheid variables*. By observing a Cepheid variable over a period of time, an astronomer can measure the period of its variability and, from the PERIOD–LUMINOSITY RELATION, can determine how much light it is throwing out into space.

Here's an analogy that may help you understand the next step: Suppose you were outside on a dark night and someone turned on a light bulb some distance away. Because you couldn't see your surroundings, you would have a hard time judging how far away the light was. It could be bright and far away, or dim and close—there would be no way of telling. Suppose, however, that you knew how much light the bulb was giving off, that it was 60, 100, or 150 watts, for example. You could then

measure the fraction of this light you were receiving at your location and, with a little geometry, figure out how far away the light was. In a sense, an astronomer measuring the period of a Cepheid variable is in the same position as you are when you read

A spiral galaxy known as NGC 4414. This image was obtained by a camera aboard the Hubble Space Telescope (HST), named for Edwin Hubble. One of the HST's goals has been to home in on the value of the Hubble constant. By identifying variable stars in this galaxy, astronomers have calculated its distance to be 60 million light years.

the wattage on the light bulb. Both of you are receiving the information you need to figure out the distance to your light source.

The first thing Hubble did was figure out the distance to the Cepheid variables—and hence to the Andromeda Nebula. The answer: 900,000 light years,* much farther away than the most distant stars in the Milky Way. With this and other results, Hubble established the basic structure of the universe as a collection of galaxies, as these distant "nebulae" now became known, rather than as one large collection of stars. This discovery alone would have been more than enough to establish his reputation.

He noticed, however, another aspect of his data, one that had been seen previously by other astronomers. It seemed that the wavelengths of light coming from atoms in the galaxies was shifted slightly with respect to light from the same atoms measured in terrestrial laboratories. Light emitted when an electron jumps from one energy level to another in an atom (*see* BOHR ATOM) will appear to have a slightly longer wavelength, and therefore appear redder, if the light comes from a distant galaxy than if it comes from the same type of atom on Earth. Hubble interpreted this as an example of the DOPPLER EFFECT, an indication that the galaxies were moving with respect to the Earth. The light from almost all the galaxies was shifted to the red, which means that they are moving away from us, the amount of the so-called *redshift* being proportional to the speed of recession.

Most importantly, Hubble was able to compare his measurement of the distance to each galaxy (from observations of Cepheid variables) with his determination of their speed (from measurements of the redshift). When he did so, he found that the farther away from us a galaxy is, the faster it is receding from us. This is what is known as Hubble's law. In equation form, it is simply

$$v = Hr$$

where v is the velocity of the galaxy, r is its distance from us, and H is a number known as the *Hubble constant*. The currently accepted value of the Hubble constant is about 70 km per second per megaparsec (a megaparsec is about 3.3 million light years). This means that a galaxy 10 megaparsecs away will be receding from us at 700 km per second, a galaxy 100 megaparsecs away will be moving away from us at 7000 km per second, and so on. And although Hubble's original determination of this law was based on observations of a handful of galaxies lying relatively close to the Milky Way, it has since been amply verified for galaxies out to the farthest reaches of the visible universe.

Hubble's law implies something that seemed quite incredible: the universe is expanding! The way I like to visualize Hubble's result is to think

* Today's more accurate distance to the Andromeda Galaxy, as it is now called, is 2.3 million light years.

of the galaxies as a bunch of raisins scattered through a clump of rising bread dough. Imagine you are standing on a raisin and can look through the dough—what would you see? Because the dough is rising, all the other raisins will appear to be moving away from you, and the farther away they are, the faster they will appear to be moving (because there is more expanding dough between you and the more distant raisins). It will also appear that you are at the center of a universal expansion, but there is nothing special about this—every raisin sees itself the same way, as the center of the dough-universe. In the same way, the galaxies are moving away from one another because the very fabric of space is expanding. All observers (including us) see themselves as being at the center of that expansion. The 15th-century philosopher Nicholas of Cusa put it best: "The universe has its center everywhere and its edge nowhere."

Hubble's law also implies something else about the universe, something truly extraordinary. The universe must have had a beginning in time. It is, in fact, a simple mental exercise to take the current expansion and "run the film backward" to show that there must have been a time when everything in the universe was crammed together into a much smaller volume than it has now. The picture of a universe that began in a hot dense state and has been expanding and cooling ever since is called the BIG BANG and is currently our best theory of the origin and evolution of the cosmos. You can use Hubble's law to get a rough estimate of the age of the universe by making a simple (though admittedly over-simplified) model. Suppose that each galaxy had been traveling at its present velocity v for all time. How long would it have taken it to get to its present position?

We know that the distance something travels is equal to its velocity multiplied by the time for which it has been traveling. Let's call the travel time t. In our simple model, because this is the time for which each galaxy has been traveling since the beginning, it is also the age of the universe. So,

$$v \times t = r \quad \text{or} \quad t = r/V$$

But from Hubble's law, r/V is just $1/H$, where H is the Hubble constant. Therefore, when we measure the current expansion of the universe to determine H, we are also getting an estimate of the length of time for which the expansion has been going on, and hence for how long the universe has existed. For the record, the current best estimate of the age of the universe is about 15 billion years, plus or minus a few billion. (In comparison, the Earth is about 4½ billion years old and the earliest life forms appeared about 4 million years ago.)

EDWIN POWELL HUBBLE (1889–1953) American astronomer. Hubble was born in Marshfield, Missouri, and grew up in Wheaton, Illinois (then a railroad suburb of Chicago). As an undergraduate at the University of Chicago, he excelled at both his studies and athletics. Hubble supplemented his college scholarship by working as a laboratory assistant for Nobel laureate Robert Millikan and by working on railroad surveying crews in the North Woods during the summers. One story he loved to recount concerned the time he and another worker were left behind to close down the camp and then missed the last train back to civilization. It took the men three days to walk out of the woods. They had no food, but, in Hubble's words, "We could have killed a porcupine or small game, but there was no need, and besides, there was plenty of water."

Receiving his BS in 1910, Hubble went to Oxford on a Rhodes scholarship. There he studied Roman and English law but, in his words, "chucked law for astronomy" and returned to Chicago for graduate work. He did most of his observational work at Yerkes Observatory, north of the city, and came to the attention of George Ellery Hale, who in the spring of 1917 invited him to join the new observatory at Mount Wilson.

Other events intervened, however. When America entered World War I, Hubble stayed up all night to finish his Ph.D. thesis, took his oral exam the next morning, and promptly volunteered for the army. His telegram to Hale read, "Regret cannot accept your invitation. Am off to war." He did not reach France until just before the end of the war, and his division saw no combat, though he was wounded by a stray shell. He mustered out of the army in the summer of 1919 and immediately went to the

Mount Wilson Observatory in California. At Mount Wilson, Hubble showed that the matter in the universe is collected into huge collections of stars called galaxies, and that the galaxies are all receding from us, according to what is now called Hubble's law.

During the 1930s, Hubble carried out an extended observational program at Mount Wilson, exploring the new world outside the Milky Way. This work quickly brought Hubble recognition not only among his astronomer colleagues, but with the general public as well. He enjoyed the limelight, and was often photographed in the company of movie stars.

Hubble's popular book *The Realm of the Nebulae*, published in 1936, added to his public recognition. Except for a stint during World War II when he served as Chief of Exterior Ballistics and Director of the Supersonic Wind Tunnel at Aberdeen Proving Ground in Maryland, Hubble spent the rest of his life at Mount Wilson. After the war, he became Chairman of the Research Committee for the Mount Wilson and Mount Palomar Observatories, and was deeply involved in the planning and design of the 200-inch (5-m) Hale Telescope on Palomar that went into operation in 1949. It was a fitting tribute to his contributions to astronomy that he was the first astronomer to use that magnificent new instrument.

Away from astronomy, Hubble was a man of wide interests. In 1938, he was elected as a trustee of the Huntington Library and Art Gallery in Los Angeles, to which he left his extensive collection of early books on the history of science. His main form of recreation was fly fishing, and there are records of many fishing expeditions to the Rocky Mountains and to the River Test in England. Edwin Hubble died suddenly on September 28, 1953, of a cerebral thrombosis.

Human Genome Project

In June, 2000, the first rough draft of the entire sequence of human DNA was published

According to the central dogma of MOLECULAR BIOLOGY, the basic blueprint for running the chemistry of any organism (including human beings) is contained in the sequence of base pairs in DNA molecules. In a sense, knowing the complete sequence of these base pairs tells you all there is to know about the chemical workings and the hereditary information associated with a given species. In 1986, a group of scientists in America initiated what came to be called the Human Genome Project, the goal of which was to map the complete sequence—the *genome*—of human DNA. In the 1980s, though, the technology for doing this was very primitive. It was believed that the project would cost billions of dollars, and would take at least until the year 2005 to be completed.

There was a great deal of opposition to this project among biologists at this time. They felt that it would introduce a kind of corporate structure, or Big Science, into their field, which had hitherto been characterized by small research groups working under the direction of a single laboratory scientist. There was a real fear that all biologists would be dragooned into performing repetitive and boring operations on human DNA. As one young post-doc told me at the time, "I don't want my life's work to be that I sequenced from base pair 100,000 to base pair 200,000 on chromosome 12." This kind of objection melted away as new technologies were developed to allow the routine work of sequencing to be done by machines.

The story of the 1990s is one of steady progression in our ability to sequence entire genomes. In 1995, for example, the Institute for Genomic Research in Rockville, Maryland, published the first complete DNA sequence of a living organism—the bacterium *Haemophilus influenzae*. It took them several years to find the whole sequence.

Other organisms followed quickly. The first genome of a eukaryotic cell (i.e., a complex cell whose DNA is housed in a nucleus), from the yeast *Saccharomyces cerevisiae,* was sequenced in 1996. This was a collaborative effort between six hundred scientists in Europe, North America, and Japan. In 1998, the first DNA sequence for a multicellular organism—the flatworm *Caenorhabditis elegans*—was published. Each of these advances required a progressively longer sequence to be elucidated, and represented a significant milestone on the way to sequencing the human genome itself.

An important player in this process was Craig Venter, then of Celeron, a private corporation which he founded. Venter introduced what has come to be known as the "shotgun" technique in DNA sequencing. The idea is that the DNA of an organism being sequenced is broken up into many small bits, and each bit is fed into a different sequencing machine. This process would be analogous to tearing up the pages of a book and giving each page to a different reader. After each bit of DNA is sequenced, massive computer programs are used to reassemble the original sequence. Because of this intense use of information technology, many scientists have begun referring to the new field of genomic research as the bio*informatic* revolution rather than the biomolecular revolution.

In June of 2000, Craig Venter and Francis Collins, head of the U.S. National Institutes of Health's genome project, announced what was called the "first assembly" of the human genome. This was, in essence, the first reconstruction of the entire human genome done by the shotgun technique. A few months later, in February 2001, the first preliminary read-through of the human genome was published. There were some surprising results.

It had been known for a long time, for example, that most human DNA plays no part in the construction of genes. The new results tell us that human beings have surprisingly few genes coded in their DNA— probably something in the neighborhood of 30,000 to 50,000. (I say "surprisingly" because scientists had expected that a complex organism such as a human being would require much more in the way of genetic structure.) These genes, however, are arranged so that they are not in a single long sequence, but consist of coding sequences called *exons,* interspersed with random sequences called *introns.* The machinery that assembles the protein from a gene whose sequence is arranged in this way turns out to have several choices as to how to put the protein together. Thus, each human gene codes for about three different proteins, rather than the single protein that one would have expected from the central dogma of molecular biology.

You can think of the completion of the first phase of the Human Genome Project as the decoding of the book of life. The next step is to figure out what all the genes are and how the proteins for which they code come together to produce the biological makeup of a human being. Scientists estimate that it will take another century for us to mine all the data and understand all the workings of the human genome.

I think this is a very pessimistic estimate—but then perhaps I have more faith in the ability of these people to deal with the complexities they are discovering than they do. Nevertheless, we are well on our way to understanding the full genetic makeup of human beings, which has enormous implications for medicine and for human well-being.

Huygens' Principle

Every point on an advancing wave front is a source of secondary wavelets

Think about a wave moving across the surface of a pond. The simplest way to describe it is to think about how the water itself moves—how the upward pressure is countered by the downward force of gravity to produce the up-and-down motion of the wave. In the 17th century, Christiaan Huygens thought about waves a little differently, and in the process gave us a powerful tool for thinking about all sorts of waves, from mundane ripples on a pond to gamma rays emitted from distant galaxies.

Huygens' principle is easy to visualize if you think of momentarily freezing the crest of a wave moving through the water. Now imagine that a small rock is dropped into the water at each point along that crest, so that each point becomes the center of a circular outgoing ripple. In most places, many of these ripples would come together and cancel one another out, leaving the water level undisturbed. Along a line parallel to the original crest, however, the wavelets will reinforce, recreating the original crest a little farther along. This, in Huygens' scheme, is how a wave moves from place to place.

Why should this somewhat contrived way of thinking about ordinary wave propagation be of any use? Consider what happens when the wave encounters an obstacle. Go back to our wave moving on water, for example, and imagine that it encounters a concrete breakwater at the entrance to a harbor. According to Huygens' principle, the parts of the wave at the breakwater will not be able to send out secondary wavelets, while those still in open water will. The disturbance from the wavelets just at the edge of the breakwater will therefore propagate into the undisturbed water behind the breakwater. In effect, the wave will be able to turn the corner when it encounters an obstacle, an ability that anyone with sailing experience will recognize immediately. (This property of waves is called DIFFRACTION).

There are many other applications –some unexpected—of this way of looking at waves. It is used extensively in optics and in communications, where waves (light and radio, respectively) routinely encounter barriers and obstacles to their progress.

One of Huygens' astronomical discoveries was made in 1655, when he became the first to see Titan, the largest moon of the planet Saturn. NASA's Cassini spacecraft reaches Saturn in 2004, and will drop a probe on a parachute to investigate the atmosphere of Titan. That probe is named *Huygens*. Thus does science honor its founders.

CHRISTIAAN HUYGENS (1629–95) Dutch astronomer and physicist. He was born in Den Haag to a family of diplomats and intellectuals. He studied at home and at the University of Leiden. He went to Paris in 1666 where he helped to found the Academy of Sciences. Most famous for his work in optics and astronomy, he improved the design of telescopes and was the first to see the rings around the planet Saturn. Huygens was also the inventor of the pendulum clock.

Ideal Gas Law

The behavior of gases can be summarized in one simple equation

The simple equation is this:

$$pV = nRT$$

It contains all the essential features of the behavior of gases: p, V, and T are, respectively, a gas's pressure, volume, and temperature (measured in degrees kelvin); R is a constant, known as the *gas constant,* which is the same for all gases; and n is related to the number of molecules or atoms in the gas (technically, it's the number of *moles* in the gas—*see* AVOGADRO'S LAW.)

To see how the equation works, imagine that we hold the temperature of the gas constant, so that the right side of the equation is constant. Then the product of the pressure and volume has to be constant as well. If the pressure goes up, the volume goes down, and vice versa. This is just BOYLE'S LAW, one of the first experimental discoveries about the nature of gases. If, on the other hand, we hold the pressure constant (e.g., by putting the gas in a balloon that is subjected to atmospheric pressure from the surrounding air), then as the temperature increases, so does the volume. This is CHARLES'S LAW, another of the experimentally discovered laws of gas behavior. AVOGADRO'S LAW and DALTON'S LAW are also consequences of the ideal gas law.

This law is an example of what scientists call an *equation of state*. It tells you how a material will respond to changes in its environment—changes in the ambient temperature or pressure, for example. Strictly, it applies only to an *ideal gas*. This is a simplified model of a gas in which the atoms are assumed to be in constant random motion, and only interact with one another during collisions. The ideal gas model makes it much easier to calculate the behavior of real gases; where it cuts corners is by ignoring the forces between molecules. Despite this, most real gases behave much as predicted by the equation—the volume occupied by nearly all common gases, such as the oxygen and nitrogen in the atmosphere, at atmospheric pressure and ambient temperature is within 1% of the value predicted by the ideal gas law. The equation therefore often crops up in first cuts of theoretical descriptions. For example, an astronomer trying to set up a mathematical model of a star will usually assume that the gas in the star obeys this equation, a procedure that allows him or her to guess how the temperature of a given region will change. (The material inside a "normal" star also obeys the ideal gas law even though its density is huge compared with the densities we encounter on Earth. This is because it consists of completely ionized hydrogen and helium—particles far smaller than atoms.) Later, as the theory improves, fancier equations of state may come into play—equations that contain more of the quirks and intricacies of real gases.

Immune System

The function of the immune system is to recognize foreign invaders in the body, communicate this information to the relevant cells, and eliminate the invader

Human beings and other organisms live in a hostile environment. All sorts of bacteria and viruses are lurking out there, just waiting to get at us, and it is the task of our immune system to protect us against them. Some of the body's defenses are purely anatomical—our skin and mucous membranes form a purely physical barrier against invasion, for example. If these outer defenses are breached, the body often counters the invasion with a generalized inflammatory response in which increased blood flow to the affected area brings in white blood cells, which, once they move through the walls of the capillaries, engulf the invaders. It is this response that produces the familiar reddening around a small cut.

The immune system proper, however, operates on a different set of principles—principles based on enlisting specific molecular agents to deal with specific targets. The most important of these molecular agents are *antibodies*. These are Y-shaped molecules. At the tips of the Y are collections of amino acids (*see* PROTEINS) of different shapes. Each of these shapes fits onto a different kind of invader, or *antigen*. In an adult human, there are as many as 100 million different shapes available on our antibodies. The immune system, in a sense, is like a large clothing store, with all sorts of sizes in stock. When an invader enters, the chances are excellent that one of the 100 million items already "on the rack" will fit it. The arrangement of amino acids in the stem of the Y determines how the antibody circulates in the body—some, for example, circulate in the bloodstream and are very efficient at eliminating bacteria and viruses, while others bind to specialized cells in the skin and intestinal linings.

The main cell that makes the antibody recognition system work is called the *B cell*, or B lymphocyte. (The name comes from the fact that these cells grow and mature in the bone marrow.) These cells are roughly spherical, and each contains many kinds of specialized antibodies in its outer coating. When the recognition process occurs—when an antigen fits itself onto the antibody of a given B cell—a process is initiated in which the B cell starts to multiply. There are two consequences of this multiplication process. One is the production of cells (called *plasma cells*) that produce large numbers of antibody molecules specific to the invader; the other is the production of memory cells that can respond to the presence of the antigen months or years after the initial invasion.

A single plasma cell can produce up to 30,000 antibody molecules each second. These molecules bind to invading bacteria, causing them to clump together so that other cells and molecules can remove them from the body. It can take several days for the plasma cells to reach maturity, however. Usually, the phenomenon of a fever "breaking" is the body's signal that the antibodies have taken over. Plasma cells live only a few days, but the memory cells last much longer—sometimes an entire lifetime. In the event of another invasion by the same antigen, these cells spring into action and produce large amounts of the antibody immediately, eliminating the time-consuming recognition process. This is what gives us immunity against a subsequent infection. The production of memory cells is the main purpose behind the use of vaccines.

B cells operate mainly to protect the body against invaders that have a "foreign" chemical signature. It is the domain of another type of immune cell—the *T cell*—to deal with body cells that have been transformed by infection or cancer. (Actually, only about half of the T cells fulfill this role; the other half play a role in regulating B cells.)

The T is for thymus, the gland where these cells grow and mature. The T cells have proteins in their outer coating that recognize specific molecules, but not specific antigens (as in B cells). Instead, T cells respond to antigens when they are combined with another type of molecule called a *histocompatibility complex,* a type of molecule found in all of an individual's cells. Think of the T cell as a sentry that moves from one place to another, challenging other cells by asking them if they have the password. If they have the right kind of histocompatibility complex on their surface, the T cell passes on. If, however, something is wrong—if the complex has become tangled up with a protein from a virus's coat, for example—the T cell attacks the cell and destroys it.

It is the ability of T cells to recognize "not self" that makes organ transplants so difficult. The T cells want to attack the transplanted organ, and have to be controlled by immunosuppressant drugs. T cells are also the target of the virus that causes AIDS, which happens to match the T cell receptors closely. Finally, it can happen that the T cells' ability to recognize self fades over time, in which case the immune system may actually attack cells within the body. This is the origin of autoimmune diseases like rheumatoid arthritis.

Inflationary Universe

Early in its life, the universe underwent a particularly rapid expansion

Astronomers have known for some time that the universe is expanding according to HUBBLE'S LAW, and that this implies that the universe began at some definite point in the past. The task of astrophysicists, then, is simple: to work their way back through the Hubble expansion, applying the appropriate laws of nature at each stage until, eventually, everything is understood.

In the late 1970s, there were several problems that had eluded solution, each dealing with a rather fundamental question about the early universe. These questions were:

— *The antimatter problem* The laws of physics treat matter and antimatter (*see* ANTIPARTICLES) on an equal footing, yet the universe is made almost completely of matter. Why should this be?
— *The horizon problem* Judging by the cosmic background radiation (*see* BIG BANG) we can detect, the universe is almost all at the same temperature, yet different parts of it could not have been in contact with one another (i.e., they were outside of each other's *horizons*), so they could not have come to thermal equilibrium. How was this equilibrium established?
— *The flatness problem* The universe seems to have almost exactly the amount of mass and energy needed to slow down and stop the Hubble expansion. Of all the possible masses the universe could have, why does it have this one?

The key to understanding how these problems were solved is that when the universe was very young, it was also very dense and very hot. All matter therefore existed as a seething mass of quarks and leptons (*see* STANDARD MODEL), which had no chance of forming into atoms. The different forces that operate in the present universe, such as the electromagnetic and gravitational forces, were all unified in a single entity (*see* THEORIES OF EVERYTHING). As the universe expanded and cooled, the forces separated from one another (*see* EARLY UNIVERSE).

In 1981, physicist Alan Guth realized that the separation of the strong force from the others, which happened when the universe was a mere 10^{-35} seconds old (that's a decimal point followed by 34 zeroes and a one!), was the crucial thing to consider. This separation is an example of a *phase transition,* an everyday example of which is the freezing of water. By analogy with the freezing of water, the "freezing" of the strong force entailed a fundamental rearrangement of the structure of the universe.

Anyone who has had to deal with pipes that burst because water freezes in them during a cold snap knows that water expands when it freezes. What Guth showed was that, in a similar way, when the strong force separated out from the others, the universe underwent a period of very rapid expansion. This expansion, which is called *inflation*, is considerably faster than the normal Hubble expansion. In a matter of about 10^{-32} seconds the universe went from something smaller than a single proton to something the size of a grapefruit—an increase in size of 50 orders of magnitude. (In comparison, the volume of water increases by only about 10% when it freezes.) This rapid expansion of the universe directly solves two of the problems mentioned above.

The easiest way to visualize the solution to the *flatness problem* is to imagine a coordinate grid drawn on a rubber sheet, and that the sheet is crumpled up in some arbitrary way. If we seize the sheet and stretch it out vigorously, the coordinate grid will wind up being flat, no matter how lumpy it was to start with. In the same way, no matter how convoluted the universe was when inflation started, it would be stretched absolutely flat by the time inflation was complete. And, since we know from the theory of RELATIVITY that the curvature of space depends on the amount of matter and energy present, this explains why the universe has exactly enough matter in it to balance the Hubble expansion.

Inflation also explains the *horizon problem,* though the explanation is a little less direct. We know from the theory of BLACK-BODY RADIATION that the radiation emitted by any object depends on its temperature. Therefore we can use the radiation from distant parts of the universe to measure the temperature of those regions. When this is done, an astonishing result emerges: To an accuracy of four decimal places, the temperature of every part of the universe is the same as the temperature of every other part. If we trace the Hubble expansion back in time, we find that different parts of the universe would have been too far apart to have come to the same temperature. With inflation, however, the universe was much more compact before 10^{-35} seconds than you would guess from winding back the Hubble expansion. During this highly compressed period, thermal equilibrium was established, and it survived through the period of inflation.

Inflation doesn't solve the *antimatter problem,* but other events that took place at the same time do. An interesting story then emerges: When particles are being forged in the maelstrom of the early universe, about 100,000,000 antiparticles are created for every 100,000,001 ordinary particles. Over the next fraction of a second, the particles and antiparticles pair up and annihilate each other in a burst of energy—essentially converting their mass into radiation. When this winnowing process is over, all that remains is the lone, leftover bit of ordinary matter. And from this bit of cosmic refuse and its fellows, the entire known universe is made.

ALAN HARVEY GUTH (1947–) American particle physicist and cosmologist, born in New Brunswick, New Jersey. He obtained his Ph.D. at the Massachusetts Institute of Technology, to which he later returned, becoming professor of physics in 1986. Guth developed his theory of inflation while he was at Stanford University, having originally studied the theory of ELEMENTARY PARTICLES. He has referred to the universe as "the ultimate free lunch."

Interference

Interference between waves can be constructive or destructive

Waves starting out from different points on a surface—in the form of ripples spreading out where droplets of water have fallen into a pool—interfere where they overlap. There are blank spots where the ripples intersect, indicating destructive interference.

Waves are one of the two ways in which energy can be carried from place to place (particles are the other). Waves are usually carried on a medium (water waves on a lake, for example), but the motion of the medium isn't the same as the motion of the wave. Think of a cork bobbing up and down as waves go by. The cork traces the motion of the water, which moves up and down, while the waves themselves go past it.

The phenomenon of interference takes place when two waves moving in different directions encounter each other. Although some waves need a medium through which to propagate, the phenomenon of interference does not depend on the medium. Light and other radiation that make up the ELECTROMAGNETIC SPECTRUM don't require a medium. Interference is therefore a property of wave motion itself. The easiest way to understand what happens is to think again about water waves and imagine that each wave carries with it an instruction for the surface. It might be "Go up 1 foot" or "Go down 6 inches." When the two waves meet, the surface will integrate these instructions—in this example, the surface will go up 6 inches (1 foot minus 6 inches).

The most striking feature of interference comes when two waves of equal amplitude arrive at a point exactly *out of phase* with each other (i.e., the crest of one wave arrives at the same time as the trough of the other). The instructions that the surface receives cancel each other out—"Go up one foot" and "Go down one foot"—so it does nothing. If the waves are water waves, there will be a still spot. For sound waves, there will be a dead spot. When the waves are light waves, there will be darkness. This is called *destructive interference*.

When the opposite happens, and two crests or two troughs arrive together, completely *in phase*, the waves reinforce each other and the amplitude is twice what it would be for a single wave. This is called *constructive interference*. Water waves will then be at their highest, sound waves at their loudest, light waves at their brightest. Of course, there is a wide range of degrees of interference that fall between totally constructive and totally destructive, leading to intermediate levels of water, sound, or light.

Destructive interference gives us a good way of telling whether something is a wave or a particle. After all, if two billiard balls collide, there is no situation imaginable in which they would both simply disappear—at worst, you'd

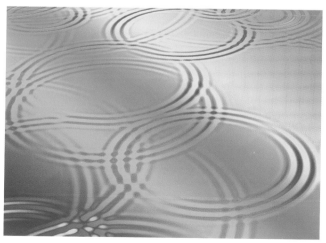

get a pile of dust. It was, in fact, the phenomenon of interference that caused scientists in the 19th century to conclude that light is a wave.

One of the simplest ways of demonstrating interference was devised by the British scientist Thomas Young. Light is shone on an opaque sheet in which two parallel slits have been cut and then allowed to fall on a screen beyond the sheet. If you directed a beam of particles at the sheet, you would expect to see those particles coming through only directly behind the slits, forming two concentrations on the screen. In particular, you would expect to see very few particles winding up between those two concentrations.

If, on the other hand, you directed a wave toward the sheet, you would see something completely different. According to HUYGENS' PRINCIPLE, each slit would act as the source of secondary waves. These waves would travel equal distances to the centerline on the screen, and therefore would arrive in phase—crest to crest, trough to trough. These are the conditions for constructive interference, and you would expect to see a maximum at that same centerline, where the light would be brightest. Thus, a wave will appear strongest precisely where the particle number is lowest. Away from the centerline, the two waves get out of phase with each other until they produce total destructive interference—a dark area. Moving still farther from the centerline, we find that the crest of the wavelet from one slit matches the crest immediately behind it from the other slit. Again, we have constructive interference. The double slit experiment, then, will produce a series of light and dark bands if a wave falls on it. We can think of this as the signature of a wave.

One of the great surprises of modern physics came when scientists directed a beam of electrons at a double slit. They saw the same alternating high- and low-intensity bands on the other side. As predicted by the DE BROGLIE RELATION, the electrons seemed to be behaving like waves, even though they had always been considered to be particles.

The principle of interference also lies behind the design of an instrument called the *interferometer*, which is widely used in physics. In this instrument, an incoming beam of light is split by a half-silvered mirror, with each split beam sent through a different arm of the apparatus (the exact configuration depending on what is being measured). Eventually, the beams are reflected back and brought together again. Differences —even very small differences—in the conditions in the two arms will show up as interference fringes when the waves recombine, because the crests of one of the waves will have moved slightly with respect to the other. Thus, this sort of apparatus can detect changes in the length of the arms of less than the wavelength of light. (In the MICHELSON–MORLEY EXPERIMENT this was especially important.)

THOMAS YOUNG (1773–1829) English polymath. He was born in Milverton, Somerset, the son of a carpenter. Young was a child prodigy who had read the Bible by age six and spoke several languages by age thirteen. Educated as a physician, he was a prominent member of London's scientific societies, and was elected to the Royal College of Physicians in 1809. He is best known for his defense of the wave nature of light, and was the first to prove that the lens in the human eye changes shape. He contributed to the science of elasticity (*see* HOOKE'S LAW). In his later years, he returned to the study of ancient languages and helped to decipher the Rosetta Stone.

Josephson Effect

An electrical current can flow through a thin layer of insulator separating two superconductors. A pair of junctions of this type can be used to make extremely accurate measurements of magnetic fields

In this machine, detectors connected to SQUIDs measure magnetic signals generated by the human brain.

The BCS THEORY OF SUPERCONDUCTIVITY explains how, at very low temperatures, the electrical resistance of certain materials falls essentially to zero, and an electrical current can pass them indefinitely. The basic mechanism for this is the formation of what are called *Cooper pairs* of electrons, which move through a conductor much more easily than the single free electrons that give rise to ordinary electrical conductivity.

In 1962, Brian Josephson, then a graduate student, realized that two pieces of superconducting material separated by a very thin layer of insulator, less than a dozen atoms thick, would behave as a single system. He applied the principles of QUANTUM MECHANICS to show that Cooper pairs would move across the layer (now called a *Josephson junction*), even when there is no voltage driving them. The existence of this current (soon verified experimentally) is called the *DC Josephson effect*.

Now, if a constant voltage is applied across the junction, the quantum mechanical prediction is that Cooper pairs will move through the junction first in one direction, then in the other. The frequency of the reversals is related to the size of the voltage—the larger the voltage, the higher the frequency. This is the *AC Josephson effect*. Because frequencies can be measured with high precision, the AC effect is used to calibrate voltages when extreme accuracy is required.

Perhaps the most widely used application of the Josephson effect comes out of yet another prediction from quantum mechanics. If we build a small loop of superconducting wire, put two Josephson junctions into it (one on each side of the loop), and then run a current through it, we have a device called a *SQUID* (in full, a superconducting quantum interference device). Depending on the size of the field, the current in the loop can go from zero (when the currents from the two junctions cancel each other out) to a maximum value (when the currents reinforce each other).

The SQUID is the most accurate device we have for measuring magnetic fields—not least because SQUIDs can be made to be very small. They are now routinely used in fields as diverse as earthquake prediction and medical diagnostics (*see* picture). One lesson to draw from the story of the Josephson effect is that even the most esoteric discoveries can have enormous practical benefits.

BRIAN DAVID JOSEPHSON (1940–) Welsh physicist. He was born in Cardiff, and attended Cambridge University, where obtained his Ph.D. in 1964, remaining there and becoming professor of physics in 1974. His theoretical derivation of the Josephson effect, made in 1962 while he was still a research student, won him a share of the 1973 Nobel Prize for Physics. The importance of the effect in computing, and its possible application in the development of artificial intelligence, led Josephson to studies of the human mind.

Kepler's Laws

The planets move in elliptical orbits around the Sun, with the Sun at one focus of the ellipse

A straight line between a planet and the Sun sweeps out equal areas in equal times

The square of a planet's period is inversely proportional to the cube of the radius of its orbit

Johannes Kepler was a man with a vision. Throughout his intellectual life, he tried to prove that the solar system was some kind of mystical work of art. First, he linked it with the five *regular solids* of classical Greek geometry. (A regular solid is a figure, each face of which is a geometrical shape whose sides are all equal. A cube, for example, whose sides are all squares, is a regular solid.) There were six known planets, each believed to be carried upon its own "crystalline sphere." Kepler argued that these spheres had to be spaced so that the regular solids each fitted snugly between adjacent pairs of spheres. The outermost pair of spheres belonged to Saturn and Jupiter; a cube just big enough to contain Jupiter's sphere would in turn just fit inside Saturn's sphere. Similarly, a tetrahedron (four sides) just fitted between the spheres that carried Jupiter and Mars; and so on. Six planets, five regular solids nested between them—what could be neater?

Unfortunately, when Kepler compared this model with the real orbits of the planets, he found that the solar system didn't behave as he thought it should. As the British biologist J.B.S. Haldane would say four centuries later, the idea of the universe (in Kepler's time, the solar system pretty much *was* the universe) as a geometrical work of art was another beautiful theory destroyed by ugly facts. The only lasting result of Kepler's youthful fling at finding a grand theory of the solar system was a model he had made for his patron Frederick, Duke of Württemburg. In this beautifully worked metal artifact, the orbits of the planets and the inscribed regular solids were all hollow, and on festive occasions each could be filled with a different liquor for the enjoyment of the duke's guests.

It was only after he went to work for the Danish nobleman Tycho Brahe (1546–1601) in Prague that Kepler came upon the ideas that have immortalized his name in the annals of science. Tycho had spent a lifetime amassing an unprecedentedly large and accurate body of observations of the motions of the planets. When he died, Kepler acquired these records. They had commercial value because they could be used to cast improved horoscopes—an aspect of early astronomy that its present-day practitioners prefer to ignore.

Working through Tycho's observations, Kepler was confronted with a problem that would be daunting even for someone equipped with a modern computer, much less for someone with no alternative but to calculate by hand. He was, of course, aware of the work of Nicholas Copernicus (*see* COPERNICAN PRINCIPLE) and, like most astronomers of his time, believed that the Earth moved around the Sun. This *heliocentric* (Sun-centered) view was the basis for the geometrical model described above. But how exactly did the Earth and the other planets move? Think about the problem this way: We are sitting on a planet that is spinning on its axis as it travels around the Sun in an unknown orbit. Streaking across our sky, we see other planets moving, also in undetermined orbits. Our task is to deduce, from what we see on our rotating, revolving globe, the shape of those orbits and the speed of the planets moving in them. And this is exactly what Kepler did, summarizing his results in the three laws stated above.

The *first law* tells us about the paths in which planets move. You may remember from your high-school geometry that an ellipse is the locus of points the sum of whose distance from two fixed points (called foci) is constant. If that sounds daunting, imagine slicing straight through a solid cone at an oblique angle. Provided your slice doesn't go through the cone's base, the cut surface is in the shape of an ellipse. Kepler's first law says that this is the shape of the orbits of the planets, provided that the Sun is at one focus of the ellipse. Each planet travels on a different elliptical path, but all the ellipses share one focus—the position of the Sun. Tycho's observations had told Kepler that the orbits of the planets are a

First law

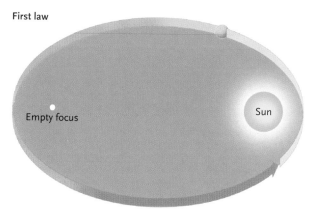

Second law

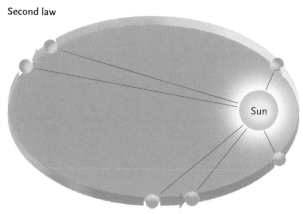

Kepler's first and second laws of planetary motion: The orbit of a planet is an ellipse, with the Sun at one focus of the ellipse; and the planet sweeps out equal areas in the same interval of time, wherever it is in its orbit. The third law specifies the mathematical relation between the planet's radius and the time it takes to complete its orbit.

KEPLER'S LAWS

nested set of ellipses. This had never even been considered by previous astronomers.

In a historical sense, the importance of the first law cannot be exaggerated. Every astronomer before Kepler had assumed that the orbits of the planets were either circular or, when that proved unable to explain observations, consisted of small circles rolling around inside of bigger ones.

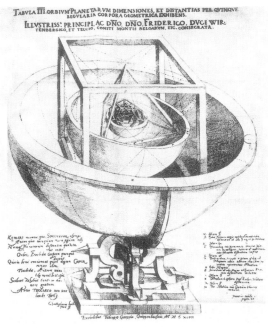

This was primarily a philosophical position, what I like to call an unexamined assumption. Philosophers argued that, unlike the Earth, the Heavens had to be perfect. Since they felt that the most perfect geometrical shape was the circle, it followed naturally that the orbits of the planets had to be circular (a presupposition I have found to be almost universal among my students). Once Kepler had access to Tycho's vast body of observations, he was able to see that this philosophical argument, like the argument that the Earth was stationary at the center of the universe, led to an incorrect description of planetary orbits.

In Kepler's geometrical model of the universe, the five regular solids, together with spheres representing the planets' orbits, were all nested inside one another.

The *second law* deals with the way planets move in their elliptical paths. I've stated the law above in its technical form, but you can visualize what it means by thinking back to your childhood. Remember when you ran toward a playground post, grabbed it, and swung yourself around? In a sense, the planets are playing this game in a playground where the Sun is the post. When its elliptical orbit carries a planet away from the Sun, it slows down, but when it comes in close it swings around the Sun at higher speed. Imagine drawing a pair of lines to the planet, one at the beginning of the day and one at the end. The area between the two lines is then the "area swept out" referred to in the law. The lines to the planet are shorter when the planet is closer to the Sun. But then the planet covers more distance in moving along its orbit (i.e., by moving faster), and the area swept out is the same as before.

The first two laws deal with the specifics of the orbit of a single planet. The *third law* compares the orbit of one planet to another. It tells us that the farther out a planet is, the longer it takes to complete its orbit—the longer is its "year." Today, we realize that there are two factors operating to produce this law. For one thing, the farther out a planet is, the longer the distance it has to travel to complete one loop. In addition, it turns out that the farther out a planet is the slower it moves.

When Kepler enunciated these laws, they were simply summaries of observations. Had you asked him why the orbits are elliptical, and why planets sweep out equal areas in equal times, he couldn't have told you. This is just what came out of his analysis. Had you asked Kepler to discuss the orbits of another set of planets around another star, he would have been at a loss to know how to proceed. He would have had to go back to the starting line and accumulate enough observations of the new system to analyze it in the same way as he had in analyzing our own. He would not have been able, in other words, to assume that the new system would operate by the same laws as he had discovered here.

One of the great triumphs of Newtonian physics was to show that Kepler's laws have a deeper foundation than this. It turns out that if you have NEWTON'S LAWS OF MOTION, NEWTON'S LAW OF GRAVITATION, and the conservation of ANGULAR MOMENTUM, then the logic of the mathematical equations leads directly to Kepler's laws. Once you know this, of course, then you know that the laws will apply to any solar system anywhere in the universe. Astronomers who search for new planetary systems (and we now know of many) routinely apply Kepler's laws to those systems even though they have never even seen the planets directly.

The third law continues to play an important role in modern cosmology. When astronomers look at distant galaxies, they detect the faint signals emitted by hydrogen atoms orbiting far from the galactic center —well outside the place where we might expect to see stars. Using the DOPPLER EFFECT, they can work out how fast the hydrogen is moving (*see also* DARK MATTER). I find it fitting that the work of the man who set us firmly on the road to understanding our own solar system should have played such a crucial, if posthumous role in establishing the structure of the wider universe.

JOHANNES KEPLER (1572–1630)
German astronomer. Kepler was born in Würtemburg, and was studying theology at Tübingen University when he received an offer to teach mathematics at a school in Graz, Austria. He made his early reputation with a series of astrological predictions about the weather in 1595. In 1598,

Kepler and other Protestants were put under severe pressure in Graz, and in 1600 he joined the Danish astronomer Tycho Brahe in Prague. Tycho's data formed the basis of Kepler's work. His later life was clouded by financial difficulties, and he died of a fever while on a journey to seek funds in Austria.

Kinetic Theory

The properties of a gas depend on the motion of the atoms or molecules of which it is composed

In a gas, the atoms or molecules are widely separated from one another and interact with one another only when they collide (to save repetition, I'll just refer to molecules from now on, which is to be understood to mean "molecules or atoms"). A single molecule, then, will travel in a straight line until it suffers a collision which changes its direction, after which it travels in a straight line until the next collision, and so on. The distance traveled between collisions is called a *mean free path*. The higher the density of a gas (i.e., the more molecules packed into a given volume), the shorter the mean free path between collisions.

In the last part of the 19th century this simple picture of the atomic nature of gases was turned into a powerful theory by a succession of great theoretical physicists. The basic idea of the new theory was to relate the kind of things we can measure in a gas—things like temperature and pressure—to the motion of the molecules. Because the molecules are always in motion, this work was given the name kinetic theory.

Take pressure as an example. At any moment, molecules in the gas are colliding with the walls of the container. Each collision produces a tiny, molecular-sized force on the wall, and the sum of the forces of all the collisions is what we perceive as pressure. When you inflate a tire, for example, you are putting more molecules from the atmosphere into the tire, so there will be more collisions against the inside of the tire than there are against the outside (the outside collisions coming from the collisions of molecules in the surrounding atmosphere). As a result, the force pushing out on the walls is greater than the force pushing in, and the tire stays inflated.

The point of the theory is that we can calculate, from information about the mean free path, how many molecules will collide with a wall. Thus, from information about the motion of the molecules, we can calculate a quantity that we can actually measure directly. Kinetic theory, in other words, provides a direct link between the world of the atom and the macroscopic world of our senses.

The same statement applies to the understanding of temperature in this theory. The higher the temperature, the faster the average speed of the molecules in the gas. The equation that relates the two is

$$\tfrac{1}{2}mv^2 = kT$$

where m is the mass of the molecules in the gas, v is their velocity, T is the temperature of the gas (measured in degrees kelvin), and k is the BOLTZMANN CONSTANT. This equation shows explicitly the connection between molecular quantities (on the left) and measurable macroscopic quantities (on the right). The average speed of molecules is directly related to the gas's temperature.

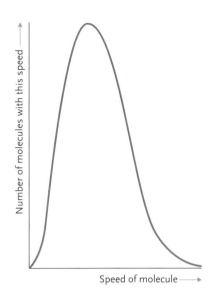

Number of molecules with this speed ⟶

Speed of molecule ⟶

Maxwell showed that the molecules in a gas have different velocities: Some move faster than average, and some slower.

The theory also says something about how the speeds of individual molecules differ from the average. Each collision between molecules has the effect of sharing out energy among the molecules in the gas: fast-moving molecules are slowed down by collisions, while slow-moving ones are speeded up (on average). At any moment, uncounted millions of molecules in the gas are changing their energy through collisions. Nevertheless, it turns out that for a gas at a given temperature, the number of molecules with a given energy (call it E) stays the same. This happens because for each of the molecules that starts with energy E and then either gains or loses energy through a collision, another molecule starting with another energy will suffer a collision and find itself with energy E. Thus, even though a given molecule will sometimes have energy E, sometimes not, the total number of molecules with energy E stays fixed. (This is similar to human populations. No one stays seventeen forever—thank God!—but the total number of seventeen-year-olds in a stable population will stay constant.)

The man who took this idea of shifted individual molecules and stable populations and turned it into a rigorous theory was James Clerk Maxwell, the same man who codified the properties of electricity and magnetism as MAXWELL'S EQUATIONS. He found that for a given temperature, the velocities of the molecules are distributed as shown in the graph below. The commonest velocity for the molecules is the average shown, but some atoms will be moving faster and some slower. A distribution of velocities like this is called *Maxwellian*, after its discoverer.

Kinship Selection

Organisms may act to benefit related individuals because those actions pass on shared genes to the next generation

The problem of altruism was traditionally a difficult one for the theory of EVOLUTION to deal with. For example, a monkey who spots an approaching leopard may cry out to warn other members of the troop, even though that action increases the risk to itself. The most naive view of natural selection is that whatever genes made that monkey give the warning cry would, over time, be eliminated from the population, since it would seem to decrease the individual's fitness. Yet altruistic behavior is observed in all sorts of animals (including human beings). How can this be?

The theory of kinship selection is one attempt to explain this and other puzzling aspects of animal behavior. The basic idea is this: Related individuals share a certain number of genes. You have half of your genes in common with each of your parents, for example, as you do with your brother or sister. What matters in evolutionary theory is not the survival of individuals, but the transmission of genes to the next generation. If the monkey who raised the warning cry had, say, three siblings in the group, then, statistically speaking, an individual sacrificing itself could well wind up with more genes in the next generation than one who does not. In the (facetious) words of evolutionary biologist J.B.S. Haldane, "I would lay down my life for two brothers or eight cousins."

Because of the way that reproduction takes place in honeybees, for example, each female bee shares all of the genes of its father and half of the genes of its mother. This means that worker bees share 75% of their genes (as opposed to the 50% shared by similarly related mammals). So working to sustain a sister as a queen bee will actually get more of the individual worker's genes into the next generation than will having daughters of her own.

Kinship selection also explains the persistence of homosexuality in many animals species, including humans. Since, by definition, homosexual behavior seems to preclude passing on genes, we might expect it to have disappeared long ago. One explanation that has been proposed is named, appropriately enough, the "helper at the nest" theory. The idea is that if an individual does not procreate, but instead works to increase the chances of relatives surviving, more of that individual's genes may make it into the next generation.

The important point about all this is that it seems to overturn our ideas about evolution. Instead of considering the fitness of individuals, as Darwin did, we are asked to consider the fitness of genes. One outcome of this way of thinking is the concept of the *selfish gene.* This is the idea that what matters is whether genes are transmitted from generation to generation, and that those behaviors that give genes advantage will survive even though they can be very harmful to the individuals involved. Or, as one wag put it, "A chicken is just an egg's way of making another egg."

Kirchhoff–Bunsen Discovery

If a particular substance emits light of a given frequency, it will also absorb light of that frequency

Robert Bunsen's enduring claim to fame was his development of the Bunsen burner, a laboratory device you may be familiar with from the school chemistry lab that burns with a very clear flame and which can therefore be used to see the colors emitted by materials heated to incandescence (*see* FLAME TEST). This was one of the first techniques used to establish the presence of various chemical elements in a material.

Throughout the middle years of the 19th century, Bunsen became the acknowledged world leader in preparing pure samples of various chemical elements. By 1859, he was taking the testing of elements by colors a step further by passing light from burning samples through a prism, which would spread the colors out. He had already discovered that for some elements—sodium, for example—the bright colors given off coincided exactly in frequency and wavelength with dark lines that appear in the spectrum of the Sun—the so-called *Fraunhofer lines*. Today, we know that these dark lines are caused by the absorption of light by sodium and other atoms as it moves up from the bright interior of the Sun through the cooler solar atmosphere. That the dark and bright lines were at the same frequency was one of many facts about the interaction of light and atoms that was known but not understood in Bunsen's time.

Also in 1859, physicist Gustav Kirchhoff, Bunsen's colleague, was using this coincidence of bright and dark lines for sodium to calibrate an instrument. He would first pass sunlight and then light from a sodium flame through a prism, then adjust the prism so that the corresponding bright and dark lines overlapped. He noticed that if he interposed the sodium flame between the incoming sunlight and his prism, the dark lines became much more pronounced than if it had just been the sunlight alone. The sodium flame, in other words, was capable of absorbing the incoming sunlight at the very frequencies at which it could emit light.

Kirchhoff, in a leap of intuition, realized that the light emitted by an atom had to be at the same frequency and wavelength as the light it absorbed. In other words, anything an atom emits it must also absorb. (In this scheme, the observed darkening of the lines corresponds to additional absorption of light by the sodium in the flame—sodium that is also emitting light of its own.)

One immediate consequence of Kirchhoff's discovery was that it showed conclusively that elements such as sodium, whose spectrum corresponded to dark lines in the solar spectrum, were actually present in the Sun. The light coming from the Sun's interior is white, which is to say that it contains all frequencies. The dark lines, in this interpretation, represent the absorption of that light by the sodium atoms in the Sun.

In terms of the BOHR ATOM it is easy to understand the basis for the Kirchhoff–Bunsen discovery. We know that light will be emitted from an atom when an electron makes a quantum jump from a higher to a lower orbit. The energy of the photon (and hence its frequency and wavelength) is determined by the energy difference between the two orbits. When light is absorbed by an atom, it pushes electrons from lower to higher orbits—and there is that same energy difference. The Kirchhoff–Bunsen discovery came about simply because the same energy is involved in moving the electron down from one orbit to another as in moving that electron up from the lower orbit back to the original one. It's the same principle that says that you gain as much energy back from the gravitational field when you go down a flight of stairs as you expend when you go up.

This discovery marked the beginning of the science of spectroscopy, an extremely important development in the history of science. For one thing, it enabled astrophysicists to determine the chemical composition not only of the Sun, but of any object in the universe that emits visible light —something that was once thought to be forever beyond our capabilities. Today, the high-tech computerized spectrometers found in laboratories all around the world cost hundreds of thousands, even millions of dollars. It's interesting, then, that Kirchhoff and Bunsen built the first spectrometers from simple glass prisms and a couple of old cigar boxes.

ROBERT WILHELM BUNSEN (1811–99) German chemist, born in Göttingen. He followed a standard career in the 19th-century German university system, gaining his doctorate from Göttingen in 1830, lecturing there, and obtaining a string of chemistry professorships, eventually settling at Heidelberg in 1852. He played a part in the development of the so-called Bunsen burner, though it was not his invention, as is often stated. Bunsen's early research was in inorganic chemistry. Not afraid to get his hands dirty, he became an expert glassblower. He came close to losing an eye in a laboratory explosion, and nearly poisoned himself with arsenic—twice! At the same time he was a great teacher, and influenced many of the next generation of German chemists.

Kirchhoff's Laws

The total electrical current flowing into any point in a circuit must be the same as the total current flowing out of that point

The total voltage throughout any loop in the circuit must be zero

The career path of Gustav Kirchhoff was in many ways typical of a physicist in 19th-century Germany. Coming late to the Industrial Revolution, Germany depended more on advanced technology for its economic well-being than did its neighbors to the west. As a result, German academics, particularly scientists, were highly valued. Shortly after his graduation, Kirchhoff married the daughter of one of his professors in the same year—"thus fulfilling," in the words of one biographer, "the two prerequisites for a successful academic career." Even before then he had, at the age of twenty one, formulated the basic laws that govern electrical circuits—laws that still bear his name.

The mid-19th century was a time when the properties of electrical circuits were being explored and exploited. The basic rules of simple circuits, like OHM'S LAW, were already pretty well worked out. The problem was that it is possible, by soldering wires and simple circuit elements together, to make very complex structures—structures that no one knew how to handle mathematically. Kirchhoff wrote down the rules that allow even the most complex of circuits to be analyzed simply, rules that are still an essential part of the toolkit of any physicist or electrical engineer even today.

There are two laws, each of which has a very simple physical interpretation. The first says simply that if you pick any point in a circuit (the usual choice being a point where wires come together), the amount of electrical charge that flows into that point has to be equal to the amount that flows out. For example, if you have a T-shaped intersection of wires, and two of the wires in the T are carrying current into the junction, then the other wire has to be carrying current out, and the amount it carries has to be equal to the amount that the other two are bringing in. The physical reason for this rule is quite simple: If it were not true, charge would accumulate at points in a circuit, and that never happens.

The second law is equally simple. It says that if you take a complex circuit you can think about the circuit as being composed of a series of loops. The current in the circuit is free to flow in any of the loops—indeed, the usual problem one faces is to figure out how much current will flow in each of them. An electron going around any loop can gain energy (e.g., from a battery) or will lose energy (e.g., in a resistor or other circuit element) anywhere along the way. The second law says that the net amount of energy that an electron gains when it goes around any loop—what you get when you add up the pluses and minuses—must be zero. This law also has a simple physical interpretation. If it were not true, then each time an electron went around a loop it would have more energy than the time before. Not only would this form the basis for a PERPETUAL-MOTION machine, it would violate the first law of THERMODYNAMICS.

The most frequent application of Kirchhoff's laws is to what are known as series and parallel circuits. In a *series circuit*, connecting a series of light bulbs, for example, electrons would leave the source, flow through the first bulb, then through the second bulb, and so on. At the end, the electrons would flow back to the source to start the circuit again. In this situation the same current flows through each bulb, and the voltage drop across each bulb is determined by its resistance according to Ohm's law.

In a *parallel circuit*, on the other hand, the wires are arranged so that the same voltage is applied across each circuit element, which means that each element can have a different current flowing through it. If you imagine the light bulbs as being on the rungs of a ladder whose uprights are the outgoing and incoming wires of the circuit, you have a good picture of a parallel circuit. The currents flowing into and away from the intersections in a parallel circuit are found by using Kirchhoff's second law.

GUSTAV ROBERT KIRCHHOFF (1824–77) German physicist. He was born in Königsberg (now Kaliningrad, Russia). By the time he graduated from the university there, in 1847, he had already formulated the basic laws that govern electrical circuits. He went on to a distinguished career at several German universities, ending up at the University of Berlin, where he was appointed professor of theoretical physics in 1875. Working with Robert Bunsen at the University of Breslau (now in Poland) from 1854, he laid the foundations for the modern science of spectroscopy (*see* KIRCHHOFF–BUNSEN DISCOVERY). He was also responsible for another set of Kirchhoff's laws, this time governing the absorption and emission of radiation by hot objects. Although he spent the second half of his life in a wheelchair because of an accident, Kirchhoff was universally described as a pleasant and optimistic man.

Lamarckianism

Evolution proceeds by the inheritance of acquired characteristics

Before Charles Darwin developed the theory of EVOLUTION by natural selection, other people had tried to work out explanations for the diversity of life forms on our planet. One of the most prominent of them was Jean-Baptiste Lamarck. Like other scientists of the 18th century, he noted that older strata of rocks generally contained simpler life forms, so that the history of living things seemed to reflect a development from simpler to more complex organisms (*see* law of SUPERPOSITION).

Lamarck took this to mean that there was a kind of evolution (he would have used the term "progression") operating in nature. He proposed that this evolution was guided by two principles. The first of these, which is the one people associate with Lamarck, is the inheritance of what are called *acquired characteristics*. In Lamarck's view, if an organism developed a particular capability during its lifetime, its offspring could inherit that capability. For example, the children of weightlifters should be more muscular, the children of intellectuals more intelligent, and so on. His most famous example was the giraffe, whose long neck he attributed to previous generations of animals stretching to reach tasty young shoots growing on the upper branches of trees. The second, lesser known

According to Lamarck, the giraffe's neck is the result of generations of them stretching to reach leaves at the top of trees. The accepted view now is that evolution favors giraffes that happen to have longer necks.

aspect of Lamarckianism was that there was supposed to exist a universal creative imperative, a kind of upward striving in which all life forms are driven to higher levels of complexity.

Today, of course, we understand that inheritance depends on genes coded in the DNA molecule, and that evolution happens because natural selection is acting on those genes. If life has become more complex, it is only because complex organisms are better at exploiting their environments and reproducing. There is no mystical "striving."

JEAN-BAPTISTE PIERRE ANTOINE DE MONET, CHEVALIER DE LAMARCK (1744–1829) French naturalist. Born in Bazantin, Picardy, the 11th child in a family of semi-impoverished French nobility. He joined the army and served in the Seven Years War, being posted to various French military establishments around the Mediterranean. His study of the plants he found on his travels brought him his first recognition from the French scientific community. After he retired from the military he worked in a bank, but remained impoverished throughout three marriages. Lamarck's interests ranged from botany (until the French Revolution, he had been keeper of the royal gardens), invertebrates—a term he coined, chemistry, meteorology, geology, and paleontology, which finally led him to formulate his theory of evolution, set out in his *Philosophie zoologique* (1809).

Lawson Criterion

If nuclear fusion is to be a source of energy, the product of particle density and the time for which the particles are held together must exceed a certain value

It is possible to get energy from fusion reactions—the bringing together of light atomic nuclei to form heavier ones (*see* NUCLEAR FUSION AND FISSION). In some reactions, the mass of the final nucleus (plus the masses of whatever other subatomic particles are created in the process) is less than those of the original nuclei, and the difference in mass is converted into energy via Einstein's familiar formula $E = mc^2$.

In stars, the primary energy source is the fusion of hydrogen (protons) into helium (*see* STELLAR EVOLUTION). The reaction takes place in three steps, the first of which is the production of deuterium (an isotope of hydrogen, a nucleus with one proton and one neutron) from the collision of two protons. Trying to control the fusion of hydrogen, the simplest fusion reaction, has also been an object of intense study by nuclear physicists since the mid-20th century. The motivation is simple: There is enough deuterium in the world's oceans to meet the energy needs of the human race for centuries to come, if only deuterium nuclei could be made to react with one another.

Attempts to harness fusion for controlled power generation bypass the steps in the stellar fusion process by reacting a mixture of deuterium, symbol ^{2}H, and tritium (an isotope of hydrogen with one proton and two neutrons in its nucleus), symbol ^{3}H. The products would be helium, a neutron, and the desired energy. The reaction can be summarized as

$$^3\text{H} + ^2\text{H} \rightarrow ^4\text{He} + \text{n}$$

To keep up the supply of tritium, it would need to be "bred" by using an isotope of lithium to capture the neutrons (the "n") emitted in the above reaction

$$^6\text{Li} + \text{n} \rightarrow ^3\text{H} + ^4\text{He}$$

The problem with getting the tritium–deuterium reaction to proceed under controlled conditions is that in order for two positively charged nuclei to come together, they have to be approaching each other at high speed to overcome their mutual electrical repulsion. In practice, this means that the tritium–deuterium mixture has to be at a temperature of millions of degrees, far too high to be held inside any material container (the mixture is a plasma—*see* STATES OF MATTER). But even reaching this temperature, which in fact is relatively easy with modern techniques, doesn't guarantee the triggering of a fusion reaction that will produce more energy than was used to get it going.

What the Lawson criterion deals with is that there has to be a certain minimum number of fusion reactions per second to maintain a reaction in any material. We can achieve fusion either by creating a very high density of interacting particles (and thereby increasing the chances of a collision) or by keeping the material together for a longer time (thereby allowing more time for the reactions to take place). It turns out that fusion can produce energy if

$$Nt > \text{about } 10^{20}$$

where N is the particle density and t is the time (the units are particles per cubic meter and seconds). This is the Lawson criterion for controlled fusion. It tells us that once the ignition temperature has been reached, there is a tradeoff between density and confinement time. We can get fusion with a lower particle density if we keep the plasma together for long enough, or if we maintain confinement for a short time but with a higher particle density.

There are two ways that engineers have tried to achieve controlled fusion, each representing a different approach to the problem of compressing a hydrogen gas, raising its temperature, and containing it while fusion occurs. The approaches are called magnetic confinement and inertial confinement, respectively.

In *magnetic confinement,* the plasma is infused with a magnetic field. As the temperature is increased, the magnetic field is squeezed together, pulling the hot plasma away from the walls of the container and supporting it in space. If the density and confinement time together satisfy the Lawson criterion, fusion will proceed. In fact, magnetic confinement machines such as the Joint European Torus (JET) at Culham in the UK have achieved this condition, but because the machines are quite inefficient none has yet extracted more energy out of the fusion reaction than was put in.

In *inertial confinement,* a tiny sphere of frozen tritium and deuterium is held in a glass shell. This droplet is then blasted on all sides with high-intensity laser beams. The outer layer of the drop boils off, setting up converging SHOCK WAVES inside the remainder. These shock waves compress and heat the hydrogen until the ignition temperature is reached, and fusion begins. A large laser installation to produce inertial fusion reactions is currently being built at the Livermore Laboratory in California. Due for completion in 2006, it will use 192 lasers to illuminate the droplet with 1.8 megajoules of energy.

JOHN DAVID LAWSON (1923–) English physicist. He was born in Coventry. He gained his BA from Cambridge University in 1943, and carried out wartime research on microwave antennas. In 1947 he joined the UK's Atomic Energy Research Establishment, experimenting with an early particle accelerator, and in 1957 he published the criterion that now bears his name. Lawson's subsequent research focused on electromagnetic problems arising in microwave tubes, particle accelerators, and radiation phenomena.

Le Chatelier's Principle

If you try to change a system that is in chemical equilibrium, that system will tend to respond so as to oppose the change

The best way to understand Le Chatelier's principle is to consider a simple chemical reaction in which two substances combine to form a third, while at the same time that product is splitting into the original reactants. This is represented schematically by the equation

$$A + B \rightleftharpoons C$$

where the double arrow sign indicates a two-way reaction. When the reaction goes from left to right, it represents the formation of substance C from A and B; when it goes the other way, it represents the splitting of C into A and B. When the system is in *chemical equilibrium,* the rates in the two directions are equal—for every molecule of C that is created at one point in the system, another molecule of C comes apart somewhere else.

Now, suppose that someone dumps some extra A into the system. For a moment, the system temporarily goes out of equilibrium because the production rate of C will go up. As the amount of C grows, however, the rate of splitting will increase until once again the rightward and leftward reactions are equal, and the rate of production of C from A and B will be equal to the rate at which C splits into A and B.

The chemical composition of rain and the fizz of an antacid tablet in water are both homely examples of Le Chatelier's principle in action. In both cases, the chemical reaction involves carbon dioxide, water, and carbonic acid:

$$CO_2 + H_2O \rightleftharpoons H_2CO_3$$

As a raindrop falls through the air, carbon dioxide is absorbed, increasing the concentration on the left side of the reaction. To maintain equilibrium, more carbonic acid is produced, eventually making the rain acidic (*see* ACID RAIN). Adding carbon dioxide, then, drives the reaction to the right. Just the opposite happens when an antacid is put into water: The sodium bicarbonate reacts with water to create carbonic acid, which has the effect of increasing the concentrations on the right of the equation. To maintain equilibrium, the carbonic acid has to dissociate to make carbon dioxide, which we see as bubbles.

HENRI LOUIS LE CHATELIER (1850–1936) French chemist. He was born at Miribel-les-Echelles into a family of scientists and educated at the prestigious École Polytechnique in Paris. He held chairs at the Collège des Mines and at the Sorbonne, and was eventually appointed Inspector-General of Mines for France, a post previously held by his father. Le Chatelier studied chemical reactions associated with mining accidents and the metallurgical industry. He designed a successful thermocouple and a water-brake for use on railroads, and invented oxyacetylene welding.

Lenz's Law

*When an electrical current
is induced in a wire by
a change in magnetic flux,
the current will flow in the
direction which makes the
magnetic field associated
with it oppose the change
in flux*

In 1831, the English scientist Michael Faraday discovered what are now
known as FARADAY'S LAWS OF INDUCTION, the first of which states that
changes in magnetic flux through an area enclosed by a conductor will
cause an electrical current to flow in that conductor, even if there is no
battery or other power source attached to it. Heinrich Lenz followed up
Faraday's discovery by determining the direction in which the induced
current flows.

Imagine a circular loop of wire with no battery or other power source
connected to it, and that the north pole of a magnet is moved toward it.
This will increase the magnetic flux through the loop, so, by Faraday's
law, a current will flow in it. Now, by the BIOT–SAVART LAW, a current
flowing in a loop will generate its own magnetic field, one which is equivalent to the field of an ordinary magnet with a north and a south pole.
What Lenz found was that the current will flow in such a direction that
the north pole of the magnetic field it produces will point toward the
magnet which is being moved toward the loop. Since north poles of magnets repel each other, the induced current in the loop will flow in such
a way as to oppose the motion of the magnet that is inducing it. This
is a general rule—all induced currents flow in such a way as to oppose
whatever is inducing the current in the first place.

Lenz's law may come to play a role in high-speed inter-city transporta-
tion. So-called maglev trains (short for magnetic levitation) already exist
in prototype form. Strong magnets are located under the train, so that
they are just above the metal track. The magnetic flux changes as the
train moves, inducing currents in the track and producing a repulsive

force between the train and the track
(just as the current in the loop discussed above creates a repulsive force
between the magnet and the loop).
This force is big enough to lift the
train a few inches off the ground, so
that the train actually flies through
the air. Levitating trains can reach
speeds well in excess of 300 miles
per hour (about 500 km/h), making
them ideal for travel between cities.

*A maglev train on a test track
in Emsland, Germany. This
prototype has reached speeds
of around 250 miles per hour
(400 km/h).*

HEINRICH FRIEDRICH EMIL LENZ (1804–
65) Russian physicist. He was born in
Dorpat (now Tartu, in Estonia), and
educated at the university there. While
still a student he took part in a round-the-
world geological expedition. He taught at
the University of St Petersburg, becoming
professor there in 1836. Lenz played a
prominent role in the Russian scientific
community. In his later career, he carried
out investigations of electrical conduction and electromagnetic phenomena.

"Like Dissolves Like"

A liquid is more likely to be able to dissolve a solid if the two are of the same polarity and have other characteristics in common

It is a well-known adage that oil and water don't mix—that is, if we put a drop of oil into a glass of water it will retain its identity and will not dissolve. On the other hand, if we put a drop of alcohol (or ethanol, as chemists know it) into water, it will quickly disappear as it dissolves, a fact that can easily be verified by adding a drop of red wine to a glass of water. Chemists explain this behavior by a rule of thumb known as "like dissolves like."

The basic point is that water is a polar molecule. The electrons in a water molecule spend most of their time near the oxygen, leaving the hydrogen side with a positive charge (even though the net electrical charge of the entire molecule is zero). Ethanol is also a polar molecule, so hydrogen bonds can form between water and ethanol molecules (*see* CHEMICAL BONDS). In a sense, water and ethanol can "get hold of" each other. Thus, when ethanol and water are mixed, the molecules interact with each other and the ethanol quickly dissolves.

Oil, on the other hand, is not made of polar molecules, so there is no bonding between it and the water molecules. The water can't "get hold of" the hydrocarbon molecules in the oil. Conversely, the oil can't bind strongly enough to any water molecule to push aside the other water molecules, as it would do if it was to go into solution. Oil is not "like" water, and doesn't dissolve.

You have used the "like dissolves like" rule many times in your life, though you probably weren't aware of it. When you use a detergent, you are using a molecule with a rather specialized structure. It has a long hydrocarbon tail (which, like oil, is not polar) coupled to a head that is strongly polar. When such a molecule encounters a layer of grease on a surface, the tail penetrates the layer, leaving the polar heads sticking out. These heads then bond with the polar water molecules, and the grease is lifted away from the surface. This is how detergents clean clothes and dishes.

Linear Momentum, conservation of

The linear momentum of an isolated system is conserved

Once an object starts to move, it has a tendency to keep on moving. This is what the first of NEWTON'S LAWS OF MOTION says: that a moving object will continue moving in a straight line unless an external force acts upon it. We call this tendency *linear momentum,* and it's something we often encounter in everyday life. The balls on a pool table continue on their way after being struck with a cue, and a javelin flies through the air for some distance at much the speed at which the thrower launched it.

Physicists define the linear momentum p of an object to be its mass m multiplied by its velocity v:

$$p = mv$$

where p and v are bold letters to indicate that these quantities have both size and direction. Thus, a car may have a speed of 20 mph, but its velocity would be stated as 20 mph due north (say). Quantities such as momentum and velocity which have direction as well as magnitude are called *vectors.*

The law deals with the total linear momentum of a system, which is the sum of the momenta of the system's individual parts. Thus, a system comprised of two pool balls of equal mass rolling directly toward each other at the same speed will have zero momentum, even though each ball has a momentum of mv. The individual momenta cancel each other—a result of the vector nature of momentum. Any quantity that doesn't change over the course of an interaction is said to be *conserved.* Several quantities in nature are conserved besides linear momentum. One other is ANGULAR MOMENTUM, the counterpart of linear momentum which applies to motion on curved paths.

When two pool balls heading toward each other on a straight line collide, several things happen. The balls rebound from each other, they flex slightly during the impact, and some of their kinetic energy is converted to heat. But whatever else may happens, we know that the total momentum of the system will be zero. If we see one ball moving off at a given speed in a given direction after the collision, we can say with certainty that the other is moving at the same speed in exactly the opposite direction.

One way of interpreting the second of NEWTON'S LAWS OF MOTION is that the rate of change of momentum is equal to the force applied to a system. Thus, in order to change the momentum of a system, a force has to act. We see this in the KINETIC THEORY of gases, in which pressure is explained as molecules hitting the wall of a container and bouncing back. Because the molecule changes direction after the collision with the wall, its momentum also changes, which means that the wall must exert a force upon it. And by Newton's third law, the molecule also exerts a force on the wall—a force we interpret as pressure.

Linnaean System of Classification

All living things can be classified in a hierarchical system whose most familiar ranks are the genus and the species

Carolus Linnaeus was a Swedish physician, a professor of medicine at the University of Uppsala. As such, he was responsible for maintaining a large botanical garden that was part of the university's research program. From all over the world, people would send him plants and seeds which would be grown in the garden. It was from his extensive studies of this large collection of plants that Linnaeus began to see a way to solve the problem of classifying all living things—what today we would call the problem of *taxonomy*. In fact, we can give him credit for inventing the categories for the quiz game of Twenty Questions, in which you are first told that the object you have to guess is animal, vegetable, or mineral. In the Linnaean scheme, everything was, indeed, either an animal, a plant (vegetable), or nonliving (mineral).

Here's an analogy that might help: Suppose you wanted to create a classification scheme for all the houses in the world. You might start by saying that houses in Europe, for example, are more similar to each other than they are to houses in North America, and then your first rough cut in classification would be to specify the continent on which a particular house is located. For each continent, you can go further by noting that houses in one country (e.g, France) are more like one another than they are like houses in another country (e.g., Norway). Thus, a second level of classification would be the country. You could carry on in this fashion down through county, city, and so forth until you came to a single street. The house number on that street would then be the final bin into which you would classify things. Every house would then be completely specified by giving continent, country, county, city, street and number.

Linnaeus saw that living things can be classified in just the same way, according to their characteristics. Human beings, for example, are more like squirrels than they are like rattlesnakes, and they are more like rattlesnakes than they are like pine trees. By going through the kind of reasoning we did for houses, it is possible to construct a classification scheme in which every living thing has its unique place.

The followers of Linnaeus have done exactly that. All living things are classified first into five *kingdoms*—plants and animals, of course, but also fungi, and there are two kingdoms of single-celled organisms (those with their DNA contained in nuclei, and those without nuclei). Next, each kingdom is broken down into *phyla*. For example, the nervous system of human beings contains a long cord down the back. This makes us members of the phylum of chordates. In most animals that possess a spinal cord, the cord is in a backbone. This large group of chordates is referred to as the *subphylum* of vertebrates. Human beings are members of this subphylum. This separates animals with backbones from those without (e.g., lobsters).

The next classification category is the *class*. Human beings are members of the class of mammals—warm-blooded animals with hair who nurture their young and give birth to live young. This division distinguishes humans from animals such as reptiles and birds. Next comes the category of *order*. We are a member of the primate order—animals with prehensile hands and feet, and binocular vision, to name a few characteristics. The classification of humans as primates differentiates us from other mammals such as dogs and giraffes.

The next two classification categories are the *family* and the *genus*. Although we are members of the family of hominids and the genus *Homo*, these distinctions mean little to us because there are no longer any other living members of either our family or genus (though others have existed in the past). For most organisms, there are many members of the same genus. The polar bear, for example, is *Ursus maritimis*, while the grizzly bear is *Ursus horibilis*. These two animals are members of the same genus but belong to different species—they don't interbreed.

The last category in the Linnaean scheme is the *species*, traditionally defined as a population of organisms that can interbreed. We are of the species *sapiens*.

It is customary to specify only the genus and species when we are talking about organisms. This is why human beings are referred to as *Homo sapiens* ("Man the wise"). This doesn't mean that the other parts of the classification aren't important—they are simply implied by giving genus and species. This so-called *binomial nomenclature* is the most lasting contribution of Linnaeus to science.

By categorizing life in this way, the Linnaean scheme gives each organism its own, unique place in the world of living things. But its success depends primarily on taxonomists correctly identifying the important physical characteristics, and people can of course make misjudgments or even mistakes—Linnaeus, for example, classified the hippopotamus as a rodent! Today, there is a move to classify according to the genetic code of individual organisms, or according to an organism's evolutionary history—its family tree—an approach known as *cladistics*.

CAROLUS LINNAEUS (1707–78) Swedish botanist and physician. He was born in Råshult and studied medicine at the University of Lund, moving to the University of Uppsala in 1728. He then began to develop a system for classifying plants and, later, animals and minerals. He recognized the affinities between diverse groups, classifying whales as mammals and putting humans and apes into the same class. Linnaeus returned to Sweden and became a professor at Uppsala in 1741. His studies of the plant collection in the university's botanical garden led him to his binomial classification of plant species. After his death, this collection and the accompanying library were bought by the English naturalist James Smith, and were later acquired by the London-based Linnaean Society.

MacArthur–Wilson Equilibrium Theory

The number of species in an isolated ecosystem will stabilize when the extinction rate is equal to the rate of immigration

There are many isolated ecosystems in nature, and they provide a particularly useful venue in which to study the comings and goings of species. Most of these ecosystems are islands surrounded by water, but there are other kinds of "islands" as well. For example, high mesas or plateaus surrounded by desert or rainforest are islands in the sky, every bit as isolated as a remote atoll in the Pacific.

The MacArthur–Wilson theory (it is sometimes called a "law") is named after ecologists Robert MacArthur and Edward O. Wilson, who formulated it in their book *Theory of Island Biogeography,* published in 1967. It deals with the number of species that will eventually inhabit such an ecosystem. Species might get blown to an island from a neighboring mainland during a storm, for example, or carried across the ocean on floating debris. Imagine that to start with the island is completely barren, with no life on it at all. At first, every new organism that arrives is quite likely to add to the number of species on the island. As time goes by, however, new arrivals are more likely to find others of their kind already there, and will not add to the diversity of the island's species. If we plotted a graph showing the immigration rate (i.e., the number of new species arriving over a given period of time) against the number of species already resident on the island, we would expect the immigration rate to be high when the resident number is low, and low when the resident number is high.

Once species arrive on the island, they will begin to go extinct. (Here, the term "extinction" simply means that they are no longer to be found on the island, not that they have disappeared from the Earth.) When the number of resident species is small, the number of disappearing species must be small as well. As the number of resident species increases, however, the number of extinctions will increase as well, both because of increased competition and because with more species there are simply more chances for things to go wrong. If we plot the number of extinctions against the number of resident species, then, we should get a curve that goes up as the number of resident species increases.

Now think about these two curves—one starting high and dropping, the other starting low and climbing. At some point, the two curves will intersect. This is the *MacArthur–Wilson equilibrium point*. If a population is at this point and a species goes extinct for whatever reason, there will always be a new immigrant species to take its place—niches do not remain unfilled for long. But if a new species arrives after equilibrium is established, then some species (perhaps the new arrival, perhaps another) will go extinct because of the added competition. The equilibrium point is thus the biological diversity that is "natural" for this particular ecosystem. The theory predicts that, over time, the number of species in an isolated system will remain roughly at this value. Observations of island ecosystems (MacArthur and Wilson did their first observations in the Bay of Florida) seem to bear out this prediction.

It is important to appreciate that the MacArthur–Wilson equilibrium is a dynamic, changing situation, not at all like a static BALANCE OF NATURE. Although the number of species may be constant over time, the particular species represented in the population are always varying as extinction and immigration constantly change the cast of characters.

The theory makes some other predictions as well. For example, if the immigration rate goes down, the number of resident species at equilibrium should decrease as well. Thus, if we look at a series of islands, those farthest from the mainland (which presumably present more of a barrier to immigration) should have lower biodiversity than those closer in. This prediction, too, seems to be borne out by observations.

ROBERT HELMER MACARTHUR (1930–72) American ecologist. He was born in Princeton, New Jersey, the son of a professor of genetics. He got his Ph.D. from Yale in 1958, and taught at the University of Pennsylvania before becoming professor of biology at Princeton in 1968. MacArthur drew together threads from ecology, genetics, and biogeography and, with Edward O. Wilson, laid the foundations for the mathematical study of populations, introducing predictive models for ecosystems (*see also* OPTIMAL FORAGING STRATEGY).

EDWARD OSBORNE WILSON (1929–) American entomologist and ecologist. He was born in Birmingham, Alabama. He graduated from the University of Alabama in 1949 and in 1955 obtained his Ph.D. from Harvard, where he became a professor nine years later. His early work was on insect societies and island populations. The publication of his book *Sociobiology* in 1972 catapulted him to international fame—and controversy, for he claimed that the same genetic imperatives apply to animal and to human societies.

Magnetic Monopoles

There are no isolated magnetic poles in nature. The magnetic flux through any closed surface is zero

A simple magnet has a north and a south pole. If you cut the magnet in half, however, you don't get separate north and south poles—you get two smaller magnets, each of which has a north and a south pole. Repeat the process and the same thing happens. However long you continued to do this (or whatever other method you chose to try) you would never get down to a single isolated magnetic pole—a magnetic monopole. In other words, there are no magnetic monopoles in nature.

This highlights a curious asymmetry between magnetic and electrical phenomena. The BIOT–SAVART LAW established that magnetic effects can be created by electrical charges, and the first of FARADAY'S LAWS OF INDUCTION showed that magnetic effects can induce electrical currents. It is possible to isolate electrical charges—a proton, for example, is an isolated positive charge. But for magnets this doesn't seem to happen.

Scientists have speculated about magnetic monopoles and hunted for them, but so far to no avail. Their search is driven largely by the BEAUTY CRITERION. To a theoretical physicist, a universe with no monopoles is like a magnificent painting with one vital piece missing. Monopoles should have been produced in abundance in the EARLY UNIVERSE, but the subsequent rapid expansion spread them out very thinly indeed. Perhaps there are only a few in the entire visible universe. My own guess is that they are more abundant than that, and someday they will show up.

If monopoles do turn out to exist, then some of our fundamental laws will need to be reformulated—for example, GAUSS'S LAW for magnetism. Imagine a single dipole magnet sitting isolated in space, enclosed by an imaginary surface (it can be any shape—something like a crumpled balloon will do). At each point on that surface there will be a magnetic field associated with the magnet, and the product of the strength of this magnetic field times the area of the little bit of the surface around the point (called a *surface element*) is called the magnetic flux. You can do this operation for each surface element, and Gauss's law for magnetism simply says that when you add up all the fluxes, the result will be zero—no magnetic monopoles.

Put another way, the lines of the magnetic field are continuous loops, so that a line leaving the surface at one point must return at another. Add up all these ins and outs, the law says, and the result is total cancellation. This result differs from Gauss's law for electrical charges because lines of electric field begin and end on charges themselves.

But if we replaced our dipole magnet with a magnetic monopole, there *would* be a net magnetic flux through the surface, equal to the strength of the magnet. Gauss's two laws would then have exactly the same form.

Magnetism

Because of the properties of atoms, materials can be ferromagnetic, paramagnetic, or diamagnetic

According to AMPÈRE'S LAW, an electrical current will give rise to a magnetic field. An electron going around an atom constitutes an electrical current, albeit a minute one. It should come as no surprise that an electron creates a magnetic field because of its motion. In fact, each electron around the nucleus of an atom can give rise to its own magnetic field, and the sum of all these magnetic fields will be the magnetic field of the atom itself. Now, in some atoms the electrons rotate or orbit in equal numbers in opposite directions so that their magnetic fields cancel out. But in other atoms, the arrangement of the electron orbitals leaves some electrons unbalanced by ones going in the other direction. Then the magnetic fields associated with the individual electrons will not cancel one another out, and the atom will possess a permanent magnetic field. You can picture such an atom as a tiny bar magnet with a north and a south pole.

It is the aggregate behavior of these tiny atomic magnets that gives materials their magnetic properties. Most magnetic materials fall into one of three broad classes: *ferromagnets*, *paramagnets*, and *diamagnets*. The other classes are *antiferromagnets* and *ferrimagnets*; in both of these the atoms are magnetic, but below a certain temperature the magnetic fields of adjacent atoms are antiparallel. For antiferromagnets the result is complete cancellation of the magnetic field. Ferrimagnets contain two or more chemical elements whose atoms have magnetic fields that differ, so that there is a net magnetic field for the substance as a whole.

Ferromagnetism

If certain materials (most notably iron, cobalt, and nickel) are at a temperature below their CURIE POINT, their atomic magnets will tend to line up and reinforce one another, and there will be a magnetic field outside of the material. We call such materials permanent magnets. Actually, the alignment of atoms in a ferromagnet does not normally extend throughout the entire material, but only over a volume containing several thousand atoms. Such a volume is called a *domain*. When iron cools, many

Iron filings are attracted to this horseshoe magnet. Iron is ferromagnetic, so the filings themselves become tiny magnets.

domains are formed, and the magnetic fields associated with different domains point in different directions. This is why iron is not normally magnetic, even though atoms are lining up inside it. If, however, the magnetic fields of the individual domains do line up—as will happen, for example, if iron is cooled in the presence of a strong external magnetic field—then the fields of the domains will reinforce one another, and there will be a large magnetic field outside of the material. This is the basic structure of *permanent magnets*.

Paramagnetism

In most materials, there is no force tending to line up the magnetic fields of neighboring atoms, so they point in random directions. The atomic magnets cancel one another out, and the material has no magnetic field. But if such a material is placed in a strong magnetic field (between the poles of a large magnet, for example), then the atomic magnets will begin to line up. The magnetic field of the material will add itself to the external magnetic field, strengthening it in the process. Materials which behave in this way are called paramagnets. When the external magnetic field is removed, however, the atomic magnets revert to their random orientations, and the material as a whole loses is magnetism. This is the basis of temporary magnetism.

Diamagnetism

In atoms which do not have a magnetic field (i.e., in which the magnetic fields associated with the electrons cancel out), another type of magnetism can arise. The second of FARADAY'S LAWS OF INDUCTION tells us that if the magnetic field is increased through a current-carrying loop, then the current in the loop will change in such a way as to oppose the increase in the magnetic field. Thus, when a material whose atoms do not normally have a magnetic field is placed in a strong magnetic field, their electrons will adjust in such a way as to create their own magnetic field opposing the external one. The magnetic field of the material then opposes the external field. Such a material is said to be diamagnetic.

The important point about the magnetic properties of matter is that they depend on the configuration of the electrons within the atom. If you take an iron magnet apart, you will find that each atom retains its magnetism. If you take the atom apart, however, you will find only the subatomic particles, and the main contributions to the atomic magnetic field will disappear. Thus, magnetism arises from the arrangements of electrical charges in an atom, not from the properties of the charges themselves.

Marginal Value Theorem

An animal will forage for food in a given area until the rate of energy consumption reaches its maximum

There are many aspects of animal behavior that are amenable to treatment by mathematics (*see,* e.g., OPTIMAL FORAGING THEORY). One of them is a problem which can be stated very simply: When an animal is foraging for food, for how long should it stay in the same general area, and when should it move on? What strategy maximizes its total energy consumption?

This problem is very typical of those dealt with in a branch of mathematics known as *game theory,* which uses probabilistic reasoning to work out strategies that maximize the chances of winning a game. The theory can be applied to ordinary games of chance, such as poker, but also to "games" such as military strategy, the deployment of sales forces, and the placement of advertising. This type of mathematics can also be used to deal with the problem of foraging strategy.

When an animal comes into a new area, it begins to consume the resources it finds there. Think of a horse moving onto a patch of tall grass, or a fox coming into an area where there are lots of rodents. As it does so, the animal begins to acquire energy. At first the amount of energy acquired will rise steeply, as the animal's energy consumption jumps from zero (during the search for the resources) to whatever it is getting from the new area. As it consumes the resources, however, the total amount of energy it has acquired will begin to level off. When all of the resources are used up, the energy consumption will drop back to zero, and the total energy acquired from the area will stay at a constant value. The question is this: What is the best moment for the animal to move on, leaving behind some known quantity of unconsumed resources and moving on to search for a new resource base? The answer is provided by the marginal value theorem, put forward in 1966 by the American ecologist Eric L. Charnov (1947–).

The theorem takes into account the time lost in transferring from one resource base to another—call this the *travel time.* There is also the time for which the animal stays in a particular resource area—call this the *staying time.* The optimum strategy is the one that maximizes the total energy consumption over time. If the animal acquires energy E after a given staying time, then the total rate of consumption for the total time t (which is the travel time *plus* the staying time) will just be E divided by t.

Let's start by supposing that all areas are allocated the same resources—that all patches of grass are equally green, for example. The mathematics of game theory tells us that the animal will maximize its total energy input over time if it leaves each area as soon as the *rate* of energy acquisition is maximum. In other words, the time to move on is just when things seem to be at their best. Hanging around does allow the animal to acquire more energy from a known source, but the amount it acquires will be less, on average, than it would get if it moved on to another area right away.

What happens in the more realistic situation in which areas vary in their resource abundance? The theorem then says that the animal's best strategy is to leave each area, regardless of its richness, as soon as its rate of energy extraction has fallen to the maximum rate obtainable from the average area. Thus, the optimum time to leave is when the marginal value of energy extraction reaches this average rate of consumption (which explains where the theorem's name comes from).

A prediction of the theorem is that foragers will spend less time in patches where there is less food. As a consequence, foragers will move on from such patches more quickly when the patches are close together than when they are far apart. Also, foragers will move on from such patches more quickly when they occur in an abundant area than when they occur in a poorer area.

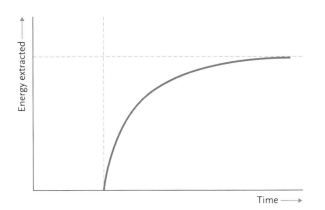

One point should be made about this approach to explaining animal behavior. No one is suggesting, of course, that the animal sits down with a calculator and works out its rate of energy consumption. Instead, in the spirit of the theory of EVOLUTION, what the marginal value theorem is saying is that, if by chance an animal stumbles upon an optimal strategy, then that animal will have more of a chance of passing its genes to the next generation. The effect will be that, over long periods of time, animals will come to their optimal strategy without ever having to think about it.

An animal finding a new food resource will gain a lot of energy to start with, but as the resource gets used up, the best strategy for the animal is to move on and seek out a new resource. Here, the vertical dashed line is when the animal begins to exploit the resource, and the horizontal dashed line is the maximum level of energy consumption.

In this respect, evolution is a lot like the "invisible hand" in 18th-century Scots economist Adam Smith's theory of economic markets in that it acts behind the scenes to produce the behavior that optimizes benefits. Classical theory's "economic man," who acts in a completely rational way with full knowledge of the market, bears no resemblance to a real human being—it's just that this behavior is favored and eventually comes to predominate. In the same way, animals who don't move from one area to another according to the marginal value theorem will, on average, be less successful than those who do. In the end, efficiency always wins.

Mass Extinctions

Several events in the history of our planet have caused the extinction of a significant proportion of all species living at the time

The extinction of the dinosaurs was a long-running mystery in paleontology. After all, dinosaurs had been around for well over 100 million years. They were the most successful group of animals that had ever evolved on our planet. Then, in a time period that could have been as long as a few tens of thousands of years or, some suggest, as short as a weekend, they apparently disappeared. So what happened?

Many explanations have been offered, ranging from the bizarre (they were hunted to extinction by little green men in flying saucers) to the credible (changes in the climate destroyed their ecological niches). My favorite explanation blamed the extinction on the appearance of flowering plants, once thought to have originated 65 million years ago, just when the dinosaurs disappeared. The idea was that until that time dinosaurs had fed on things like pine needles, full of natural oils, but when they had to switch to spinach they all died of constipation!

Actually, paleontologists seldom focus solely on the extinction of the dinosaurs, for the fact of the matter is that when the dinosaurs bit the dust 65 million years ago, fully 70% of all species on the Earth did the same. The event, whatever caused it, was what scientists call a mass extinction. We know of many such events, and the mass extinction that did for the dinosaurs was neither the most severe nor the most recent of them. In fact, depending on how you define a "mass" extinction, there have been anywhere from five to over a dozen mass extinction events in the last 500 million years. The biggest occurred about 280 million years ago, and the most recent about 13 million years ago. Although some scientists have suggested that they all share a common cause, explanations tend to concentrate on changes in the Earth's climate.

In 1980, a father-and-son team at the University of California at Berkeley came across something that led to the best current theory of the extinction of the dinosaurs. Nobel laureate Luis Alvarez and his son Walter carried out some delicate analysis of sediments that had been laid down just at the time of the extinction. In the sediment they found extremely high concentrations of iridium, a heavy metal rather like platinum. Iridium is very rare on the surface of the Earth, because the Earth went through a molten phase early in its history, and heavy metals tended to sink toward the center. However, iridium is much more abundant in some types of asteroids. Their suggestion, sometimes called the *Alvarez hypothesis,* was that this material had been deposited by the impact of an asteroid about 7 miles (11 km) across. The major killing mechanism was a dust cloud that blanketed the Earth for several years, blocking out sunlight and killing most of the living things on the planet.

At first, scientists responded to this claim with a high degree of skepticism, even hostility. But over the years, evidence in its favor began to accumulate. For example, geologists sifting through sediments laid down at the time of the hypothesized impact found a mineral called shocked quartz, which could only have been formed at the high temperatures and pressures associated with an impact. Gradually, opinion began to swing over to the Alvarez explanation. Then, in 1992, the smoking gun was found—a crater more than 100 miles (170 km) across, largely buried under ocean-floor sediments next to the Yucatán peninsula, Mexico. Chicxulub crater (it was named after a nearby fishing village) is one of the

An artist's impression of the day that the Alvarez asteroid struck the Earth, marking the end of what is known as the Cretaceous period of the Earth's history. The apprehensive-looking dinosaur in the foreground is the familiar Tyranno-saurus rex.

largest craters known on Earth, and it is now widely accepted as the site of the asteroid impact that finished off the dinosaurs. The recent discovery that isotopes characteristic of asteroids are present in certain other sediments suggest that the mass extinction of 280 million years ago may have had a similar cause.

The debate on asteroid impact has now largely shifted to the other mass extinctions—were they also caused by impacts or were there other causes, such as massive volcanic eruptions or sudden changes in sea level? These are the sorts of questions that evolutionary scientists are trying to answer today.

LUIS WALTER ALVAREZ (1911–88) American physicist. He was born in San Francisco and educated at the University of Chicago. During World War II he helped to develop ground-approach radar and worked at Los Alamos on the atomic bomb project. At the University of California at Berkeley he improved the bubble chamber, a device for recording the interactions of high-energy particles, for which he received the 1968 Nobel Prize for Physics. The Alvarez hypothesis resulted from work he carried out in 1981, with his son, geologist Walter Alvarez (1940–).

Maximum Sustainable Yield

There is a maximum number of organisms that can be harvested from a given population in a sustainable manner

Populations of organisms such as fish or game are not static, but vary according to the conditions of their environment and the population density. A standard way of representing this is to use what is called a *recruitment curve,* obtained by plotting a graph showing how the net increase in population (births minus deaths) varies with the population density. For low densities, there is little addition ("recruitment") to the population, simply because there are few organisms to give birth. At high densities, though, there is intense competition for resources, and net recruitment is again low because the death rate is high. In between these two extremes, the net recruitment rate will rise to some maximum value, then fall as the density increases.

The maximum point on the recruitment curve represents the maximum number of individuals that can be added to the population by natural processes. If more individuals than this are removed from the population, the population will decline and the species will eventually be driven to extinction. The maximum number that can be harvested in a sustainable manner is given by the maximum point on the recruitment curve, and that is what is called the maximum sustainable yield (or the *optimum yield*). This number is extremely important in the conservation and management of wildlife—it is, for example, used to set the maximum harvests for commercial fisheries around the world, as well as to set quotas for hunting seasons.

The concept of maximum sustainable yield is not always easy to apply in practice, however. For one thing, biologists don't always have enough data to make a clear determination of the recruitment curve. For another, the assumption that there is only a single maximum on the recruitment curve may not be true, and that can complicate matters. To take account of this, biologists routinely build in safety factors when they set up yield strategies.

The quantities of fish we take from the sea have to be controlled if fish populations are to be sustained.

Maxwell's Demon

Can the second law of thermodynamics be violated?

In science, as in fiction, some of the most interesting characters are imaginary ones. Nowhere is this truer than in discussions of the second law of THERMODYNAMICS in which a being called Maxwell's Demon makes his appearance. The "Maxwell" is James Clerk Maxwell, the author of MAXWELL'S EQUATIONS. The second law states that the orderliness of an isolated system cannot increase. A gas made up of molecules moving at different speeds, for example, cannot spontaneously separate into two sections, one of which has all the faster-than-average molecules and the other has all the slower ones.

Most physical processes are *reversible*. For example, water can be frozen and the resulting ice melted back to water, iron can be magnetized and demagnetized, and so on. Processes governed by the second law of thermodynamics, however, are not reversible—the orderliness of a closed system is either constant or decreasing, never increasing. This behavioral imbalance greatly puzzled physicists in the late 19th century, and Maxwell devised a means whereby the second law could, apparently, be reversed. As a "thought experiment," he pictured an arrangement in which a container is split into two parts connected by an atomic-sized trapdoor in the wall that separates them. The upper chamber contains an ordinary gas, while the lower one starts out containing a vacuum.

Now imagine a tiny creature standing next to the trapdoor, watching the molecules whiz around him. Whenever he sees a fast molecule heading for the trapdoor, he opens it and lets the molecule through. When he sees a slow molecule he leaves the trapdoor closed, so the molecule just bounces back into the upper chamber. If the creature kept this up for long enough, it's obvious that the gas would separate into two sections—

slow molecules in the upper chamber, fast molecules in the lower. Thus, the creature would have produced a violation of the second law of thermodynamics. Not only that, but the hot and cold sections of the chamber could now be used to drive a heat engine (*see* CARNOT'S PRINCIPLE), and work or energy could be extracted from the system. If the creature never tired of operating the trapdoor, the process could continue forever, and thus we would have a PERPETUAL-MOTION machine.

A portrait of James Clerk Maxwell, whose research into the theory of electromagnetism and the behaviour of molecules in a gas made him one of the greatest physicists of the 19th century.

The fanciful operator of the trapdoor, named Maxwell's Demon by Maxwell's fellow physicists, has enjoyed a peripheral existence in the folklore of thermodynamics ever since. A perpetual-motion machine would be pretty useful, but it turns out that the Demon has to use energy, in the form of photons, to observe the molecules. Also, the Demon and the trapdoor would lose orderliness to the gas (in technical parlance, they would *gain entropy*). Thus the total orderliness of the system, which includes both Maxwell's Demon and the trapdoor, still decreases, and the Demon is no longer considered to be even a theoretical threat to thermodynamics.

The first counterargument surfaced soon after the development of QUANTUM MECHANICS. If the demon is going to make a measurement of the velocity of incoming molecules, then we have to take account of HEISENBERG'S UNCERTAINTY PRINCIPLE. Along with the measurement of the velocity will come an uncertainty in the position, and if that uncertainty is too big, the molecule may well miss the trapdoor even if it's open. So, the argument goes, Maxwell's Demon is nothing more than a classical, macroscopic monster trying to live in the quantum world, where the rules are different from what he's used to. As soon as you incorporate quantum mechanics into the Demon's life, he is no longer a threat.

Another powerful argument against the Demon emerged from the study of information and computers. If we represent the Demon by a computer which takes in information about approaching molecules and changes bits of information in its processing unit as a result, then after the trapdoor has been either opened or allowed to remain closed, the computer has to reset itself. But that takes energy—in fact, exactly the amount of energy required to compensate for the increase in orderliness produced by the Demon. In this scheme, the side effects of the Demon's operation cancel out his intentional work, and the second law of thermodynamics is safe.

Too bad—the Demon was such an engaging character!

Maxwell's Equations

Four equations describe all phenomena associated with electricity and magnetism

By the mid-19th century, scientists had discovered a number of things about the behavior of electricity and magnetism and about the connections between the two. Among these discoveries were:

—COULOMB'S LAW, which describes the force between electrical charges,

—GAUSS'S LAW for magnetism, which says that there are no isolated magnetic poles in nature,

—the BIOT–SAVART LAW (*see also* AMPÈRE'S LAW, OERSTED DISCOVERY), which says that moving electrical charges can produce a magnetic field, and

—FARADAY'S LAWS OF INDUCTION, which say that changing magnetic flux can produce an electric field, or cause a current to flow in a conductor (*see also* LENZ'S LAW).

These four laws form the basis of Maxwell's equations, named for James Clerk Maxwell. You may wonder why we should refer to "Maxwell's equations" if the underlying laws were discovered by several of his predecessors. The first reason is that Maxwell derived and expressed the laws in a generalized mathematical form (Faraday, for example, never wrote his laws in the form of equations at all). Second, he introduced many new ideas of his own, including adding concepts missing from the original laws. Third, he put the whole subject of electromagnetism onto a sound theoretical basis. And fourth, he then went on to make numerous predictions and discoveries on the basis of his equations, including predicting the existence of the entire ELECTROMAGNETIC SPECTRUM.

Let's start with the second of those reasons. The Biot–Savart law tells us that when there is an electrical current flowing in a wire, there will be a magnetic field around that wire. But what would happen if the current flowed through a wire onto a large plate, separated by a small gap from another plate connected to another wire? As electrons run onto the first plate, they exert repulsive forces that push electrons from the second plate into its wire. The net effect is that negative charges flow along the wire into the first plate, and out of the second plate into the wire connected to it, even though no current actually flows across the gap.

Maxwell realized that Ampère's law as it stands cannot be applied to this situation unambiguously, since it's not clear whether a current is flowing in the system. He also realized that, even though no charges were flowing across the gap between the plates, the electrical field (the force that would be experienced by an imaginary electrical charge placed there) was increasing. He therefore postulated that in the world of electromagnetism, a changing electrical field could play the role of an electrical current in producing magnetic fields. Maxwell's addition to the equation, representing what is now known as a *displacement current,* has since been well justified by experimental evidence.

With this missing piece of the puzzle in place, Maxwell was able to manipulate the equations mathematically to make a rather astonishing prediction. The equations seemed to predict the existence of waves made up of electrical and magnetic fields, with a velocity related to the forces between charges or between magnets. When he put the numbers in, Maxwell found, much to his amazement, that the speed of these waves was precisely the same as the speed of light! In other words, the familiar phenomenon of light was actually related to electricity and magnetism. Moreover, Maxwell was able to predict the existence of the entire electromagnetic spectrum, from radio waves to gamma rays. Thus a theoretical inquiry into the nature of electricity and magnetism has produced enormous benefits for humanity. In a sense, it has led to everything from microwave ovens to dental X-rays.

JAMES CLERK MAXWELL (1831–79) Scottish physicist, perhaps the greatest theoretical physicist of the 19th century. He was born to an old and prominent family in Edinburgh. He attended Edinburgh University and Cambridge, having published his first scientific paper (on a technique for drawing a perfect oval) at the age of 14. He reached the rank of professor of experimental physics at Cambridge before succumbing to cancer in his 48th year.

Clerk Maxwell, as he is often called, did his early work on the theory of color and color vision. He showed that all colors can be understood as a mixture of three primary colors, red, green, and blue, and explained colorblindness as a defect of the receptors in the eye. He also was the first to project an actual color photograph (of a tartan ribbon) to an audience, at a meeting of the Royal Institution at London in 1861. Almost as an aside, he carried out the first rigorous calculations of the structure of the rings of Saturn,

showing that they could not be a fluid, but must be made up of small particles.

Maxwell made important contributions to many fields of science. Most importantly, he put the science of electricity and magnetism on a firm mathematical footing. He began his work on electromagnetic theory in the mid-1850s. It is rather ironic that Maxwell was a firm believer in the luminiferous ether, and cast much of his theory in terms of that nonexistent medium. He didn't live to see the MICHELSON–MORLEY EXPERIMENT do away with the need for an ether. Fortunately, his theory is still valid, ether or no ether.

Finally, he made fundamental contributions to the field of statistical mechanics, working out the energy distribution of molecules in a gas of colliding atoms, a cornerstone of the KINETIC THEORY. Last but not least, he introduced a note of serious whimsy into the field with his introduction of MAXWELL'S DEMON.

Mendel's Laws

Inherited traits are passed from parent to offspring through the action of genes

Offspring receive one gene for each trait from each parent

Gregor Mendel is generally considered to be the founder of the modern science of genetics, and the pea plants with which he experimented have joined Newton's apple as key icons in the folklore of science. Motivated by an interest in agriculture and the orchards around his monastery in the town of Brünn (now Brno in the Czech Republic), Mendel embarked on a long, painstaking series of experiments in plant hybridization that eventually led him to the notion that heredity is governed by genes.

His technique was simple if painstaking: Starting with pea plants, he would cover the flowers so that only carefully selected pollen could fertilize each plant. He could then compare the properties of the parent plants to those of their offspring to try to deduce the laws of inheritance.

In a classic set of experiments he cross-bred two sets of peas—one tall, one short. All offspring in the first generation were tall (not short or medium sized, as might be expected). When these first-generation plants were crossed with one another, however, only three-quarters of the resulting plants were tall, and the other quarter were short. The explanation that Mendel developed for this (and many other) results depended on two postulates:

— there is a "unit of inheritance," which he called a "factor" but which we now call a *gene,* and the offspring receives one gene from each parent; and
— if the offspring receives genes for conflicting characteristics, one gene will be *dominant* and be *expressed* (i.e., the characteristic coded for by the gene will appear in the individual), while the other will be *recessive* (i.e., not expressed).

What this means for the peas is that each first-generation offspring received a "tall" gene and a "short" gene—one from each parent. The fact that all the offspring were tall tells us that the tall gene is dominant. Each offspring, however, still carried a "short" gene as an unexpressed passenger in its hereditary material. In the next generation, one offspring (on average) will have two "tall" genes; two will have a "tall" and a "short," and hence will be tall plants; and one will receive two "short" genes, and hence will be a short plant. By this scheme, Mendel could explain many characteristics of inheritance that had long been a mystery—why some diseases (e.g., hemophilia) will skip generations, or why a brown-eyed parent can have a blue-eyed child.

Gregor Mendel, the Moravian monk who founded the science of genetics.

As has happened from time to time in the history of science, Mendel's work, which was completed in 1865, did not immediately receive due recognition. He summarized the work in a paper that was read at the Natural Sciences Society of Brünn, and it was printed in the society's journal, but at that point he seems to have become discouraged. Although the journal was distributed to over a hundred scientific organizations around the world, the issue containing his revolutionary work appears to have gathered dust in libraries for thirty years. It was only toward the end of the 19th century that scientists working on problems of inheritance rediscovered Mendel's work and were able to give him (posthumously) the recognition he so richly deserved.

This is not to say that his ideas were accepted uncritically. There had been a long history of discussion in the life sciences about the doctrine of *preformationism*—the notion that somehow the sperm and the egg already contained a miniature of the adult organism. For example, Anton van Leeuwenhoek (1632–1723), the first scientist to make systematic use of the microscope, believed that each sperm contained a miniature human being, and the function of the egg was to provide the nourishment that would allow it to grow. The question now was whether the development of the embryo was driven by internal, inherited factors as Mendel reasoned or was determined by external, environmental factors that might, for example, affect the nutrients in the egg. Scientists are now starting to look in detail at the way entire organisms develop from single fertilized eggs, and are coming to understand that external factors, such as substances to which the embryo is exposed in utero, can affect how genes are turned on, and thus how the organism develops.

Today we understand that Mendel's genes are actually segments of the DNA molecules carried in cells, and the central dogma of MOLECULAR BIOLOGY tells us that they exert their influence by coding for PROTEINS, which in turn act as enzymes to control chemical reactions in living things (*see* CATALYSTS AND ENZYMES).

GREGOR JOHANN MENDEL (1822–84) Moravian monk and plant geneticist. He was born Johann Mendel in Heinzendorf (now Hynčice in the Czech Republic), the son of a peasant farmer. He took the name Gregor in 1843 when he entered the monastery at nearby Brünn (now Brno). In 1851 he was sent by his abbot to the University of Vienna to study science, including botany, after which he taught natural sciences at a local school.

In 1856, perhaps motivated by childhood memories of assisting his father, a peasant farmer, he began to investigate hybridization in plants, in particular the edible pea, which was grown in the monastery garden. Mendel's results, the bedrock of modern genetics, sparked little interest when they were published in 1865. Three years later he became abbot of Brünn and largely abandoned research for administrative duties.

Michelson–Morley Experiment

Light does not require the presence of a "luminiferous ether" in order to move through space

It is very hard to imagine nothing—the absence of everything. The human mind seems to want to fill empty space with some kind of material, and for most of history that material was called the ether. The idea was that the emptiness between celestial objects was filled with a kind of tenuous Jell-O. When MAXWELL'S EQUATIONS first predicted the speed of electromagnetic waves, it was assumed that those waves traveled through the ether the way that waves travel on the surface of the ocean. In fact, in the last quarter of the 19th century, physicists constructed elaborate models of what the ether would look like, some complete with gears and axles whose turning pushed the wave along.

In 1887, two American physicists, Albert Michelson and Edward Morley, put together an experiment designed to detect the presence of the *luminiferous ether*—the material that was then believed to permeate space and provide a medium for light and other radiation to travel through. Michelson was well known for his skill at constructing optical instruments, while Morley had a reputation as an exceptional and patient experimenter. Their experiment is simple to describe, if rather more difficult to carry out.

The pair used an instrument called an *interferometer*, in which a single beam of light is split into two by a half-silvered mirror (a plate of silvered glass that reflects half the light that hits it, and transmits the other half). These two beams are sent off at right angles to each other, and each is reflected back from a mirror. The two beams are then brought back together, and any small difference in the distance traveled by each beam

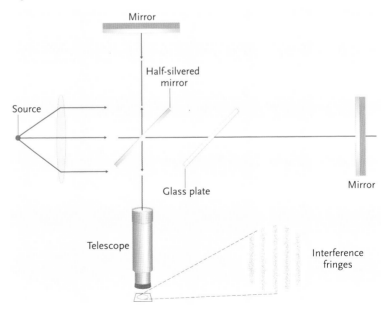

A schematic diagram of the setup used by Michelson and Morley. If the ether existed, the interference fringes produced by the optical system would have changed position when the equipment was pointing in a different direction, but they did not.

can be detected by looking at the pattern of the recombined beam (*see* INTERFERENCE).

The principle of the Michelson–Morley experiment was based on the idea that if there really is an ether, then as the Earth goes around the Sun we should experience an ether "wind" as a result of our motion through the ether. If the ether wind is in one direction now, then in six months' time it will be coming from a different direction because the Earth will have completed half an orbit around the Sun and will then be heading in the opposite direction. By carefully monitoring their apparatus, Michelson and Morley were able to make a very important statement: there is no evidence for any ether wind. (Modern experiments with lasers and improved interferometers have confirmed and tightened up this conclusion significantly.)

With the necessity for an ether removed, the conflict between Newtonian mechanics (in which there was a preferred frame of reference) and Maxwell's equations (which predicted a universal speed of light in all reference frames) that led to the theory of RELATIVITY was sharpened. The experiment removed the ether as a "preferred" frame of reference. Although Einstein later claimed that the experiment played little part in his development of relativity, it clearly helped to persuade other scientists to accept that theory.

ALBERT ABRAHAM MICHELSON (1852–1931) German-American physicist. Michelson was born in what is now Strzelno, Poland, in what was then a disputed region between Germany and Russia. He moved to America with his family at age two. He grew up in California, where his father was a merchant in gold-rush towns. He was admitted to the Naval Academy in Annapolis at the special request of his congressman, and after a tour of active duty, he was appointed an instructor in physics there. Michelson now began working on precise optical determinations of the speed of light.

In 1881, he resigned from active service and joined the Case School of Applied Sciences (now part of Case-Western Reserve University) in Cleveland, Ohio, to pursue his inquiries. He received the 1907 Nobel Prize for Physics for his work in determining the length of the standard meter and his precision measurement of the speed of light.

EDWARD WILLIAMS MORLEY (1838–1923) American physicist and chemist. He was born in Newark, New Jersey, the son of a Congregational minister. Educated at home because of poor health, he was trained for the church but chose instead to study chemistry and natural history. Morley was an exceptional experimenter, establishing a highly accurate value for the relative masses of hydrogen and oxygen in pure water. This quality proved invaluable in his famous collaboration with Albert Michelson.

Milankovič Cycles

Because of variations in the Earth's orbit, the planet goes through periodic ice ages

One of the great surprises of 19th-century geology was the realization that large glaciers had once moved down from the Arctic and covered most of Europe and North America. Two geological features in particular pointed to this glaciation. If you think of an advancing glacier as something like a bulldozer, it's not hard to see that it will move dirt and rocks in front of it. When the glacier reaches its maximum extent and starts to recede, the pile of rocks left behind becomes a line of hills—the so-called *glacial moraines.* In addition, when a glacier moves, the ice within it flows (albeit slowly) and carries rocks along with it. If you look at the floor of a mountain valley that has been carved out by a glacier, there is a good chance that you will find deep parallel grooves in the rock. The origin of these *score marks* is easy to understand if you picture the glacier as dragging along rocks imbedded in its underside, acting rather like a file or a piece of sandpaper. Moraines and score marks provide dramatic evidence for the past existence of glaciers.

Soon after the discovery that an ice age had existed, it became quite clear that it was not a single event, but that ice ages seem to have come and gone with some regularity in the Earth's past. Why this should happen remained a puzzle until a rather remarkable scholar decided to look into it early in the 20th century. In his memoirs, Milutin Milankovič tells how he came to think about the origin of ice ages. He and a friend, a poet, were in a coffee house celebrating the publication of a volume of patriotic verse that his friend had written (as junior faculty members at the University of Belgrade, coffee was all they could afford). A wealthy businessman seated near them was so taken with the verses that he bought ten copies of the book on the spot. The pair began to drink wine and celebrate in earnest. After the first bottle, Milankovič "looked back on his earlier achievements which now seemed narrow and limited." By the end of the third bottle, the poet was planning to write an epic, and Milankovič had resolved to "grasp the entire universe and spread light to its farthest corners."

During World War I Milankovič was on the general staff of the Serbian army. Captured by Austro-Hungarian forces, he became a prisoner of war in Budapest. Happily for Milankovič (and for science), colleagues at the Hungarian Academy of Science arranged for him to have a desk

The Gilkey Glacier in Alaska. The curved bands show the annual movement of the ice, which travels more slowly along the edges than in the center. During ice ages, glaciers like this spread southward from the Arctic.

where he could work provided he gave his word that he would not try to escape. He agreed, and spent most of the war working out a theory of periodic ice ages.

The explanation of what are now called Milankovič cycles has to do with the orbit of the Earth. According to NEWTON'S LAW OF GRAVITA-TION, a planet going around the Sun should follow an elliptical orbit (this is also the first of KEPLER'S LAWS of planetary motion). In addition, the conservation of ANGULAR MOMENTUM means that if the Earth is spinning on its axis, the direction of that axis must remain fixed in space. But in the real solar system the Earth does not orbit the Sun in lonely splendor. It is subject to the gravitational forces of the Moon and the other planets, and these forces have small but significant effects on both the orbit and the rotation of the Earth. There are three such effects:

— *Precession* The Earth's axis does not in fact stay pointing in the same direction, but twists slowly around. This effect is called precession, and you may have seen it at work in a gyroscope. When a gyroscope is set in motion it spins rapidly about its axis, but the axis itself traces out a cone. The Earth's axis does the same thing, over a time period of about 26,000 years. Right now, for example, the Earth is tilted so that the northern hemisphere (where most of the land is) is tilted away from the Sun in January, when the Earth is closest to the Sun. In 13,000 years' time, the situation will be reversed, and the northern hemisphere will be tilted toward the Sun during January, which will then be in the middle of the northern hemisphere's summer.

— *Nutation* In addition to the lazy precession of the Earth, the axis of the Earth itself wobbles slightly. Right now, the Earth's axis is tilted at 23° to the plane of its orbit. It can, over a period of 41,000 years, wobble back and forth between 22° and 23°, under the influence not only of the Moon but also of Jupiter (distant, but massive).

— *Change in shape of orbit* The shape of the Earth's orbit itself changes over time due to the influence of the gravitational pull of the planets. It goes from being an ellipse that's elongated in one direction, to being a circle, to being an ellipse that's elongated in a direction perpendicular to the first, then back to being a circle, and so on. This particular cycle lasts about 93,000 years.

What Milankovič realized is that each of these factors affects the amount of sunlight received by different regions of the Earth. For example, the precession of the Earth's axis affects the character of the northern hemisphere's winters and summers (I am concentrating on the northern hemisphere because that is where most of the land, and hence most of the glaciers, are to be found).

Milanković understood that the Earth's long-term climate is unstable (*see* EQUILIBRIUM). In a period when the amount of sunlight falling on the northern hemisphere is decreasing, you would expect snow to stay on the ground a little longer each year. Since snow is a good reflector, the extra snow cover will reflect more sunlight and cool the Earth further. This, in turn, will lead to more snow the next winter, more ground cover, more reflection, and so on. Over time, the snow will build up and the glaciers will move south. The Earth will have entered an ice age. At the other end of the cycle, when the Sun is adding energy to the northern hemisphere, things will go the opposite way—when a little ice melts, it opens up some bare ground which absorbs sunlight, which warms the Earth, and the same sort of instability drives the ice sheets back.

Milanković realized that there are three cycles operating, each of which corresponds to a different astronomical effect. His idea was that when they all reinforce one another, you could get a period of cooling and an ice age. Under normal circumstances, however, the three would not be pulling in the same direction and we would soon return to a normal climate. Ice ages, then, happen when all three orbital effects act together to pull the Earth toward a colder climate, and so can be expected to have occurred often in the planet's history.

Over the last 3 million years there have been at least four major periods of glaciation, and there have been other such periods in the past. For the record, I should point out that the last ice age reached its maximum about 18,000 years ago, and that scientists talk about our current situation as being an *interglacial*—a chilling thought indeed.

MILUTIN MILANKOVIĆ (1879–1958) Serbian climatologist. He was born in Dalj (now in Croatia) and was educated in Vienna, where he began a career as a civil engineer. In 1904 he joined the University of Belgrade, where he spent the rest of his academic life. During World War I he was a prisoner of war in Budapest, where Hungarian colleagues allowed him to continue his research. Milanković strove for many years to reconstruct the Earth's past climate. His findings are regarded as questionable now that better dating techniques are available.

Miller–Urey Experiment

Molecules necessary to life could have been formed by chemical reactions on the early Earth

When the Earth formed 4½ billion years ago, it was a hot, lifeless ball in space. Today, of course, it teems with life. The question, then, is how we got from there to here, and in particular how the molecules that comprise living things arose on an Earth without living things. In 1953, what is now considered to be a classic experiment was performed at the University of Chicago. It started scientists on the road to answering this fundamental question.

In 1953, Harold Urey was already a Nobel laureate; Stanley Miller was his graduate student. Miller's idea for the experiment was simple: In a basement laboratory, he would put together a small microcosm of what scientists believed the Earth's early atmosphere was like, and then stand back and watch what happened. With Urey's encouragement he assembled a simple apparatus from a glass bulb and some tubing that allowed substances evaporated from the tube to make a circuit, be cooled, and be returned to the bulb. He filled the bulb with gases that Urey and the Russian biochemist Aleksandr Oparin (1894–1980) believed were present in the early atmosphere—water, hydrogen, methane, and ammonia. To simulate the effect of the Sun, Miller heated the bowl with a Bunsen burner, and to simulate the effect of lightning bolts he installed two electrodes in the glass tubing. The idea was that material would evaporate from the bulb and pass up into the tube, where it would be subjected to electrical sparks. The material would then be cooled and returned to the bulb to begin the cycle again.

After the system had been running for a couple of weeks, the liquid in the bulb was turning a dark reddish-brown. Miller analyzed this liquid and found that it contained amino acids, the basic building blocks of PROTEINS. The way was now open for scientists to investigate the origin of life in terms of basic chemical processes. Since 1953, sophisticated versions of the Miller–Urey experiment, as it has become known, have yielded all sorts of MOLECULES OF LIFE—including complex proteins essential for cellular metabolism, and fatty molecules called lipids that make up cell membranes. Energy sources other than electrical sparks (heat and ultraviolet radiation, for example) seem to produce the same result. There seems little question, then, that all of the parts needed to assemble a cell could have been made in chemical reactions that took place on the early Earth.

The rationale of the Miller–Urey experiment is that, for several hundred million years, lightning bolts in the early atmosphere sparked the production of organic molecules that rained down on the primeval ocean, creating what is called the *primordial soup* (*see also* EVOLUTION). As yet undetermined chemical reactions in this "soup" would then have led to the first living cell. In recent years, serious questions have been raised about this scenario, particularly about whether ammonia was present in the early atmosphere. In addition, a number of alternate routes to the first cell have been advanced, from the enzymatic action of the biochemical molecule RNA to simple chemistry at deep ocean vents. Some scientists have even suggested that the origin of life is related to the new science of COMPLEX ADAPTIVE SYSTEMS, and that life itself might be an emergent property of matter. Today, this field is in a state of splendid turmoil as different sets of hypotheses are advanced and tested. Out of this turmoil should emerge the story of how our most distant ancestors came to be.

STANLEY LLOYD MILLER (1930–)
American chemist. He was born in Oakland, California, and was educated at the University of California at Berkeley and the University of Chicago. He spent most of his career, from 1960, as a professor of chemistry at the University of Californian at San Diego. His work on the Miller–Urey experiment earned him a research fellowship at the California Institute of Technology.

HAROLD CLAYTON UREY (1893–1981)
American chemist. He was born in Walkerton, Indiana, into a family of clergymen. He studied zoology at Montana State University and took his doctorate in chemistry at the University of California at Berkeley. He was a pioneer in applying the methods of physics to chemistry, and was awarded the 1934 Nobel Prize for Chemistry for his discovery of deuterium, a heavy isotope of hydrogen. Much of his subsequent work was concerned with the differences in the rates at which chemical reactions proceed when different isotopes are used.

Millikan Oil-Drop Experiment

The charge on the electron is 1.6 × 10⁻¹⁹ coulombs

After the discovery of the ELECTRON, scientists knew that this particle was one of the fundamental constituents of matter. Measuring its properties, however, remained a formidable technical challenge. It was Robert Millikan who made the first precision measurement of the electrical charge on the electron. His experimental apparatus consisted of a chamber with metal plates at the top and bottom. These plates could be connected to an electrical potential (voltage). A fine spray of oil droplets was then introduced into the chamber, and the droplets were monitored one by one as they fell. Because the droplets were so small, in the absence of a voltage on the plates they soon reached TERMINAL VELOCITY, where the downward force of gravity was balanced by the upward force of air resistance. From the measurement of this velocity, the radius and the mass of the droplet could be determined. At this point, the electrical field was turned on and adjusted so that the droplet remained stationary. The downward force of gravity on the drop was then balanced by the upward electrical force.

In the normal course of events, each oil droplet would have acquired an electrical charge on moving through the air, just as items of clothing in a dryer acquire a static charge by rubbing against one another. Thus, the voltage required to halt the fall of the droplet would be determined by the total charge on the droplet. This charge had to be an integer (whole-number) multiple of the charge on the electron, because a whole number of electrons must have contributed to this charge—the total charge, measured in terms of a charge on the electron, couldn't have a fractional part.

The next step was to irradiate the droplet with X-rays, a process that changed the ionization of the molecules in the oil droplet, and thus the total electrical charge. The electrical field was again changed to balance gravity, and the balancing electrical voltage was measured. Then the droplet was irradiated again to change the charge, and so on. Eventually, a large body of data was accumulated, consisting of many values of electrical charge. Each value was found to be an integral multiple of the charge on the electron. From this collection of data, the charge on a single electron was deduced.

The experiment was extremely painstaking. In the course of it, Millikan had to remeasure quantities such as the viscosity of air, a process that would take him five years of meticulous effort. He published his value of the electrical charge in 1913, and was awarded the Nobel Prize for Physics in 1923.

Millikan's procedure has not remained a dusty museum exhibit in the history of science. During the 1960s, when the idea of the existence of quarks was first suggested (*see* STANDARD MODEL), high-tech versions of his apparatus were built to search for this new particle. The idea was that if there are such things as free quarks (i.e., not bound into subatomic particles), their charge (predicted to be one-third or two-thirds of the charge on the electron) would show up clearly in the data. The fact that no such particles were seen is taken as evidence that quarks cannot be separated as individual entities from the particles in which they reside.

ROBERT ANDREWS MILLIKAN (1868–1953) American physicist. He was born in Morrison, Illinois; his father was a Congregationalist minister, and his mother a former dean of women at a small college in Michigan. He attended Oberlin College in Ohio, where his skill in Greek landed him a job as an instructor in physics in an affiliated preparatory school. He quickly developed a love for this new field, and pursued it on his own.

Admitted to Columbia University as the sole graduate student in physics, he took his Ph.D., did the customary year in Europe to explore the frontiers of his field, and then took up a faculty position at the University of Chicago. There he became a recognized authority on the teaching of physics (he authored the standard introductory textbook of his time). It was at Chicago too that he performed his oil-drop experiment, which not only yielded the first precision measurement of the electrical charge on the electron, but propelled him to a position of prominence in American science. He also began to turn into something easily recognizable to the modern reader—a public scientific intellectual.

During World War I he was appointed to the rank of colonel, directing research in the U.S. Army Signal Corps. He spent a great deal of time directing scientific organizations, and in 1921 became, in effect, president of the newly formed California Institute of Technology in Pasadena, California. He stayed active in research, doing pioneering work in the new field of cosmic ray physics. Eventually he became the best-known scientist of his era, following in the tradition of John Tyndall and Michael Faraday in England and prefiguring such American scientist-popularizers as Carl Sagan.

Mimicry

Evolution can lead to some organisms mimicking others, either to send out a common danger signal to potential predators or to produce a resemblance to a species that is to be avoided

According to the theory of EVOLUTION, organisms will tend to develop traits and characteristics that contribute to fitness—that is, to their ability to pass on their genes to the next generation. A few species have evolved a striking physical resemblance to other species, a phenomenon known as mimicry.

Many species develop a kind of early-warning system to warn off would-be attackers—think of the white stripes on a skunk, or the distinctive banding on stinging insects such as wasps and bees. The reason for the evolution of these signals is clear. An attacker that encounters a skunk or a wasp and suffers the consequences is likely to avoid the members of that species in future. Members of the species that carry the genes for the distinctive markings are therefore likely to survive longer and have more offspring—a classic case of natural selection at work. However, for the warning system to be effective, encounters between potential attackers and members of the species must be sufficiently frequent. It is these warning signs that other species have developed an ability to mimic.

The first kind of mimicry was described in 1852 by Henry Bates. The best example of so-called *Batesian mimicry* is seen in butterflies. The caterpillars of monarch butterflies store a toxic chemical known as glycoside as a byproduct of their metabolism, and this chemical is passed on to the adult butterfly. As a consequence, the butterfly has an unpleasant (and even poisonous) taste to birds, its main predators. After one encounter with a monarch butterfly, then, birds tend to leave them well

alone. A number of butterflies that do not taste bad and are not poisonous have taken advantage of the chemical warfare that the monarch wages against predators. The viceroy butterfly, for example, has the same orange and black markings as the monarch, and the two can easily be confused. As a result, the viceroy butterfly is also avoided by its predators, and is therefore "fitter" in the Darwinian sense.

The second kind of mimicry was first described in 1878 by the German naturalist Fritz Müller (1822–97). The best illustration of *Müllerian mimicry*, as it is now known, is provided by stinging insects such as wasps and bees. All

The viceroy butterfly (below) has the same coloration as the monarch butterfly (above). The viceroy's body lacks the toxin present in the monarch, and this mimicry helps to protect it from discerning predators.

of these have a distinctive pattern of bands on their bodies. The idea is that by increasing the pool of distinctly marked insects, it is more likely that potential attackers will go through the learning process outlined above. A bird that has been stung by one species of wasp, for example, is likely to avoid other species of wasp. In Mullerian mimicry, each mimicking species is capable of delivering the unpleasant experience—a sting in the case of wasps and bees.

With Batesian mimicry there is a delicate balance between the mimicker and the mimicked. If there are many viceroy butterflies in a region, for example, then a bird is likely to encounter the tasty viceroys before it sees a monarch. In this case, the value of the monarch's defenses (and of the viceroy's mimicking) is lost.* Consequently, Batesian mimicry works only if the number of mimickers is smaller than the number of mimicked. As often happens in the natural world, there can be too much of a good thing.

The key difference between these two kinds of mimicry, then, is this: The wasp can really sting, so Mullerian mimicry is based, in some sense, on reality. Batesian mimicry, though, is all bluff, and there is no real adverse consequence for predators that ignore the warning signs.

* In the case of the viceroy and the monarch, this balance was upset in January 2002 when a severe storm killed an estimated quarter of a billion monarchs in Mexico, where they had gathered for the winter.

HENRY WALTER BATES (1825–92) English naturalist and explorer. He was born in Leicester, the son of a factory owner. Bates lived an unusual life. Forced to end his formal education at age 13, he was apprenticed to a local hosiery manufacturer. He was an avid entomologist and published his first paper, on beetles, when he was 18. Later, he traveled around the Amazon Basin collecting insects for collectors in England, discovering 8000 new species. At the urging of Charles Darwin, Bates published several books on his travels and the insects he had studied. Eventually he became assistant secretary at the Royal Geographical Society in London and was recognized as a noted scientist by his peers.

Molecular Biology, central dogma of

One gene on the DNA molecule codes for one protein which runs one chemical reaction in the cell

One of the greatest discoveries of 19th-century biology, given ample confirmation in the 20th, was that life is based on chemistry. There is no VITAL FORCE, no intrinsic difference between materials in living systems and those in nonliving systems. In fact, the best analogy for a living system is to compare it to an industrial chemical plant in which an enormous number of chemical reactions are taking place. Materials arrive at the loading bays and finished products are shipped out. Somewhere in the factory office—perhaps stored in a computer as programs—are instructions on how to make the entire system work. In just the same way, the chemical business of cells (*see* CELL THEORY) is governed by instructions in the cell's nucleus, its "front office."

The last half of the 20th century saw an enormous elaboration of this insight. We now understand how the information for carrying out the chemical reactions in cells is transmitted from one generation to the next, and how it is used to keep a living cell running. The place where information is stored in cells is the DNA (deoxyribonucleic acid) molecule, which has the well-known double-helix or "twisted ladder" structure. The important working information is carried on the rungs of the ladder, each of which is made of two base molecules (*see* ACIDS AND BASES). Their full names are adenine, guanine, cytosine, and thymine, usually referred to simply as A, G, C, and T. If you read down one side of the DNA molecule, you will be following a sequence of bases. Think of this sequence as a message written in an alphabet that has only four letters. It is this message which determines how the cell runs its chemistry, and which therefore determines the characteristics of the living thing of which it is part.

The gene that Gregor Mendel discovered (*see* MENDEL'S LAWS) is actually just a sequence of base pairs along a DNA molecule. It appears that the human *genome,* the sum of all human DNA, contains about 30,000 to 50,000 genes (*see* HUMAN GENOME PROJECT). In advanced organisms such as human beings, genes are often interrupted by

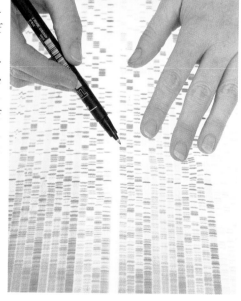

An autoradiogram—an image consisting of dark bands which indicate how far various segments of DNA have moved through a gel. The sequence of bands allows geneticists to reconstruct the code written in the DNA molecule.

bits of "junk" DNA, but in simple organisms the gene may be uninter-rupted. In any case, the cell knows how to read the information con-tained in the gene. In human beings and other advanced organisms, the DNA is wrapped around a molecular core to form a structure called a *chromosome*. It takes 46 chromosomes to hold all of the DNA for a human individual.

Just as information on a hard drive in a factory's office needs to be translated into widgets on the factory floor, information contained in DNA needs to be translated by cellular hardware into chemical processes in the body of a cell. The molecule most intimately connected with this chemical translation process is *ribonucleic acid,* RNA. If you imagine tak-ing a DNA ladder, cutting it in half, and then replacing each T with another similarly shaped molecule called uracil (represented by U), then you have RNA. When it is time for a gene to be translated, the molecules in the cell come along and "unzip" the section of DNA that contains the gene. Bits of RNA that are floating around the cell can then attach them-selves to the exposed ends of the DNA bases. As is the case for DNA itself, only certain linkages are allowed. Thus, if there is an exposed C in the DNA, only a G from an RNA building block will be able to bind to it. When the bits of RNA have lined themselves up, other enzymes come along and assemble the entire RNA molecule. The message in the bases of the RNA, then, bears the same relationship to the original DNA as a pho-tograph does to a negative. In this process, the information in the gene on the DNA has been transferred to the RNA.

This particular kind of RNA is called *messenger RNA,* or mRNA. Because it is much shorter than the entire DNA molecule in the particu-lar chromosome on which it is found, the mRNA molecule can move out of the cell nucleus through pores in the membrane and into the cell body itself. This is how mRNA carries the information from the nucleus (the "front office," in our analogy) out into the body of the cell.

In the body of the cell, two other kinds of RNA are found, each of which plays a crucial role in the final assembly of the protein for which the gene codes. One of these is called *ribosomal RNA,* or rRNA, and makes up part of a structure in the cell called the ribosome. Think of this as the analog to a workbench in a factory—it provides a structure on which the assembly can take place.

The other kind of RNA found in the body of a cell is called *transfer RNA,* or tRNA. This molecule has three base pairs at one side and a spot where an amino acid can be attached on the other side (*see* PROTEINS). The three bases at the end of the tRNA molecule will attach themselves to the RNA provided they match three bases in the RNA itself. (There are 64 tRNA molecules all told—four times four times four—and each of them can attach to one and only one triplet of exposed bases in mRNA.) Thus, the assembly process consists of a particular tRNA molecule attaching to the mRNA, dragging its specific amino acid with it. In the end, the entire mRNA will have tRNA molecules attached to it, and there will be a string of amino acids at the other end of the tRNA arranged in a certain order.

The order of amino acids, of course, is the primary structure of the protein. Other enzymes come along and clean up the assembly, but the end product is a protein whose primary structure is determined by the message written in the gene on the DNA. This protein folds up, assumes its final shape, and acts as an enzyme (*see* CATALYSTS AND ENZYMES) which governs one chemical reaction in the cell.

Although every organism in nature has a different message written in its DNA, all organisms use the same genetic code—the same connection between a triplet of DNA bases and an amino acid in the resulting protein. This similarity of living things is one of the strongest pieces of evidence for the theory of EVOLUTION, since it strongly suggests that human beings and all other organisms evolved from a single biochemical ancestor.

Molecular Clock

*The longer it has been
since two species have
shared a common ancestor,
the more differences there
will be in their DNA*

According to the central dogma of MOLECULAR BIOLOGY, the chemical identity of every living organism is determined by the sequence of base pairs in that organism's DNA. According to the theory of EVOLUTION, species develop over time, and this development is paralleled by changes in a species' DNA. There are many ways in which DNA can change. There can be, for example, a slow accumulation of mutations in individual bases, wholesale copying errors, and injections of strings of nucleic acid sequences by viruses. One thing is clear about changes in DNA, however—the more time that has elapsed since two species shared a common ancestor, the more time there has been for such changes to take place, and consequently the more differences there will be between the sequences of DNA in the two species.

There are several points to be made about this statement. For one thing, counting differences in DNA sequences allows us to construct a family tree of all living things. Humans and chimpanzees, for example, share 98% of their DNA, which implies that we shared a common ancestor fairly recently. In the same sense, we share much less of our DNA with frogs, which implies that our branch of the tree of life split off from the branch occupied by amphibians much further back in time. The theory of evolution predicts that the tree constructed in this way must match what has been put together over the past century from studies of the fossil record. The fact that the two trees match is, to my mind, one of the strongest pieces of evidence in favor of evolution. It also shows that the theory of evolution is *falsifiable* (one of the main requirements of any scientific theory, as discussed in the Introduction), since it *could* have turned out that humans are more closely related genetically to frogs than they are to chimpanzees.

The notion of the molecular clock takes the use of DNA evidence to a deeper level. If there is some average rate at which changes in DNA occur—if there is a steady "tick" of the molecular clock—then by counting the number of differences between base pairs in the sequences of DNA between two species we can estimate when their last common ancestor lived. If there is a steady rate of change in DNA, analysis of modern DNA can tell us about the timescales involved in the development of the tree of life.

When this idea was first advanced in the 1980s, there was hope that changes in DNA would be found to proceed at a common rate—that there would be a single tick of the clock. As it turned out, however, there are many different molecular clocks, each ticking at a different rate. For example, base pairs in an important gene cannot change much without crippling the whole organism, so the clock for base pairs in genes ticks relatively slowly. Most segments of DNA, however, do not affect the chemistry of the organism, so the clock associated with these segments can tick more rapidly.

Perhaps the aspect of the molecular clock that has attracted the most attention is its application to recent human evolution. To understand these events, you have to know that inside of every cell of an advanced organism are tiny organelles called *mitochondria*. This where the cell's fuel is burned—where the basic function of metabolism takes place. We believe that mitochondria first entered complex cells billions of years ago in an act of SYMBIOSIS. Two cells that had been evolving independently found that they could do much better in partnership if one lived inside the other. One piece of evidence that this happened long ago is that mitochondria contain their own small loop of DNA (DNA that in humans contains some 26 genes).

Sperm cells contain no mitochondria, so all of the mitochondrial DNA in your body comes from the ovum that was contributed by your mother. Mitochondrial DNA, in other words, is passed through the female line. It also appears that the molecular clock in mitochondrial DNA ticks about ten times faster than the clock in DNA from the cell's nucleus. This makes the analysis of mitochondrial DNA easier, since the number of changes associated with a specific span of time will be greater for it than for nuclear DNA.

The first use of mitochondrial DNA that caught the public's attention was a study in which samples from 147 individuals around the world were collected and sequenced. The first analysis seemed to indicate that all modern humans could trace their ancestry back to a single woman who lived in Africa about 200,000 years ago. This woman was promptly christened Eve (or, for the more scientifically inclined, Mitochondrial Eve), and was even featured on the cover of a major news magazine.

Unfortunately, this dramatic result didn't stand up to a more complete analysis, and scientists no longer talk about Eve (she succumbed to a critical analysis of the computer program that had been used to analyze the DNA data). Today, scientists believe that the DNA data show that all modern human beings are descended from a rather small breeding population—some 5000–10,000 individuals—who lived in Africa around 100,000 to 200,00 years ago.

Molecular Orbital Theory

The bonding of atoms in molecules depends on the overlap of the atomic wave functions

When atoms come together to make molecules, the CHEMICAL BONDS that are formed between them involve the outermost electrons of all the atoms. There are several ways of describing the bonding process theoretically. One is *valence bond theory,* according to which bonds form between atoms when pairs of electrons from overlapping orbitals are exchanged. Molecular orbital theory is another.

These sorts of approximate theories are useful in the sciences because they give us a simple, intuitive way of thinking about physical processes. With modern computers we can perform detailed calculations of the energies involved in the bonding process, but such calculations give us precious little insight into what's actually happening as atoms come together. The role of approximate theories is to give us precisely that sort of insight.

Molecular orbital theory begins by noting that electron orbitals in atoms can be described in terms of what are called their *wave functions* (*see* SCHRÖDINGER'S EQUATION). It then describes how the atomic orbitals are transformed into molecular orbitals when a chemical reaction takes place. Like the more familiar types of waves, the wave functions of electrons in orbitals undergo INTERFERENCE, and it turns out that the orbitals in molecules can, to a good approximation, be represented by the orbitals that result from the interference of the wave functions of the atoms.

For example, consider what happens when two of the innermost atomic orbitals in neighboring atoms come together. If the wave functions in the region of overlap interfere constructively, electrons are likely to appear in the region between the two atomic nuclei, pulling the two atoms together. If the interference in the region of overlap is destructive, on the other hand, there will be no electrons in the region of overlap, and there will be a net repulsive force between the atoms. Thus, the two atomic orbitals combine to form two molecular orbitals: one which tends to bond the atoms, and one which tends to push them apart. It is the interplay between the two that determines whether a stable molecule will be formed.

To see how this picture works, we can look at why hydrogen forms molecules containing two atoms each, but helium does not. With hydrogen, there are two electrons available—one from each atom—and space for two electrons in the lowest (binding) molecular orbital. So electrons are likely to be found between the nuclei, and molecular hydrogen will form. With helium, though, there are four electrons, so both the binding and the non-binding atomic orbitals will be occupied. Numerical calculations indicate that in this case the non-binding effects win out, and helium molecules, should they ever form, would be highly unstable. That is why helium gas is composed of single atoms.

Molecules of Life

The molecules of life are modular. Important classes of molecules are proteins, carbohydrates, lipids, and nucleic acids. There are many other molecules that serve as "energy currency" inside the cell

Life—mysterious, complex, enigmatic—is built around quite large molecules and quite simple chemical reactions. If you were a designer of large molecules, there are two ways you could go about it. One way, analogous to hand-crafting jewelry, would be to build each molecule from scratch to do a specific job. The other way, analogous to modern construction techniques, would be to make a series of simple modules from which a wide variety of larger molecules could be assembled by putting the modules together in different ways. It turns out that the molecules of life are modular in just this way. The theory of EVOLUTION would suggest that this is, indeed, the simplest way to get to large molecules, since very complex molecules don't have to be constructed at the beginning of an evolutionary process. Instead, new modules can be added over time to extend the range of large conglomerates, a process very much in keeping with the spirit of evolution.

Proteins

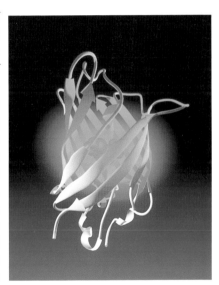

The basic building block of PROTEINS is a molecule called the *amino acid*. Think of an amino acid as a collection of atoms with a hydrogen sticking out of one side, an oxygen and a hydrogen pair sticking out of the other side, with various other components in between. These amino acids are assembled into proteins, rather like beads on a string, by having the hydrogen ion (H^+) from one amino acid combine with the hydroxyl ion (OH^-) from another to form a molecule of water (think of the two as squeezing out a drop of water between them). The most important role of proteins is as *enzymes* (*see* CATALYSTS AND ENZYMES) which govern the chemical reactions of cells, but they are also important components in the structure of living things. Your hair and fingernails, for example, are made from proteins.

A computer graphic showing the molecular structure of a protein. This one, known as green fluorescent protein (GFP), is used by medical researchers to trace the spread of malignant tumors.

Carbohydrates

Carbohydrates are combinations of oxygen, hydrogen, and carbon in the ratio of 1 : 2 : 1. These molecules serve as the energy source for many living systems. The most important carbohydrate may be the six-carbon sugar glucose ($C_6H_{12}O_6$), which is the end product of PHOTOSYNTHESIS and which therefore serves as the basis for the entire food chain in the biosphere. Combining glucose molecules together as basic modules can make complex carbohydrates. As with proteins, carbohydrates can also play a subsidiary role in cells by forming parts of structures. Plant fiber, for example, is made from cellulose, which is a string of glucose molecules hooked together in a specific way.

Lipids

Lipids are organic molecules which do not dissolve in water. Think of blobs of fat floating on top of gravy, and you have a good picture of lipids. In living things, lipids play two important roles. One class, the *phospholipids,* consist of a small head containing a phosphate group (i.e., a phosphorus atom connected to four oxygen atoms) and a long tail made of hydrocarbons. In this molecule, the hydrocarbon tail is *hydrophobic,* which is to say that the molecule attains its lowest energy state when this end is not in the presence of water. The phosphate head, on the other hand, is *hydrophilic*—the lowest energy state will be attained when it is in contact with water. If these molecules are put into solution, they will immediately seek their lowest energy state and align themselves in such a way that the tails of the molecules are together and the heads are out in the water. This type of *bilayer* is very stable because the heads will be in contact with water, while water will be excluded from the region around the tails. Moving any lipid molecules requires an input of energy, either to take the hydrophilic part away from water or to expose the hydrophobic part to it. In fact, cell membranes and the membranes that separate one part of the cell from another are all made from these lipid bilayers. They are flexible, tough, and do the job of separating life from non-life.

In addition, lipids play a role in the storage of energy. Roughly speaking, a given weight of lipids will store twice as much energy as the same weight of carbohydrates. This is why, when you overeat and your body wants to store energy as a contingency for the (now postponed) time when food will no longer be available, it will store that energy in the form of fat. On this simple fact the multibillion-dollar diet industry is built.

Nucleic Acids

DNA and RNA (*see* MOLECULAR BIOLOGY) are the molecules that carry information on how to run the chemistry of a cell and facilitate the transfer of that information from the DNA of a cell to the cell's body. The proteins that serve as enzymes for all the chemical reactions in an organism are coded for in that organism's DNA.

Energy Currency Molecules

The basic business of life requires energy. In particular, it requires processes by which energy can be created in one place and used in another. There is a veritable army of molecules in the cell to carry out this function. Probably the most important are two molecules called adenosine triphosphate (ATP) and adenosine diphosphate (ADP). Each consists of a group of carbon, hydrogen, and nitrogen atoms known as adenine bound to a molecule of ribose (a sugar), with the whole thing attached to a tail consisting of a string of phosphates. As the names imply, ADP has a two-phosphate tail, while ATP has a three-phosphate tail. The energy transfer takes place as follows: When a chemical process such as photosynthesis is going on inside of a cell, the energy generated is used to add a third phosphorus to the tail of ADP. The resulting ATP molecule will then move to other parts of the cell, and other chemical processes can gain access to this energy by removing the last phosphate of the ATP, converting it back to ADP.

There are, as suggested above, other molecules that can move energy around a cell. Think of this set of molecules as being analogous to the ways in which you can pay your bills. You can use cash, bank transfers, credit cards, and so on, whichever is most convenient for you. In the same way, the cell can use ATP (the equivalent of hard cash) or any of a large selection of other, more complex molecules to carry out its work.

Moore's Law

Every index of computer performance improves by a factor of two every two years

During the 1960s, when the information revolution was in its infancy, Gordon Moore, who later went on to be one of the founders of Intel, noticed something interesting about the development of computers. It seemed that the memory available for computers was doubling every two years or so. This realization became a kind of rule of thumb in the computer industry, and it quickly became obvious that it wasn't just memory that obeyed this rule, but almost any index of performance for a computer—the size of circuits, the speed of processing, and so on.

The development of computers has continued to follow Moore's "law." What is astonishing is that the intervening decades have witnessed several complete revolutions in technology. We have gone from computers with individual transistors to those with integrated circuits to those with microchips, yet in the background Moore's law just seems to keep proving its validity. No one in Silicon Valley in the 1960s, for example, could have foreseen today's manufacturing techniques with their ability to put millions of circuit elements on a chip of silicon the size of a postage stamp, yet when Moore's law demanded that kind of integration, it showed up. If anything, Moore's law seems to be speeding up—over the last few years the doubling time seems to have shrunk to something more like 18 months rather than 2 years.

Sooner or later, though, the laws of nature will put an end to the dominance of Moore's law. Take the size of circuit elements, for example. The law predicts that by the year 2060 they will have to be the size of a single atom—something not allowed by quantum mechanics!

GORDON EARLE MOORE (1929–) American computer engineer and businessman. He was born in San Francisco, and received his Ph.D., in physical chemistry, from the California Institute of Technology. Moore worked for a while under William Shockley (1910–89), co-inventor of the transistor, and began to research semiconductors. Shockley was beginning to show signs of the eccentricity that made him a controversial figure, and Moore and several colleagues departed. With one of them, Robert Noyce (1927–90), Moore founded the Intel Corporation in 1968 (where he still serves as Chairman Emeritus) and they set about designing and manufacturing complex integrated circuits—the "chips" that power modern personal computers. Moore's "law" comes from a comment he made in 1965 in an article for *Electronics* magazine about how the technology of integrated circuits would bring down the cost of computers.

Murphy's Law

If something can go wrong, it will go wrong

There are many versions of Murphy's law—the toast always falls buttered side down, the person in front of you in line at the store can't find the right change, and so on. I had always assumed that Murphy's law was just one of those folk sayings that express a kind of cynical but humorous view of the world, and that Murphy was just a fictional character. I was amazed to learn, therefore, that not only was Murphy a real person, but that he was a U.S. Air Force engineer who actually said something like the standard version of the law that carries his name.

"Murphy" was actually Captain Edward Aloysius Murphy (1917–), a graduate of West Point and former bomber pilot, who, in the mid-1940s, worked on some of the early experiments to test the response of the human body to extreme acceleration. In the experiments, carried out at Edwards Air Force Base in the California desert, a volunteer was strapped into a kind of sled, and rockets were used to accelerate the sled along a railroad track. The highest acceleration—or rather, deceleration—came at the end of the journey, when the sled was slowed down (violently) by a pool of water on the tracks.

Needless to say, this was a system in which lots of things can go wrong. As the designer of one of the instruments on the sled, Murphy is supposed to have gotten fed up with constant questions about why his systems weren't working properly. What he actually said—the first statement of Murphy's law—was, "If there is more than one way to do something and one of those ways won't work, somebody is going to come along and try it." I suppose it's a vindication of Murphy's law that this original phrasing has become, through repeated misquoting, the familiar statement given above. Incidentally, the problem with Murphy's instrument was that a technician had installed it backward—another beautiful example of the law in action.

Of course, Murphy's "law" is not a law in the sense in which that word is used elsewhere in this book. It has not, in other words, been put through the rigorous testing required by the scientific method. Nonetheless, it is a comforting bit of folk wisdom we can all use to help us get through those times when life just doesn't want to cooperate with us.

It also, however, illustrates something important about the engineer's outlook on life. Engineers know that the first (and second and third) time a complex system is tested, it won't work. It's not supposed to. The whole point of the test is to find out what bugs there are in the system so that they can be removed. There is a fundamental mismatch between the way engineers and the general public look at this issue. For example, the widespread ridicule to which the American space program was subjected in the 1960s, when rocket after rocket exploded on the launch pad, simply showed that the public didn't understand the purpose of the tests.

Once the bugs were eliminated, of course, the ridicule disappeared in the success of the Apollo program. The same phenomenon could be seen in the public discussion of the American missile defense system in the early years of the 21st century.

I suppose it is Murphy's law that leads designers to build large safety factors into their structures and machines, "just in case." Most buildings, for example, are designed to withstand at least 50% higher stresses than they are ever likely to encounter in real life—just because the architects know something will go wrong.

The fact of the matter is that engineers like to think about systems failing. I remember a seminar I attended at the University of Virginia in the early 1970s, before the launch of the first space shuttle. The speaker was the NASA engineer in charge of designing the shuttle engine, and he spent a full 90 minutes explaining in gory detail all the reasons why the engine would never work. I have never seen such a rapt audience—those guys were reveling in the contemplation of a system in which so many things could go wrong. I think that every engineer has to have this kind of Calvinist streak to succeed. The fact that they then go out and design systems that work flawlessly is, of course, beside the point.

Mutuality, principle of

There are many examples of species of plants and animals providing essential services to one another

We are used to thinking of plants and animals as coming in separate, distinct and independent species, and indeed they usually do. But there are also many instances of separate species that have evolved together so that each depends on the other for some benefit, or even survival, or in which each species is in some other way essential to the other species. The theory of EVOLUTION allows for this through the principle of COEVOLUTION.

Sometimes this mutualism manifests itself as a pattern of behavior. The clown fish, for example, lives near sea anemones. When threatened, the fish retreats into the anemone's tentacles, gaining shelter. In return, the clown fish attacks other fish that like to eat anemones. Both organisms thus derive mutual benefit from the arrangement. A variation of this sort of mutualism is when one species nourishes another—humans raising crops and cattle, for example, or ants raising crops of fungi.

A more intimate form of mutualism is when one organism lives inside another. Perhaps the most striking example is found in the digestive system of cows and other ruminants. These animals, like humans, cannot digest cellulose, the main structural molecule found in plants. Instead, these animals have a chamber known as a *rumen* in which a variety of microbes live. After vegetation is chewed up, it passes to the rumen where these microbes break down the cellulose. (The partially broken-down material can be brought up and chewed again—this is what a cow is doing when it chews its cud.) In a sense, the rumen of a cow is a complete micro-ecosystem, made up of many different organisms whose function is to digest cellulose for its host. In the same way, higher plants have roots that are made up of root tissue and fungi intertwined, the fungi supplying the plant with various minerals.

Mutualism also plays an important role in the NITROGEN CYCLE. Most nitrogen in the environment is in molecular form, N_2. In order to use nitrogen, however, plants need to have it *fixed*—converted into atomic nitrogen. The main mechanism for fixing nitrogen is through the action of certain species of bacteria which live in mutual relationships with plants. The best-known example of this is the root nodules found on the roots of legumes. The standard farming practice of planting fields with legumes and then plowing them under adds fixed nitrogen to the soil and encourages the growth of crops planted later.

Nebular Hypothesis

The solar system formed from a contracting cloud of gas and dust

This idea about the formation of the solar system, a field that is part of *cosmogony,* is one of the oldest surviving theories in astronomy. It was first put forward in a general way by the German philosopher Immanuel Kant (1724–1804); Pierre Simon de Laplace later made the first attempt at a scientific analysis using NEWTON'S LAW OF GRAVITATION.

The scenario starts with a cloud of gas and dust floating in space. By chance, some parts of the cloud will be denser than their surroundings, and hence will have a greater mass. The force of gravity then comes into play, causing surrounding matter to fall into these denser regions, which increase their mass further. The cloud will thus collapse around the points of highest density, and each of those points will eventually become a star. Astronomers see many such regions of star formation in the Galaxy.

In general, the gases in the cloud will be swirling around in all directions but, again just by chance, more will be swirling in one direction than in another. The cloud will have a net rotation, clockwise or counterclockwise. As the cloud condenses under the influence of gravity, the conservation of ANGULAR MOMENTUM causes its spin to increase. So when the cloud collapses, most of the mass becomes concentrated in a ball at the center (this ball becomes the star), and the rest of the mass is

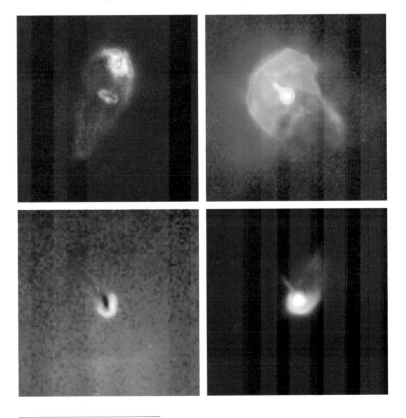

These photographs taken with the Hubble Space Telescope show planetary systems in the process of formation in the Orion Nebula. The nebula is a huge stellar nursery where stars and planets condense out of huge clouds of gas and dust. About 5 billion years ago, our own solar system would have looked like this.

spread out in a spinning disk in the protostar's equatorial plane. It is from this disk that planets form.

In the disk, grains of solid material collide and start to stick together. They form bodies called *planetesimals,* ranging in size from boulders to mountains, and these bodies come together to form planets. The types of planet depend on the distance from the newborn star. Close in, the temperatures are kept high by radiation from the star, in which nuclear reactions have now started (*see* STELLAR EVOLUTION). Materials with low melting and boiling points have no chance to solidify, and the resulting planets are small, rocky bodies—in our solar system, these are the so-called *terrestrial planets*—Mercury, Venus, Earth, and Mars.

This period in the history of the solar system, at least as predicted by some current computer models, was a strange one. While the planets were forming by the sort of accretion process described above, there may have been a dozen or more Mars-sized objects orbiting the Sun. These objects are thought to have engaged in a kind of highly protracted game of celestial billiards, some crashing into each other and spawning fragments that were eventually incorporated into the planets, others being slung out of the solar system by the dominating gravitational influence of Jupiter. There may still be bits and pieces of our solar system floating around in space, far from the star that gave them birth.

The Earth's Moon, which astronomers often class as a terrestrial planet, is believed to have formed later when one of those Mars-sized objects collided with the Earth (*see* GIANT IMPACT HYPOTHESIS). Such impacts between large objects must have been common in the final stages of formation of the planets. This explains another puzzling feature of the solar system. The rates at which planets spin on their axes (the length of their "day") varies widely. In the case of Venus, the rotation is *retrograde*—in the opposite sense to the spins of its planetary neighbors. It is hard to reconcile these differences with a smooth, orderly process. Once you realize that the rotation that a planet ends up with will have been strongly affected by the collisions between several protoplanets, it becomes easier to understand.

Farther from the young Sun, where temperatures are lower, different kinds of planets form. Here it is cool enough for ices to condense, and large, planet-sized objects made of rock and ice form. Their strong gravity allows them to capture hydrogen and helium from the solar nebula (something the small terrestrial planets with their weaker gravity could not do), and grow to great size. Although these *gas giants*—in our solar system, Jupiter, Saturn, Uranus, and Neptune—are large, they have a low density. Saturn, for example, has a density less than that of water, so it would float if you could find an ocean large enough to drop it into. We believe that all of the gas giants have small, rocky cores, assembled by the same process that produced the terrestrial planets.

The outermost planet, small and icy Pluto, was long regarded as an oddity. In the 1990s, astronomers discovered the first members of what is known as the Kuiper Belt, a huge reservoir of large comets orbiting beyond Neptune. Pluto is probably just the largest known member of the Kuiper Belt—one of perhaps several million remnants left over from the formation of the planets.

This picture of the formation of planets explains the main features of the solar system—the fact that the inner planets are small and rocky, the outer planets are large and predominantly gaseous, and that all the planets orbit the Sun in the same direction. In 1995, astronomers obtained the first good evidence of planets in orbit around other stars (in essence, they saw the star being pulled around by the gravitational attraction of the planets in their orbits). We now know of many more planets outside of our own system than are in it: as of this writing, a confirmed 83 planets in 71 solar systems (certain to be more by the time you read this). However, only one of these systems looks anything like ours. In all the others, the planets appear to have highly elliptical orbits, whereas in our own system the orbits are much more circular. And in many of them, the planets orbit closer to the star than Mercury does to the Sun—indeed, some of these planets revolve around their star in a matter of a few days.

Astronomers have also found many *circumstellar disks*—flattened clouds of matter circling young stars. So it appears that the nebular hypothesis can explain the presence of many planetary systems, even though few of the systems found so far resemble our own.

PIERRE SIMON, MARQUIS DE LAPLACE
(1749–1827) French mathematician and astronomer. He was born in Beaumont-en-Auge, the son of a peasant farmer, but his precocious ability in mathematics quickly gained him a place in the French academic world. He developed the mathematical physics of Isaac Newton, working out the consequences of NEWTON'S LAW OF GRAVITATION for the solar system. His monumental five-volume work *Traité de mécanique céleste* established the field known as celestial mechanics. Laplace rose to a position of some prominence under Napoleon, serving as his minister of the interior.

Nerve Signals, propagation of

Nerve signals are carried by the movement of ions across membranes in the nerve cell, and by neurotransmitters from one cell to the next

The nervous systems of humans and other animals have evolved into complex networks which depend for their operation on chemical reactions. The key member of the nervous system is a specialized cell called the *neuron*. Neurons consist of a compact cell body containing the nucleus and other organelles. From this cell body proceed a number of branch-like appendages. Most of these appendages, called *dendrites,* serve as points of contact for signals from other neurons. One appendage, generally the longest, is called the *axon* and serves to carry signals to other neurons. At its tip, the axon may split into many branches, each of which is capable of connecting to another neuron.

The outer layer of the axon is a complex structure, with many molecules that serve as channels through which ions can move, both inward and outward. In these molecules, one end is pushed down and clamps over its target atom. Energy from other parts of the cell is then used to push that atom to the outside of the cell, while the reverse process brings some other molecule into the cell. The most important of these molecular pumps are those that move sodium ions out of the cell and potassium ions into it.

When the cell is resting and no nerve signals are being transmitted, the sodium–potassium pump operates to move potassium ions to the inside of the cell and sodium ions outside—think of the cell as having fresh water inside and salt water outside. Because of this imbalance, the axon membrane has a voltage of 70 millivolts across it (about 5% of the voltage across a typical AA battery).

When the state of the cell changes, however, and an electrical signal triggers the axon, the membrane distorts and the sodium–potassium pump, for a short time, operates in reverse. Positive sodium ions pour into the interior of the axon while potassium ions are pumped out. For a moment, the inside of the axon becomes positively charged. This distorts the channels in the sodium–potassium pump in a different way, blocking further sodium inflow. Potassium ions then flow out, restoring the previous voltage. Meanwhile, the sodium ions inside the axon spread out, distorting the "down-axon" side of the membrane. This changes the downstream pumps and causes the signal to propagate further. The sudden change of voltage, caused by the sudden flows of sodium and potassium ions, is called the *action potential.* Once it has passed a given point on the axon, the pumps go back into action and restore the resting state.

The action potential actually moves quite slowly—no more than a fraction of an inch per second. To speed the signal up (because, after all, it wouldn't do for a signal from your brain to take a minute to reach your hand), axons are sheathed in a substance called myelin which blocks the flow of sodium and potassium. The myelin sheathing has gaps in it at

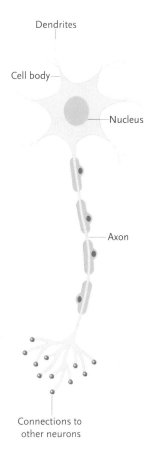

Dendrites

Cell body

Nucleus

Axon

Connections to
other neurons

*The structure of a neuron.
Neurons are the fundamental
components of the nervous
system. These elongated cells
transmit nerve impulses.*

specific intervals, and the nerve impulse jumps from one gap to the next, speeding up the propagation process.

When a signal reaches the end of the main body of the axon, it has to be transmitted, either to the next downstream neuron or, in the case of neurons in the brain, through the many branchings to many other neurons. This transmission uses a completely different process than the one which gets the signal down the axon. Each neuron is separated from its neighbor by a small gap known as a *synapse*. The action potential cannot jump this gap, so some way has to be found to get the signal to the next neuron. At the end of each branch of the axon are tiny sacs called *vesicles*, each holding a specific type of molecule known as a *neurotransmitter*. When the action potential arrives, these sacs are distorted and produce a spray of molecules that cross the synapse and attach themselves to specific molecular receptors in the membrane of the downstream neuron. This attachment distorts the membrane of that neuron. We'll discuss in a moment whether that distortion initiates another action potential (it remains one of the great unsolved questions of neurophysiology).

Once the neurotransmitters have done their job of transmitting the nerve impulse from one neuron to the next, the molecules can simply diffuse away, or they may be broken down chemically, or the neurotransmitters may simply be moved back into their vesicles (a process known by the awkward name of *re-uptake*). One of the astounding scientific findings of the late 20th century is that human mental states can be profoundly affected by drugs that affect the emission and re-uptake of neurotransmitters. Prozac® and similar antidepressants block the re-uptake of the neurotransmitter serotonin. Parkinson's disease seems to be related to a deficiency of the neurotransmitter dopamine in the brain. On the frontiers of psychiatric medicine, researchers are seeking a better understanding of how these chemicals affect the human psyche.

There remains the fundamental question of what causes a neuron to initiate an action potential or, in the jargon of neurophysiologists, what makes a neuron "fire." This is especially interesting when it comes to neurons in the brain. A neuron in the brain may receive neurotransmitters from up to a thousand of its neighbors. Just how all of these signals are processed and integrated is largely unknown, although many research groups are working on it. All we know is that there is a process in the neuron that does, in fact, integrate these signals and decide whether or not to initiate an action potential (fire) and pass the signal on. On this fundamental process rests the entire functioning of the brain. How fitting it is that this deepest mystery of nature should remain, for the moment at least, a mystery of science as well!

Newton's Law of Gravitation

There is an attractive force acting between all bodies in the universe

A painting of Sir Isaac Newton by an unknown artist. Arguably the greatest scientific mind the human race has ever produced, Newton was responsible for much of our current picture of the operation of the universe.

In his later years, Isaac Newton remembered it this way: He was walking in the apple orchard on his parents' farm when he happened to notice the Moon in the daytime sky. At the same time, an apple falling from a nearby tree caught his eye. From his work on motion (*see* NEWTON'S LAWS OF MOTION), he knew that it was the force of gravity that made the apple fall to the Earth. He also knew that, since the Moon was moving in an orbit around the Earth, there had to be a force acting on it as well—otherwise it would simply fly off into space. He wondered whether the same force that pulled the apple to the ground could be the one keeping the Moon in its orbit.

You have to know a little of the historical background to appreciate the brilliance of this insight. When scientists such as Galileo studied the ACCELERATED MOTION of bodies falling to the surface of the Earth, they believed that they were seeing a strictly terrestrial phenomenon—one that existed only near the surface of our planet. When other scientists such as Johannes Kepler (*see* KEPLER'S LAWS) studied the motions of objects in the heavens, they assumed that whatever it was that governed motion out there had to be different from what governed motion down here. For most of history, in fact, scientists had argued that heavenly objects must follow circular paths, since the heavens were perfect and the circle is the perfect geometrical figure. Thus, to use the modern term, there were two kinds of gravity in people's minds. There was terrestrial gravity, operating on the imperfect Earth, and celestial gravity, operating in the perfect heavens.

What Newton's insight did was to unify these two gravities. From this time forward, the false division between the Earth and the rest of the universe could no longer be maintained.

The result of Newton's calculation is what we now know as Newton's law of gravitation, or sometimes the *law of universal gravitation*. It states that there is attractive force, called gravity, between each and every pair of objects in the universe. Like all physical laws, it is encapsulated in a mathematical equation. If the masses of two objects are denoted by

m and M, and the distance between them is D, then the magnitude of the gravitational force is

$$F = GMm/D^2$$

In this equation, G is a number known as the gravitational constant. In SI units, it has a numerical value of about 6.67×10^{-11}.

There are several points to be made about this law: First, it explicitly applies to all bodies in the universe. Right now, for example, you are exerting a gravitational force on this book and the book is exerting an equal and opposite force on you. Of course, these forces are very small—much too small to be detected even with the most sensitive equipment—but they are real (and calculable) nonetheless. In the same way, you are exerting a force on the most distant quasar, more than 10 billion light years away, and it is exerting a force on you. Again, this force is much too small to be measured.

The next point is that the law applies to objects at the surface of the Earth (as well as any place else). Right now, the Earth is exerting a downward force on you, a force that can be calculated from the above equation. We call this force *weight*. If you were to drop something, it is this same force that would cause it to accelerate downward. Galileo showed how to measure the rate of ACCELERATED MOTION of any object dropped at the Earth's surface. This acceleration was given a special letter, g.

For Galileo, g was just a number to be measured—it could have had any value. For Newton, however, g could be calculated directly from the equation by inserting the mass of the Earth for M and the radius of the Earth for D, and remembering that according to the second of NEWTON'S LAWS OF MOTION, the force on an object is equal to its mass times the acceleration. Thus, what was a number that had to be measured for Galileo had become for Newton a number that can be calculated or predicted.

Finally, the law explains the working of the solar system, and KEPLER'S LAWS of planetary motion can be derived from it. For Kepler, his laws were simply summaries of data on the motion of planets, with no explanation about why they should be what they are—no theoretical underpinning. In the grand Newtonian scheme, however, they become consequences of the laws of motion and the law of universal gravitation. Again, what is arbitrary at one level becomes a logical consequence of laws stated at a deeper level.

The picture of the solar system that emerges from this law, the picture that unifies earthly and heavenly gravity, can be understood from a simple example. Suppose you were standing at the top of a sheer cliff with a pile of cannonballs and a large cannon. If you just dropped a cannonball over the cliff edge it would fall straight down, accelerating as it went. Its fall would be described by Newton's laws of motion as applied to an object being accelerated by gravity. If you now used the cannon to shoot a second cannonball outward, it would fall in an arc. Again, its trajectory would be described by Newton's laws, this time as they are applied to a body falling under the influence of gravity that has also been propelled horizontally. You could keep shooting cannonballs outward, increasing the charge each time, and each time the cannonball would fall farther from the base of the cliff.

Now imagine that you put enough powder in your cannon to send the cannonball all the way around the world. If we ignore air resistance, when the cannonball had circled the Earth once it would come back to its original position moving at exactly the same speed as when it originally left the cannon. It's obvious what would happen next—the cannonball would make another circuit, then another, then another, and so on. It would, in other words, behave exactly like a satellite, or the Moon. In this progression we have gone from a falling body (Newton's apple) totally under the influence of "terrestrial" gravity to a satellite in orbit (the Moon) without making any transition from one kind of gravity to another. This was the insight that allowed Newton to unify the two gravities.

One last question remains: Did events really happen as Newton said they had? There is no contemporary record of his work on gravity, but documents can get lost. On the other hand, we know that Newton was particularly un-pleasant and relentless in establishing his scientific priority, and it would have been completely in character for him to shade the truth if he felt his priority was

Although the famous apple tree at Newton's childhood home in Woolsthorpe, Lincolnshire, no longer exists, grafting has produced several generations of descendants. This one stands in the grounds of Babson College in Wellesley, Massachusetts.

threatened. By pushing the discovery back to 1666, rather than have it recorded as 1687, when it was finally published, he gave himself two decades' advantage on the priority front.

I have to admit that, although some historians wax apoplectic on this issue, I can't get excited. Newton's apple makes a great story and a wonderful metaphor for the mysterious process of human creativity. Whether it is also historically true seems to me to be a secondary issue.

Newton's Laws of Motion

An object moving in a straight line will continue to do so unless acted upon by an external force

The acceleration of the object is then proportional to the applied force and inversely proportional to the object's mass

For every action there is an equal and opposite reaction

Depending on how you look at them, Newton's laws are either the end of the beginning or the beginning of the end. They mark a crucial turning point in the history of science, a brilliant compilation of everything you need to know about the motion of physical objects—what has traditionally been called the science of mechanics. Together with NEWTON'S LAW OF GRAVITATION, the laws of motion opened up an entirely new vista on how the solar system operates. They can be thought of as the beginning of modern science.

But Isaac Newton did not pull his laws out of thin air. In fact, they represent the culmination of a long historical process. For centuries, philosophers and mathematicians such as Galileo (*see* ACCELERATED MOTION) had tried to work out the rules that governed the motion of material objects, but they were always hampered by what I like to call unspoken assumptions—ideas about the way the world is that are so powerful and apparently so obvious that they were never questioned. For example, philosophers felt that celestial objects had to move on circular orbits, or at best on circles rolling within circles. Why? Because, according to a notion going back to the ancient Greeks, planetary orbits had to conform to heavenly perfection, and the only perfect geometrical shape was the circle. It took the genius of Johannes Kepler to look at the data honestly and show that the orbits are, in fact, ellipses (*see* KEPLER'S LAWS).

The first law

Against this historical backdrop, Newton's first law of motion can be seen as a truly revolutionary statement. It says that if you leave something alone, it will just continue doing whatever it's doing. If it is moving in a line, it will continue to move in a line. If it's standing still, it will continue to stand still. To get it to do anything else, you have to intervene by applying a force. An airplane will move down the runway, for example, only if its engines are on. That may seem pretty obvious, but things become a little more subtle when we consider motion in a circle, for which Newton's analysis becomes extremely important.

Imagine taking a ball attached to a length of string and twirling it around your head. As long as you keep twirling, the ball moves in a circle. It doesn't move in a straight line, so according to Newton's first law you must be applying a force. The ball wants to move off in a straight line, according to the law, and you have to keep pulling it back into the circular track. You can actually feel the force you are applying in your hand. If you let go of the string, the force will no longer be acting, and the first law says that the ball should then move off in a straight line. This, of course, is exactly what it does (or at least it is what it would do in space—on the

Earth, it flies off, but is then pulled toward the ground by the Earth's gravity). If you were to try this out, you would find that the ball continues with its instantaneous velocity at the moment that you let go, and goes off at a tangent to the circular motion, perpendicular to the radial direction.

Substitute a planet for the ball and the force of gravity for the string, and you have the Newtonian model of the solar system.

This analysis of circular motion sounds simple, but you have to remember that it had eluded some of the best minds in the history of scientific thought (Galileo, to name one). The problem is that when something is moving around a circle at a constant speed, it looks as if it is in a kind of equilibrium. In point of fact, although the speed of the object may not change, the direction of its motion does. This constitutes an acceleration, and therefore qualifies as the type of "change of motion" Newton was talking about.

The first law has another role to play in our investigation of the natural world. It tells us that if we see a change in the motion of an object, there must be a force causing that change. So when we see iron filings jump up and cling to a magnet, or pieces of clothing sticking together when they come out of a dryer, we can be sure that we are seeing the action of forces in nature. (The forces in these examples are magnetism and static electricity, respectively.)

The second law

If the first law tells us how to recognize when a force is acting, the second law tells us what happens when it does act. It tells us that the bigger the force, the bigger the acceleration. Furthermore, for a given force the acceleration will be less for a massive object than for a less massive one. Both of these statements seem intuitively right.

In equation form, Newton's second law is written as

$$F = ma$$

where F is the force, m is the mass of the object, and a is the acceleration. This equation may well be one of the most useful (and most used) equations in science, for if we know the forces acting on a system and have a knowledge of the objects that comprise it, then we can calculate how that system will change over time.

It is the second law that gives Newtonian physics its mechanistic, clock-like flavor. Tell me the positions and velocities of everything in the system, Newton is saying, tell me the forces acting, and I will tell you every future state of that system. It was only with the advent of QUANTUM MECHANICS that scientists had to modify their view of how the universe works.

The operation of a rocket illustrates all three of Newton's laws of motion. Once launched, the rocket will keep going (first law), it will keep accelerating as long as it burns its fuel (second law), and its forward motion is a reaction to the gases blasting out in the opposite direction (third law).

The third law

This is probably the one aspect of Newton's work that has gained the most recognition outside of the sciences. It is often borrowed for use outside of its scientific context because of its useful parallels in real life, and crops up in discussions of topics as diverse as interpersonal relationships and global politics. For Newton, however, the third law is a precise statement about the nature of physical forces. It says that if body A exerts a force on body B, then body B will exert an equal and opposite force on body A. For example, when you are standing up you are exerting a force equal to your weight downward on the floor or the ground. The third law then says that the floor or ground exerts an upward force equal to your weight on you. You can actually feel this force on the soles of your feet—a continual experimental verification of this particular law of nature.

It is important to recognize that the two forces Newton talks about in this law act on different objects. When an apple falls from a tree, the Earth is exerting a force on the apple (which is why it accelerates downward), but the apple is also exerting an upward force on the Earth. The reason the apple moves perceptibly and the Earth doesn't has to do with the second law. The apple has a small mass, and the gravitational force exerted by the Earth causes it to accelerate perceptibly. The Earth, on the other hand, has a much larger mass, so, according to the second law, its acceleration is correspondingly much smaller. (In fact, while the apple falls from the tree, the Earth moves upward much less than the distance across the nucleus of an atom.)

Taken together, then, Newton's three laws of motion provided scientists with the tools they needed to begin their comprehensive investigation of the universe. And despite all of the advances since Newton's time, if you want to design a car or send a spaceship to Jupiter, you will still use these three laws.

ISAAC NEWTON (1642–1727) English scientist, who many consider to be the greatest scientist who ever lived. He was born to a family of small landowners near the town of Woolsthorpe in Lincolnshire. His father died three months before he was born, and when his mother remarried two years later he was left in the care of his grandmother. Many scholars ascribe Newton's eccentric behavior later in life to the fact that he was deprived of his mother's company until the death of his stepfather nine years later.

For a brief period, young Isaac was trained in the business of farming. As often happens with people who achieve greatness in later life, legends about his early idiosyncrasies abound. According to one story, when he was sent out to watch cattle one day he sat under a tree with a book while his charges wandered off. In any case, it was soon recognized that he was someone who belonged in school, so he was sent back to grammar school in Grantham. In 1661 he entered Trinity College, Cambridge.

As an undergraduate, he quickly mastered the standard curriculum and moved on to reading the leading scientists of the time, particularly the followers of French philosopher René Descartes (1596–1650), who advocated a mechanical view of the universe. He received his bachelor's degree in the spring of 1665, and what happened next is one of the most incredible sequences of events in the history of science. That year marked the last visit of the bubonic plague to England, and amid the mounting death toll, Cambridge University was closed. Newton returned to Woolsthorpe for almost two years, with only his new books and his own intellect for company.

By the time the university had reopened in 1667, Newton had (1) developed the branch of mathematics known as calculus, (2) developed the basis for the modern theory of color, (3) derived his law of universal gravitation (*see* NEWTON'S LAW OF GRAVITATION), and (4) solved some mathematical problems that no one had been able to solve up to that time. In his own words, "In those days I was in the prime of my age for invention, and minded Mathematics and Philosophy more than at any time since." (As I often ask my students when I recount these accomplishments, "What did *you* do last summer vacation?").

Shortly after his return to Cambridge, Newton was elected to a fellowship at Trinity, where his statue still graces the chapel today. He gave a series of lectures on the theory of color, arguing that colors arise from the basic characteristics of the light wave (what we now call the wavelength), and that light was corpuscular in nature. He also invented the reflecting telescope, a feat which brought him to the attention of the Royal Society. Eventually, his work on light and color led to the publication of his *Opticks* in 1704.

Newton's championship of the "wrong" theory of light brought him into conflict with Robert Hooke (*see* HOOKE'S LAW), a leader of the Royal Society. This conflict continued until Hooke's death, and was marked by charges of plagiarism and disputes over priorities of discovery. So stressful was this to Newton that he withdrew from intellectual life for a period of six years, an act that some modern psychologists suggest may be attributed to a nervous breakdown exacerbated by the death of his mother. During this period he dabbled in alchemy, numerology, and the writing of religious tracts.

In 1679 Newton returned to the work that was to make him famous, working on the orbital motions of planets and satellites. This work, again accompanied by controversies over priority with Hooke, led to the enunciation of the law of universal gravitation as well as to what we now call NEWTON'S LAWS OF MOTION. This work was summarized in the book *Philosophiae naturalis principia mathematica*, presented to the Royal Society in 1686 and published the year later. This work, which is the text for the modern scientific revolution, brought Newton international recognition.

His religious interests, and his staunch Protestantism, also brought the newly famous Newton to the attention of a wider circle of English intellectuals, most notably the philosopher John Locke. Spending more and more time in London, Newton entered the stimulating political life of the capital, and was appointed Master of the Mint in 1696. Although this position was traditionally a sinecure, Newton took it quite seriously, seeing through a change of coinage and instituting rigorous measures to stop counterfeiting. It was during this period that he became embroiled in another famous priority dispute, this time with the German philosopher Gottfreid Leibniz (1646–1716) over the invention of calculus. In his later years he brought out new editions of his major works and served as president of the Royal Society, as well as maintaining his position at the mint.

Nitrogen Cycle

Nitrogen is continually cycled through the Earth's biosphere by a variety of chemical and other processes, and human beings now add more fixed nitrogen to the biosphere than does nature itself

Nitrogen is one of the commonest elements in the *biosphere,* the narrow shell around the Earth that supports life. Right now, for example, almost 80% of the air you're breathing is made up of this element. Most of the nitrogen in the air is in a very tightly bonded form (*see* CHEMICAL BONDS), in which two nitrogen atoms are locked together as a nitrogen molecule, N_2. Because the bond between the two nitrogen atoms is so strong, living things cannot make use of nitrogen in its molecular form—first, the nitrogen has to be "fixed." *Fixation* is a process in which nitrogen molecules are split apart, leaving the individual atoms free to combine chemically with something else, such as oxygen, thus preventing the nitrogen atoms from recombining with one another. The bonds between nitrogen and other types of atom are sufficiently weak that the nitrogen can be incorporated into living things. The fixation of nitrogen is therefore an extremely important part of the processes of life on our planet.

The nitrogen cycle is an interlocking series of circular pathways by which nitrogen is circulated through the biosphere. We can start by thinking about organic material decomposing in the soil. Various microorganisms use the nitrogen in decomposing material to make molecules necessary for their own metabolism, and release the rest as ammonia (NH_3) or the ammonium ion (NH_4^+). Other microorganisms then fix this nitrogen, usually in the form of nitrates (NO_3^-). Taken into plants (and ultimately into all living things), this nitrogen is used to form the MOLECULES OF LIFE. When the organism dies, the nitrogen is returned to the soil and the process starts again. Nitrogen is lost through the cycle when it is incorporated into sediments or released by certain kinds of

Nitrogen is continuously cycled through the Earth's biosphere around a network of inter-linking pathways. To the natural processes is added the extraction of nitrogen for artificial fertilizers.

bacteria, and added through volcanic eruptions and other forms of geological activity.

One way of thinking about the biosphere is to contrast the huge reservoir of nitrogen in the atmosphere and oceans with the smaller reservoir in living things. Between these two reservoirs is a bottleneck in which nitrogen is fixed by various means. Under normal circumstances, nitrogen from the environment is fed through this bottleneck into living systems, and then returns to the environment when the living systems die.

Now for some numbers. There are about 4 million billion tons of nitrogen in the atmosphere, and about 20,000 billion tons in the Earth's oceans. A very small fraction of this—about 100 billion tons—is fixed and incorporated into living things each year. Of the 100 billion tons of fixed nitrogen, only about 4 billion tons is found in living plants and animals—the rest is stored in decomposing organisms and will eventually be returned to the atmosphere.

In nature, the most important method of fixing nitrogen involves the operation of bacteria. The numbers aren't as well known as we would like them to be, but bacteria fix somewhere between 90 and 140 million tons of nitrogen. The nitrogen-fixing bacteria that are most familiar to us live in nodules on the roots of legumes. This explains the custom in agriculture of planting peas or other legume crops and then plowing them under—the idea is that the fixed nitrogen is then available for use by future crops grown in that field.

A smaller fraction of nitrogen is fixed by lightning. You may be surprised to learn that lightning flashes are quite common—about a hundred strikes a second worldwide, or about five hundred strikes since you began reading this paragraph. Lightning heats the atmosphere around it, in effect "burning" the nitrogen by combining it with oxygen, creating various nitrogen oxides. Although this is a rather spectacular form of fixing, it converts only 10 million tons of nitrogen per year.

Thus, all told, somewhere between 100 million and 150 million tons of nitrogen are fixed by natural processes each year. Human activities also fix nitrogen and deliver it to the biosphere (the planting of legumes is just one such activity, accounting for about 40 million tons of nitrogen fixed each year). Furthermore, when fossil fuel is burned, either in an electrical generator or in an internal-combustion engine, it heats the air around it, just as lightning does. Every time you drive your car you are adding fixed nitrogen to the biosphere. Roughly 20 million tons of nitrogen a year are fixed by the burning of fossil fuels.

By far the greatest addition of fixed nitrogen from human activity, however, comes from the use of fertilizers. As with many technological advances, the ability to fix nitrogen on a large scale was driven by military needs. In pre-World War I Germany, a way of producing ammonia (a form of fixed nitrogen) was developed for use in munitions production. The supply of nitrogen is often a limiting factor in plant growth, and farmers benefited from the availability of commercially fixed nitrogen in the form of artificial fertilizers. Each year, we now create a little over 80 million tons of fixed nitrogen for agricultural use (it should be pointed out that not all of this goes to growing food crops—suburban lawns and gardens are fertilized by the same means).

Adding up the human contributions to the nitrogen cycle, we come up with about 140 million tons a year—a number that is in the same ballpark as the amount fixed naturally. Thus, in a relatively short period of time, humans have come to dominate this particular natural cycle. What will the consequences be? Every ecosystem can absorb a certain amount of nitrogen, and its effect is relatively benign—it simply makes the plants grow faster. Once an ecosystem is saturated, however, the nitrogen is washed out and carried elsewhere. Probably the worst ecological problem associated with nitrogen is the *eutrophication* of lakes. Nitrogen fertilizes algae in the lake, and when the algae die their decomposition uses up most of the available oxygen in the water, thereby crowding out other life forms.

Nevertheless, we have to understand that, in the grand scale of things, modifications to the nitrogen cycle are not the worst problem facing humanity. In the words of Stanford plant ecologist Peter Vitousek, "We are heading for a green and weedy world, but it's not a disaster. It's very important to be able to distinguish between disaster and degradation."

Nuclear Fusion and Fission

It is possible to get energy from some atomic nuclei, either by splitting large nuclei into smaller pieces (fission), or by making small nuclei join together to make a larger one (fusion)

According to the theory of RELATIVITY, mass is a form of energy, as indicated by Einstein's equation $E = mc^2$. This raises the possibility of converting mass into energy or energy into mass. Such conversions proceed via reactions which take place inside the atomic nucleus and between elementary particles. In particular, some of the mass of the atomic nucleus can be converted into energy in two ways: Large nuclei can be split into smaller ones, a process called *fission,* or small nuclei can be combined into larger ones, a process called *fusion.* Fusion reactions are quite common in the universe—they are the means by which most stars generate their energy. Fission reactions are one of the means by which we humans generate electricity. In both fission and fusion reactions, the mass of the products is less than the mass of the reactants. The difference in mass is converted into energy according to $E = mc^2$.

Fission

There are a few naturally occurring isotopes (uranium-235 is the commonest) for which energy is released when their nuclei are split into smaller pieces. For example, if a nucleus of uranium-235 is hit by a neutron which is travelling fast enough, the nucleus will split into two largish segments and a spray of other particles. This spray of particles will typically include two or three neutrons. If you add the masses of the large segments and the particles together, however, you find that the sum is less than the sum of the masses of the original uranium nucleus and the bombarding neutron. The difference in mass appears as energy shared among the final products, primarily as kinetic energy. The fast-moving particles move away from the site of the fission and collide with other particles in the material, heating them up. Thus, the energy generated in nuclear fission ultimately appears as heat in the surrounding materials.

Natural uranium, as it is dug out of the ground, contains about 99.3% uranium-238 and 0.7% uranium-235. But only the rarer isotope undergoes fission when hit by a neutron *and* produces more neutrons. The uranium-238 will absorb the neutrons in non-fission reactions. So, to produce a working nuclear reactor, the proportion of uranium-235 has to be increased from its natural value to around 5%. This process is called *enrichment.*

In a nuclear reactor, the uranium-235 isotope in enriched natural uranium undergoes fission as a result of being hit by a neutron. On average 2.5 additional neutrons are produced during each such fission reaction. These neutrons in the fission products go on to strike other uranium-235 nuclei and cause further fissions. This is the so-called *chain reaction.* As long as more neutrons are produced during each fission than are lost via other processes, the reaction will continue and energy will be generated.

In a nuclear bomb, the chain reaction is allowed to "run away" so that vast amounts of energy are generated in a tiny fraction of a second, resulting in a colossal explosion. In nuclear power generators, the energy has to be released in a more manageable fashion. The reaction is controlled by introducing material that absorbs some of the neutrons and so reduces the rate of fission reactions. Cadmium absorbs neutrons strongly and is therefore often used to control nuclear reactors. It is made into rods that may be lowered into the reactor to absorb more neutrons and reduce the energy generation, or pulled out so that fewer neutrons are absorbed and more energy generated. That energy is then used to heat water and to drive relatively conventional turbines to generate electricity.

Fusion

Fusion is the opposite of fission—it is the coming together of small nuclei to make larger ones. The commonest kind of fusion reaction, which occurs in stars, starts with hydrogen nuclei. The net effect of the reaction is to combine four hydrogen nuclei (protons) to form one helium nucleus (two protons and two neutrons) plus a few other particles. As in fission, the masses of the particles produced add up to less than the masses of the particles we started with. This difference in mass appears as kinetic energy in the final particles and is the ultimate source of heat and energy in a star.

In a star, the fusion reaction does not take place all at once, by having four nuclei collide, but takes place in three steps. In the first step, two protons come together to form a nucleus of deuterium (one proton and one neutron) plus some other particles. In the second step, another proton hits a deuterium nucleus and forms helium-3 (two protons and one neutron) plus some other particles. Finally, two helium-3 nuclei come together to form helium-4, two protons, and some other particles as well. The net effect of the entire chain of reactions is that four protons are converted to helium. (*See also* STELLAR EVOLUTION.)

Although fusion reactions take place in stars, and uncontrolled fusion reactions occur in hydrogen bombs, we have not yet been able to control fusion reactions and produce usable energy (*see* LAWSON CRITERION). However, sometime in the future it is possible that "fusion power" will be available on a commercially viable scale, fusion taking place either within tiny spheres of deuterium and tritium bombarded by intense laser beams, or within extremely hot plasmas contained within magnetic "bottles."

Null Hypothesis

In any statistical analysis, the possibility that there is no result must be considered

The reason we analyze data in any scientific experiment is to be able to choose between *hypotheses*. If, for example, you believe that nature should behave in a certain way in a given situation, and you perform an experiment to see whether you are right, then you want to be able to claim that the data support your hypothesis and not someone else's. In other words, we expect the data to show us that the results of the experiment depend in one way on the variables and not in another way. In most situations it is not possible for there to be a single "clean" experiment, so we must repeat measurements many times to insure that the result is reliable. This often requires us to make statistical analyses of the data we obtain. Also, it often happens that there are many factors involved in coming to a conclusion, and it is important to separate out what is important and what is not—the wheat from the chaff, as it were.

For example, when scientists started to look for a relationship between smoking and lung cancer, it wasn't enough to find one smoker who contracted (or failed to contract) the disease. Massive amounts of data on people who smoked and didn't smoke had to be gathered and analyzed to show that there *is* a relationship between smoking and lung cancer. In this sort of analysis, the null hypothesis plays a crucial role. The null hypothesis is basically the proposition that the effect being searched for doesn't exist. As far as the sought-for correlation between smoking and lung cancer goes, the null hypothesis would say that no such correlation exists. The question then becomes one of deciding whether the data are good enough to rule this hypothesis out.

With smoking and lung cancer, the null hypothesis was ruled out long ago—no reputable scientist would invoke it now. But there was a time when there simply weren't enough data to rule it out, and researchers could not prove that the difference in the incidence of lung cancer between smokers and non-smokers wasn't simply a matter of chance. Only by collecting a large body of data, thereby reducing the probability that the result is down to chance, can the null hypothesis be ruled out.

In this example, we accumulate a lot of data—what scientists would call assembling a large *sample*—to rule out the null hypothesis. In others, such as the work of Tycho Brahe that led to KEPLER'S LAWS of planetary motion, the same end was reached by making more accurate measurements in all cases—though the data have to be good enough to show that the effect you are claiming can be separated from the null hypothesis.

So the next time you read a study that purports to establish a correlation between a disease and a cause, ask yourself whether the researchers have really looked at enough cases to rule out the null hypothesis.

Occam's Razor

The simplest explanation is likely to be the right one

William of Occam was one of the most prominent philosophers of the 14th century, but today he is known for one thing—an offhand remark he made in one of his books that has acquired the name Occam's razor because it provides a way of "shaving away" complexity in arguments.* What Occam actually said was, "*Non sunt entia multiplicanda oracter necessitatem,*" which translates roughly as, "Hypotheses should not be multiplied without reason." It is a warning against advancing convoluted arguments when simple ones will do.

For example, suppose that someone sees a bright and unexpected light in the night sky, a UFO. One explanation is that this person has seen the lights of a spacecraft piloted by extraterrestrials. This explanation requires many of Occam's superfluous "hypotheses"—the existence of extraterrestrials, their ability to built interstellar ships, their interest in Earth, their inability to avoid detection (despite their advanced technology), and so on. But there are many other, simpler explanations for lights in the sky—airplanes, the planet Venus (the number one explanation for "UFOs"), the (in)famous weather balloons, and so on. Each of these explanations requires a relatively small number of hypotheses. Thus, even though no one will ever be able to prove that the light wasn't from an extraterrestrial spaceship, most people will (either consciously or unconsciously) apply Occam's razor and reject that possibility.

I have to say that, although scientists often talk about Occam's razor and even use it when dealing with pseudoscientific topics such as UFOs, I don't recall ever having heard it invoked in serious scientific discussions. The reason, I think, has to do with the fact that scientists feel uncomfortable advancing philosophical arguments, and will hardly ever do so if solid experimental evidence is available. Forced to make a choice between theory A and theory B, in other words, scientists will turn to experiment and observation as the ultimate arbiters, rather than to philosophical principles such as Occam's razor. In this, the razor is like the BEAUTY CRITERION—scientists feel comfortable with it, and even think it is right, but they seldom use it.

* Interestingly, the multi-page entry for Occam in the *Dictionary of Scientific Biography*, normally the last word on such matters, makes no mention of his razor.

WILLIAM OF OCCAM (1285–1349) English theologian and philosopher. He was born in Occam (also spelt "Ockham"), a hamlet in Surrey. He became a Franciscan monk and studied theology at Oxford. In 1324 he was charged with heresy and became caught up in a dispute between the Franciscan order and the papacy. William ended up in Bavaria, writing tracts on conflicts between civil and papal authority. He introduced his "razor" originally to simplify theological arguments.

Octet Rule

Atoms tend to lose or gain electrons until their outer shell holds eight electrons

The arrangement of the chemical elements in the PERIODIC TABLE is explained by the way in which electrons fill the available energy levels, or *shells,* in atoms. In particular, the *noble gases,* such as neon, xenon, and argon, have eight electrons in their outermost shell, which is why they do not easily engage in chemical reactions. In energy terms, a filled outermost electron shell is the lowest state that most atoms can assume. This is the basis of the octet rule.

This rule explains how atoms form ions. Take sodium as an example. It has 11 electrons: 2 in the innermost shell, 8 in the next, and 1 in the outermost shell. The outermost electron is easy to remove, so that when energy is added to the sodium atom (perhaps by collisions with other atoms), a sodium ion with one positive charge is easily formed. It takes ten times as much energy to remove an electron from an inner shell, however, so a sodium ion with a double positive charge is rarely seen. In the same way, calcium, an atom with 2 electrons in its outermost shell and 8 in the next lower one, forms an ion by losing 2 electrons. Thus, when atoms become ions, they tend to resemble noble gas atoms.

The octet rule is a useful guide when you are beginning to understand CHEMICAL BONDS, but it is no more than that, and there are many elements to which it does not apply. Tin, for example, has 14 electrons outside of its closed shell, but it loses only 2 or 4 electrons before the cost of removing more electrons becomes prohibitive, and therefore forms ions with a positive charge of 2 or 4.

The octet rule is one of those rules that were originally discovered as seemingly arbitrary regularities by chemists through experiment and observation, but which are easy to understand in terms of the ATOMIC THEORY.

Oersted Discovery

An electrical current can give rise to a magnetic field

Electricity and magnetism, though they appear to be totally different, are actually closely related to each other. The credit for melding these two phenomena goes to James Clerk Maxwell, whose work on the electromagnetic theory of light began in the 1850s and lasted until his premature death in 1879. But there was a chain of discoveries that led to MAXWELL'S EQUATIONS stretching back through the 19th century that was started by a Danish physicist named Hans Christian Oersted.

Oersted had two qualities that are usually considered to be severe impediments to a research career: a passionate interest in philosophy and a strong desire to explain science to the general public. In his early days as an itinerant scholar in Paris, for example, he severely dented his scientific credibility by his intemperate defense of some obscure German philosophers. His arguments that there ought to be a connection between electricity and magnetism also had, to the modern ear at least, an uncomfortable ring of the mystical to them. He argued, for example, that inevitable conflict between the positive and negative aspects of electricity would lead to magnetism.

For whatever reason, in 1820 Oersted set up a lecture demonstration at the University of Copenhagen, using the newly invented battery to supply a current. He showed that, by turning on an electrical current near a compass, he could deflect the compass needle. This was the first unambiguous demonstration of a fundamental connection between electricity and magnetism. It served as the inspiration for other scientists, most notably Ampère (*see* AMPÈRE'S LAW) and Biot and Savart (*see* BIOT–SAVART LAW), to put some mathematical flesh on the details of that connection, and pave the way for Maxwell's theory of electromagnetism.

Because of his devotion to the public understanding of science, and because his discovery was made in the course of a lecture demonstration, the annual prize awarded by the American Association of Physics Teachers is named the Oersted Medal in his honor.

HANS CHRISTIAN OERSTED (1777–1851) Danish physicist. Oersted was born in Rudkøbing, the son of an apothecary. His early education consisted of learning German from a family he lived with for a while, and serving as his father's assistant from the age of 11. After the family moved to Copenhagen he entered the university there, obtaining a degree in pharmacy in 1797 and a doctorate two years later. He completed his education with a tour of major European laboratories, where he learned about the new studies of electricity and magnetism. After supporting himself for several years as a public lecturer in science, he was appointed to the faculty of his alma mater in 1806. In 1820 he made the fundamental discovery that demonstrated the unification of electricity and magnetism. He was appointed director of Copenhagen's Polytechnic Institute in 1829.

Ohm's Law

The electrical resistance of a conductor does not depend on the voltage across that conductor

The phenomenon of electrical resistance can best be understood in terms of an analogy with water flowing through a pipe. The moving water can be compared to an electrical current, which is a flow of electrons in a conductor, and voltage as the counterpart of the pressure that pushes the water through the pipe. Resistance is a process by which the energy of motion (of water or electrons) is converted into heat. It was this phenomenon that Ohm set out to understand.

Increasing the pressure of water flowing through a pipe leads to an increase in the proportion of energy lost as the flow becomes more turbulent. Ohm carried out a series of measurements indicating that what happened with an electrical current was rather different. He found that, over a wide range of voltages, the resistance of the material didn't change at all. This is what we know as Ohm's law. In equation form, it is simply

$$V = IR$$

where V is the voltage across a conductor, I is the current flowing through it, and R is the resistance.

Today, we understand the mechanism of resistance in terms of the moving electrons that constitute the current colliding with the stationary atoms in the conductor (*see* FREE ELECTRON THEORY OF CONDUCTION). Each collision transfers a fixed fraction of energy to the atom—energy that we perceive as heat—and the fact that other collisions are occurring doesn't affect this fraction. Increasing the voltage or the current affects only the number of electrons flowing in a wire, and not the fraction of energy transferred in individual collisions, which is why resistance is largely independent of voltage.

When Georg Ohm was performing the experiments that led to what we call Ohm's law, the ATOMIC THEORY had been around for a little over a decade and the discovery of the ELECTRON lay well into the future. For him, $V = IR$ was simply an experimental result. We now recognize it as an approximation to a more complex picture of electrical conduction, but one that is good enough to allow us to build the electrical circuits and appliances that play such important roles in our lives. Ohm has been honored by having the unit of electrical resistance named after him.

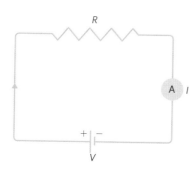

For this circuit, Ohm's law tells us that the voltage V equals the current I indicated by the ammeter (A) times the resistance R.

GEORG SIMON OHM (1789–1854)
German physicist. He was born in Erlangen (according to some sources, in 1787), and studied at the university there. Ohm was an instructor in mathematics and science during the early part of his career.

He didn't obtain a university chair until five years before his death, becoming professor of physics at Munich in 1849. Recognition had come slowly for Ohm, who had published his law in 1827. He also studied sound and hearing.

Olbers' Paradox

Why is the sky dark at night?

Perhaps the real paradox here is how this scientific riddle became so attached to the name of Wilhelm Olbers. In fact, this is one of several instances where posterity has given us an eponym that fails to credit the true originator. As far as historians of science can tell, it was the English scientist Edmond Halley (1656–1742) who first remarked on the problem, in 1720, and 22 years later the Swiss astronomer Philippe de Chéseaux (1718–51) posed the question again, supplying essentially the same answer as Olbers would in 1823.

Olbers' argument was simple: If the universe is infinite and unchanging (something that would have appeared obvious to astronomers of his time), then no matter which way we look, our line of sight should eventually encounter a star. This means that, no matter which way we look, we should see a bright point. The entire sky, in other words, should be bright with starlight. Yet the sky is dark.

Olbers tried to resolve the paradox by arguing that light from distant stars was absorbed by intervening matter, such as dust clouds. When the first law of THERMODYNAMICS was discovered, however, this explanation failed. If a dust cloud absorbs some of the energy in starlight, it will heat up. Over time, it will become luminescent and give off its own light.

The paradox was resolved by the findings of 20th-century astronomy. We now know (*see* HUBBLE'S LAW) that the universe has been around for a finite time. If the big bang happened 15 billion years ago, then, when astronomers use their various instruments to probe the depths of the universe, they can detect radiation only from objects less than 15 billion light years away. Thus, the number of stars in the sky, while large, is not infinite, so there need not be a star located along each line of sight. Similarly, we know that stars do not shine forever (*see* STELLAR EVOLUTION), but stop shining when their nuclear fuel is used up. Even if there is a star along a particular line of sight, it does not follow that we should be seeing its light right now—it could have stopped shining a long time ago. Either of these explanations resolves the paradox, although obviously neither could have been known to Olbers and his contemporaries.

HEINRICH WILHELM MATTHÄUS OLBERS (1758–1840) German astronomer, born in Arbegen. He was educated as a physician and had a successful practice in ophthalmology in the north German town of Bremen. His great intellectual love, however, was always astronomy. Over the years he built up one of the best astronomy libraries in Europe (it was later bought by Pulkovo Observatory to form the core of its own collection). Olbers discovered the asteroids Pallas and Vesta, and introduced several new methods for calculating the orbits of comets.

"Ontogeny Recapitulates Phylogeny"

During its development, a fetus traces out the evolutionary path that has been followed by its species

During the 19th century, scientists studying the development of the human embryo *in utero* noticed that during the first months of development the embryo bore a striking resemblance to other kinds of vertebrates. For example, at about a month the human embryo bears a series of striations in the neck region that look for all the world like primitive gills. Later on, the embryo bears a resemblance to reptiles, then to birds, and then to other mammals. This gave rise to the slogan above, which derives from a statement made by the German naturalist Ernst Haeckel (1834–1919) in his book *Riddle of the Universe,* published in 1899. It says that during the development of an organism, *ontogeny* (the genesis of an individual) retraces the path of *phylogeny* (the genesis of *phyla*—*see* LIN-NAEAN SYSTEM). The human embryo, for example, is first identical to a fish, then to a reptile, and so on, until it finally breaks away to assume its individual human identity. This is one of those ideas that is clear, beautiful, reasonable—and completely wrong.

In fact, the human embryo never possesses gills or any other appendages that might be required by this supposed development of the human embryo. The gill-like slits that appear are called *pharyngeal arches.* In fish, these cells do, indeed, develop into gills, but in humans they are precursors for parts of the head and neck. Just as the theory of EVOLUTION is not suggesting that humans descended from the apes, but shared a common ancestor with them, embryology does not imply that the human embryo retraces our evolutionary steps, only that it develops different structures from the same primordial cells. (There is some similarity between "ontogeny recapitulates phylogeny" and the equally false theory of the TRIUNE BRAIN.)

The extraordinary thing about this idea, which was even elevated to the status of a *law of biogenetics,* is that almost as soon as it was first put forward, evidence was found to disprove it—yet it has persisted (astonishingly, it is still to be found in some textbooks). There are connections between ontogeny and phylogeny, but there is no embryological recapitulation. It often pays to be skeptical of the obvious!

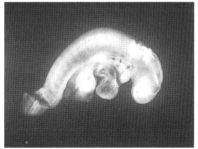

The embryo of a Japanese eel, at age 36 hours (above) and a human embryo at age 29 days (below). Their resemblance led to the false belief that the development of an organism recapitulates the evolutionary history of the species.

Optimal Foraging Theory

A predator's choice of prey depends on how long it takes to find the prey and on how long it takes to capture and consume it

Predators such as this salmon-hunting brown bear must balance energy return against handling time.

Sometimes abstract mathematics comes together with real-world phenomena in a particularly compelling way. Optimal foraging theory, developed by Robert MacArthur (of the MACARTHUR–WILSON EQUILIBRIUM THEORY) and Eric Pianka in 1966, is one of those cases. Most animals are capable of eating more kinds of prey than they actually consume. So, what are the principles that guide their choice of prey? This problem is typical of those dealt with in a branch of mathematics known as *game theory*.

We start by saying that a given type of prey will yield a fixed amount of energy—call it E. To obtain this energy, the predator must spend time doing two things: It must search until it finds the prey, and it must then capture and consume the prey—ecologists call this the *handling time* for the prey. The predator's rate of energy consumption will then be the energy E divided by the sum of the search and handling times. Optimal foraging theory predicts that animal behaviors will evolve to favor those strategies that yield the highest rate of energy consumption.

There are many consequences of this prediction. If, for example, the rate of return for searching and handling a new kind of prey is less than for going after a type already in the repertoire, the animal will restrict its choice of diet. This explains why animals have a narrow diet.

The theory also predicts how animals will behave in certain situations. For example, it is possible for the search time to be long, but the handling time short—think of a bird hopping around a tree looking for insects, or a bear going through a forest turning over logs looking for termites. Once a prey is found, little is added to the total time by consuming it—essentially, all the cost is in the search. Animals in this situation will be generalists, consuming a wide variety of prey.

If, in contrast, the search time is short but the handling time is long, you can expect different behavior. A lion on the Serengeti plains of Africa, for example, lives within sight of antelope herds, so its search time is essentially zero, but actually catching an antelope may take a lot of time and energy. In this case, the choice of prey will be narrow. The lion will target old, lame, and immature animals in order to minimize the handling time.

Animals can be forced to abandon their optimal foraging strategy. For those not at the top of a food chain, there is the ever-present risk of predation. Faced with this risk, an animal may have to forage for low-energy food in a safe location rather than follow what in the absence of predators would be its optimal foraging strategy—otherwise it may end up as the E in someone else's equation.

Ozone Hole

The ozone layer over the South Pole is depleted during the southern-hemisphere spring

It is important to be clear that the ozone hole, despite its name, is not a gap in the atmosphere. Ozone is made from three oxygen atoms hooked together, rather than two as in the standard oxygen molecule. The ozone in the atmosphere is concentrated in what is called the *ozone layer,* roughly 20 miles (30 km) up in the stratosphere. There it absorbs ultraviolet light from the Sun—light that would otherwise be damaging to life here at the surface. Any threat to the ozone layer must therefore be taken seriously. In 1985, British scientists working at the south pole established that levels of ozone in the atmosphere were abnormally low during the Antarctic spring. The ozone has become depleted at the same time every year, sometimes by more, sometimes by less. A similar but less pronounced ozone hole appears over the north polar region in the northern spring.

In the years following the discovery of the southern ozone hole, scientists worked out the sequence of events that causes it to form. When the Sun goes down for the long Antarctic night, temperatures plummet and high-altitude clouds of ice crystals form. These crystals provide sites for a series of complex chemical reactions that build up a supply of molecular chlorine (i.e., two atoms of chlorine locked together). When the Sun rises to herald the Antarctic spring, the ultraviolet radiation breaks the bonds that hold the chlorine molecules together, releasing a flood of chlorine atoms into the atmosphere. These atoms act as catalysts for the conversion of ozone into ordinary oxygen via the twin reactions

$$Cl + O_3 \rightarrow ClO + O_2 \quad \text{and} \quad ClO + O \rightarrow Cl + O_2$$

The net effect of these reactions is to break down ozone molecules (O_3) into ordinary molecules of oxygen (O_2), leaving the original chlorine atom free to repeat the process (each chlorine molecule can break up a million ozone molecules before other chemical reactions remove it from

This image of the ozone hole over Antarctica was assembled from data gathered in October 1999 by an orbiting instrument known as the Total Ozone Mapping Spectrometer.

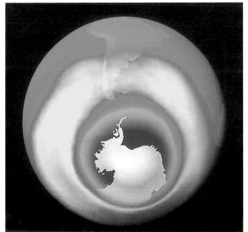

the atmosphere). The result of this chain of events is that ozone starts to disappear over Antarctica, thereby creating the ozone hole. Soon, however, the Antarctic vortex breaks down and fresh air (containing new ozone) floods into the region and the hole disappears.

In 1986 an international meeting on the threat to the ozone layer was held in Montreal, and the industrialized countries agreed to cut and ultimately stop production of chemicals known as *chlorofluorocarbons* (CFCs). By 1992, the replacement of these chemicals was going so well that the date for the final elimination was set in 1996. Today, scientists believe that the ozone layer will recover completely in about 50 years. (*See also* GREENHOUSE EFFECT.)

Pauli Exclusion Principle

No two electrons can occupy the same state

The Austrian physicist Wolfgang Pauli was one of the small group of European scientists who established QUANTUM MECHANICS in the 1920s and 1930s. The principle to which his name is now attached is one of the main ideas of this branch of physics. The easiest way to think about it is to compare electrons to cars in a parking lot. Only one car can go into each parking space, and once the spaces are all filled no more cars can be stored in the lot—extra cars have to go to another parking lot. In just the same way, for electrons in atoms there are only so many "parking spaces" in each orbital, and once those spaces are filled, any extra electrons have to go to the next higher orbital.

Now, electrons behave as though they are spinning (by convention, their spin is either $+\frac{1}{2}$ or $-\frac{1}{2}$), and two electrons, provided they have opposite spins, can occupy each space. It is as though a left-hand-drive car and a right-hand-drive car could occupy the same parking space simultaneously, whereas two left-hand-drive cars or two right-hand-drive cars could not. This is why the first row of the PERIODIC TABLE has two elements (hydrogen and helium) in it—there are only two states in the lowest orbital, reserved for electrons with opposing spins. In the next orbital there are eight states available (for four $+\frac{1}{2}$ spin electrons and four $-\frac{1}{2}$ spin electrons), which explains why there are eight elements in the second row of the periodic table. And so on.

Inside old stars, the temperature is so high that the atoms are mostly ionized and the electrons roam freely. The Pauli exclusion principle applies there, but in a slightly modified form. It now says that only two electrons, with opposite spins, can occupy a certain volume of space and at the same time have a certain range of velocities. But things are different if the density of a star's interior exceeds about 10^7 kilograms per cubic meter (10,000 times the density of water, or about 100 tons in a matchbox). Here the principle results in an increase in pressure. This additional pressure is called *electron degeneracy pressure,* and it means that there is a stable size for older stars that is similar to the size of the Earth. Such stars are called white dwarfs and they are the end point in the lives of stars with masses similar to the Sun's (*see* CHANDRASEKHAR LIMIT).

Although I have described the Pauli exclusion principle as it applies to electrons, it can be extended to any particle which, like the electron, has a spin with a half-integer value (i.e., whose spin is $\frac{1}{2}$, $\frac{3}{2}$, $\frac{5}{2}$, …). The electron has a spin of $\frac{1}{2}$, and so does the neutron. This means that neutrons, like electrons, need a certain amount of "elbow room" around them. If a white dwarf star's mass exceeds about 1.4 times the mass of the Sun (*see* CHANDRASEKHAR LIMIT), then gravity forces the protons and electrons within it to combine into neutrons. But the neutrons, like the electrons in white dwarfs, then generate a pressure, called *neutron degeneracy pressure,*

The Pauli Effect

Scientists such as Isaac Newton and Michael Faraday carried out experiments to investigate aspects of the physical world, then turned to developing a theory to explain the results of their experiments. No longer. Since the start of the 20th century, the sort of specialization that has plagued every other area of human endeavor has infected the sciences. Almost every scientist is now either an experimentalist or a theoretician. It is almost impossible to be both.

Wolfgang Pauli was clearly a theorist, and had the theorist's cultivated contempt for the "plumbers" who got their hands dirty in experiments. He became something of a legend for this disdain, as well as for his total inability to make any equipment work. If he so much as walked through a lab, people expected the equipment would break down. A catastrophic explosion in a laboratory at the University of Leiden in the Netherlands is supposed to have taken place at the exact moment Pauli stepped off the train from Zurich.

Whether any of this is true or not, the Pauli effect—the supposed ability of the man to destroy experiments by his very presence—entered the folklore of physicists. But like the BOHR EXPLANATION, it probably doesn't bear too close a scrutiny.

enabling the star to stabilize as a *neutron star* with a radius comparable to the size of a city. However, gravity can overcome even neutron degeneracy pressure if the star has more than two or three times the mass of the Sun, in which case it will collapse right down to a BLACK HOLE.

The Pauli principle is an example of a type of law of nature that is destined to become increasingly important in the age of computers. It's not like NEWTON'S LAWS OF MOTION—it doesn't predict what will happen to a given system, given its initial state. Instead, it tells us what kinds of things *can't* happen. It is what biologist and complexity theorist Harold Morowitz (1927–) calls a "pruning rule": It tells us that in very complex problems (like the problem of working out the details of the orbitals of an atom that has many electrons) there are some possibilities we don't have to ask our computer to consider. Thus, the law prunes away the dead wood in the problem, leaving only the possible solutions, hopefully allowing the computer to come to a solution in a reasonable amount of time. Rules like the Pauli exclusion principle will therefore become more important as we come to depend more and more on computers to solve difficult and complex problems.

WOLFGANG PAULI (1900–58) Austrian-Swiss physicist. He was born in Vienna, the son of a distinguished professor at the University of Vienna; Ernst Mach (*see* SHOCK WAVES) was his godfather. While he was still at school he mastered both special and general RELATIVITY. He studied theoretical physics at the University of Munich, along with such distinguished fellow students as Werner Heisenberg (*see* HEISENBERG'S UNCERTAINTY PRINCIPLE), and obtained his Ph.D. in 1922.

Pauli was one of the pioneers of QUANTUM MECHANICS and made several important contributions to the new science, the most striking of which

was the exclusion principle, in 1924, and for this work he was awarded the 1945 Nobel Prize for Physics. His idea of a spin quantum number was confirmed experimentally two years later. Pauli also explained an apparent loss of energy in beta decay (*see* RADIOACTIVE DECAY) by suggesting that another particle, later identified as a neutrino, was emitted along with an electron.

During World War II Pauli worked in America at the Institute of Advanced Study at Princeton. After the war he returned to Europe, resuming his position as professor of experimental physics at the federal Institute of Technology, Zurich, and assuming Swiss citizenship.

Penicillin, discovery of

Penicillin, the first antibiotic, was discovered by accident. It works by blocking the growth of bacterial cell walls

In 1928, Alexander Fleming was starting another bread-and-butter experiment in his long-term research into how the human body fights off bacterial infections. After starting some cultures of the bacterium *Staphylococcus*, he noticed that some of the culture dishes had become contaminated by a common mold called *Penicillium*—the stuff that turns bread blue if you leave it lying around too long. Around each spot where the mold was growing, Fleming noticed a zone where bacteria could not grow. He concluded that the mold was producing something that was killing the bacteria, and eventually isolated the molecule we now call penicillin, the first modern antibiotic.

An antibiotic works by inhibiting a chemical reaction which is essential to a bacterium's life. Penicillin works by blocking molecules that build new cell walls for the bacterium, something like chewing gum stuck on a key preventing it from opening a lock. (Penicillin has no effect on humans or other animals because our cells do not have the same outer coverings as those of bacteria.)

Throughout the 1930s, people tried to perfect penicillin and other antibiotics, but were unable get the drugs in pure enough form. In fact, early antibiotics were like many of our modern anticancer drugs—it was a close decision as to whether the drug would kill the invader before it killed the patient. It wasn't until 1938 that two scientists at Oxford

Sir Alexander Fleming at work in his laboratory.

University, Howard Florey (1898–1968) and Ernst Chain (1906–79), managed to isolate penicillin in its pure form. Under intense pressure from the medical needs of World War II, the drug was brought into mass production in 1943. In 1945, Fleming, Florey, and Chain shared a Nobel prize for their work.

Penicillin and other antibiotics have saved countless lives since their discovery. Penicillin was also the first drug for which the evolution of resistance to ANTIBIOTICS was seen.

ALEXANDER FLEMING (1881–1955) Scottish bacteriologist. He was born in Lochfield, Ayrshire. He graduated from St Mary's Hospital Medical School in London and worked there all his life, except during World War I when he served in the Royal Army Medical Corps and became interested in controlling wound infections. His accidental discovery of penicillin in 1928, the year he became professor of bacteriology, earned him a share of the 1945 Nobel Prize for Physiology or Medicine.

Periodic Table

The chemical elements show a regular pattern of chemical properties when they are arranged in order of increasing atomic number

Dmitri Mendeleyev said that it came to him in a dream. Like many chemists in the mid-19th century, he was trying to come to terms with the flood of new chemical elements that were being discovered. He was writing an introductory book on chemistry. Surely, he thought, there had to be some sort of order to all of these substances—some way of arranging the elements that made them more than just a random list of names. It was this ordering, this arrangement, that was the subject of his dream.

Called today the periodic table of the elements, Mendeleyev's scheme was to rank the chemical elements in rows in order of increasing mass, then arranging the length of the rows so that chemical elements in the same column have similar chemical properties. Thus, for example, the column on the far right of the table is made up of helium, neon, argon, krypton, xenon, and radon. These are the *noble gases*—substances that are very reluctant to combine with other elements or take part in chemical reactions.* By contrast, the elements in the leftmost column—lithium, sodium, potassium, and so on—react violently, often explosively, with other substances. Similar sorts of statements can be made about the other columns in the table—they all contain elements with similar chemical properties, but those properties vary from one column to the next.

We shouldn't underestimate the intellectual courage that Mendeleyev showed in publishing his result. For one thing, when he put his original table together it contained gaps—elements we now know to exist, but were then yet to be discovered. (In fact, the discovery of these elements, which included scandium and germanium, was one of the great predictive triumphs of the periodic table.) For another, Mendeleyev had to assume that the weights of several elements had been incorrectly measured, because otherwise they wouldn't fit into the pattern. As it turned out, he was right again.

When the periodic table was first published, it was simply a statement of an observed pattern in nature. As with KEPLER'S LAWS of planetary motion, the table contained no inkling about why things should be that way. It was not until the advent of QUANTUM MECHANICS and, in particular, the PAULI EXCLUSION PRINCIPLE that the arrangement of the elements in the periodic table began to make real sense.

Today, we think of the periodic table in terms of how electrons occupy shells in atoms (*see* AUFBAU PRINCIPLE). The chemical properties of an atom—which tell you what kinds of associations it will make with other atoms—are determined by the number of electrons in its outermost shell. Hydrogen and lithium each have one outer electron, so they will behave similarly in chemical reactions. In the same way, helium and neon both have filled outer shells, so they, too will behave similarly to each other but very differently from hydrogen and lithium.

---

* They were once called *inert gases*, but the name was changed in 1962 when it was found that xenon could be made to react with fluorine.

1 H																	2 He
3 Li	4 Be											5 B	6 C	7 N	8 O	9 F	10 Ne
11 Na	12 Mg											13 Al	14 Si	15 P	16 S	17 Cl	18 Ar
19 K	20 Ca	21 Sc	22 Ti	23 V	24 Cr	25 Mn	26 Fe	27 Co	28 Ni	29 Cu	30 Zn	31 Ga	32 Ge	33 As	34 Se	35 Br	36 Kr
37 Rb	38 Sr	39 Y	40 Zr	41 Nb	42 Mo	43 Tc	44 Ru	45 Rh	46 Pd	47 Ag	48 Cd	49 In	50 Sn	51 Sb	52 Te	53 I	54 Xe
55 Cs	56 Ba	57 to 71	72 Hf	73 Ta	74 W	75 Re	76 Os	77 Ir	78 Pt	79 Au	80 Hg	81 Ti	82 Pb	83 Bi	84 Po	85 At	86 Rn
87 Fr	88 Ra	89 to 103	104 Rf	105 Db	106 Sg	107 Bh	108 Hs	109 Mt	110 Uun	111 Uuu	112 Uub		114 Uuq		116 Uuh		118 Uuo

	57 La	58 Ce	59 Pr	60 Nd	61 Pm	62 Sm	63 Eu	64 Gd	65 Tb	66 Dy	67 Ho	68 Er	69 Tm	70 Yb	71 Lu
Lanthanides															
Actinides	89 Ac	90 Th	91 Pa	92 U	93 Np	94 Pu	95 Am	96 Cm	97 Bk	98 Cf	99 Es	100 Fm	101 Md	102 No	103 Lr

The table that makes sense of chemistry. The first column contains the alkali metals, and the second column the alkaline earth metals. On the far right are the noble gases. To the left of them, nonmetals are above the "staircase," below which are other metals. Between these and the alkaline earths are the transition metals, with the lanthanides and actinides as offshoots. Hydrogen (H) has a place to itself, at the top.

Chemical elements up to uranium (92 protons and 92 electrons) are found in nature. Beyond this come artificial elements that have been created in laboratories. The highest claimed so far is number 119, with scientists in Germany leading the charge to create new elements.

DMITRI IVANOVICH MENDELEYEV (1834–1907) Russian chemist. He was born in Tobol'sk, Siberia, the youngest of 17 children. Mendeleyev's early life was not easy. His father, a schoolteacher, became blind, and his mother managed a glass factory to support the family. His father died when Mendeleyev was 13, and then the factory burnt down, after which his mother died. All he knew of science was what he had picked up from a brother-in-law.

One of his mother's last acts was to get Dmitri enrolled in the Pedagogical Institute at St Petersburg. There, Mendeleyev took a degree in chemistry, and went on to study in France and Germany. In Karlsruhe he met the Italian chemist Stanislao Cannizaro (1826–1910), whose careful distinction between atomic and molecular weights impressed the Russian. Back in St Petersburg, Mendeleyev became professor of chemistry at the Technical Institute in 1864.

Mendeleyev's periodic table, which he slowly put together from the late 1860s, was slow to gain acceptance, but it event-

ually made him Russia's most famous scientist. However, in 1890 he spoke out in support of students who were agitating for social reform. He was immediately removed from the university, and after 1893 was denied any official position. As a final injustice, he lost out on the 1906 Nobel Prize for Chemistry by a single vote. The prize went to Henri Moissan (1852–1907), who had isolated fluorine—one element, against Mendeleyev's classification of them all.

Period– Luminosity Relation

The longer the period of a Cepheid variable star, the more energy it is emitting

When Keats wrote, "Bright star, were I as steadfast as thou art," he didn't have Cepheid variables in mind. Most stars, and fortunately for us the Sun is one of them, produce a more or less steady output of light and other forms of radiation (*see* ELECTROMAGNETIC SPECTRUM). There are, however, several classes of stars whose radiation varies over time— stars that are called, appropriately enough, *variable stars,* or simply "variables." The class of Cepheid variables played an essential historical role in helping to establish the scale of the universe. (Their name comes from the fact that the first star known to be of this type was Delta Cephei—a fairly bright star in the northern constellation of Cepheus.)

If you monitor a Cepheid variable, you will find that its brightness increases severalfold, then it dims, rather more slowly than it brightened. This pattern repeats over and over again, on a timescale that can vary from a few days to many weeks, depending on the particular star. A Cepheid's *period*—the time between successive maxima (or minima) in brightness—and its *range*—the difference in light output between maximum and minimum—remain constant.

Cepheid variables became astronomers' first *standard candle*. This is an object whose light output is known—a 100-watt light bulb, for example, is a splendid standard candle. If you can locate a standard candle in space, you can measure the amount of light you actually receive from it and compare your measurement with the amount of light you know it is emitting. From the difference you can deduce how far away the source is. Standard candles enable astronomers to add a third dimension—distance—to the two-dimensional display of the night sky.

Early in the 20th century, the American astronomer Henrietta Leavitt began to study Cepheid variables. By 1912 she had established that the longer it takes a Cepheid to go through its cycle, the brighter it appears to be. Edwin Hubble tweaked this result by relating the period to a Cepheid's intrinsic *luminosity*—the amount of energy the star is pouring into space. Thus was established the period–luminosity relationship. It was Hubble's identification of Cepheid standard candles in what is now called the Andromeda Galaxy that established the existence of other galaxies, which in turn led to his discovery of HUBBLE'S LAW.

HENRIETTA LEAVITT (1868–1921) American astronomer. She was born in Lancaster, Massachusetts. Leavitt graduated from what is now Radcliffe College in 1895 and was appointed by astronomer Edward C. Pickering to help classify the stellar spectra that were being amassed at the Harvard College Observatory. It was her studies of Cepheid variable stars in the Small Magellanic Cloud, a small companion galaxy to the Milky Way, that led to her discovery of the period–luminosity relationship.

Perpetual Motion

It is possible to make a machine that will run forever, or, better still, provide a limitless source of energy

There is something about the laws of nature that brings out the contrariness in human nature. Nowhere is this more obvious that in the enduring belief that it is possible to build a perpetual-motion machine—a machine that will run forever without any outside assistance. As a scientist who has some contact with the public, I receive about one letter a year informing me of plans for such a machine. Sometimes the letters offer me a percentage of the earnings that could be made from such a machine if I will bring it to the attention of the appropriate authorities.

There seem to be two kinds of perpetual-motion machines—those that violate the first two laws of THERMODYNAMICS and those that violate only the second law. Here is an example of the first kind of machine: A metal ball rests between the north and south poles of a magnet. A heavy piece of metal shields the ball from the north pole, so that when the ball is released it moves toward the south pole. As it approaches that pole, the metal shield near the north pole is lifted while another shield descends between the ball and the south pole. The ball reverses direction, rolling back toward the north pole. At just the right moment, the shield is dropped at the north pole and the ball starts rolling back toward the south pole. How is it supposed to work? Energy is extracted from the rolling ball, and by putting the shields on a kind of balance beam, no energy is expended in raising and lowering them.

The problem with this particular machine is that if a metal shield is moved in a magnetic field, FARADAY'S LAWS OF INDUCTION guarantee that an electrical current will flow in the metal. This means that energy will be drained from the system through the operation of OHM'S LAW. It is fairly easy to show that if the magnets are strong enough to make the ball move, they will also be strong enough to generate large resistive losses in the metal shields as they fall, so that the machine, which looks so attractive on paper, simply will not work.

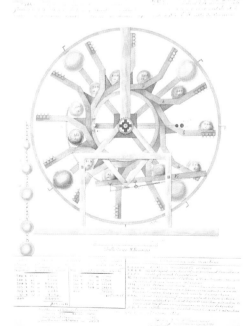

A perpetual-motion machine proposed in 1834. The weight of the balls moving outward along the arms to the right was supposed to keep the wheel turning forever.

Some proposed perpetual-motion machines are more sophisticated than this one, and it requires more subtle thought to see the flaw in their design. But there are always flaws, which is why no such machine has ever been demonstrated to work. This fact was recognized, in fact, by the U.S. Patent Office in the mid-20th century. Harassed by a flood of patent applications for perpetual motion machines, the Office declared that in future any such application would have to be accompanied by a working model. Since then they have not been bothered by applicants.

Phlogiston

There is a substance called phlogiston that comprises the flammable part of any material

This old chemical theory was based on the idea that there was something incorporated into any material which would burn that represented its flammable portion. This something was called phlogiston, from the Greek word for "burned." The idea was that when a material burned, the phlogiston was given off and was carried away in the air. Wood, for example, was thought of as a mixture of ashes and phlogiston, so that when wood burned the phlogiston was given off, leaving the ashes behind. In the same way, metals were thought to be mixtures of phlogiston and materials called calxes.

A major problem with this theory is that, whereas ashes are usually lighter than the original wood, calxes (or, as we would say today, metallic oxides) are often heavier than the original metal. Today we know that this is because carbon dioxide and water, the major combustion products of wood, are gases which go off into the atmosphere, whereas when metals combine with oxygen (when iron rusts, for example) the resulting oxide is a solid which stays put.

The final nail in the coffin of phlogiston theory was supplied by Antoine Lavoisier, who showed that it was chemical combination with the newly discovered element oxygen that explained both the apparent weight gain and the apparent weight loss in the chemical reactions of combustion.

ANTOINE-LAURENT LAVOISIER (1743–94) French chemist. He was born in Paris and received a broad scientific education at the Collège Mazarin. Investment in a tax-collecting company provided him with the funds to establish a laboratory, where he proceeded to lay the foundations of modern chemistry. He introduced rigorous experimental methods that included the careful weighing of reactants and products. As well as disposing of the phlogiston theory (though he still counted light and heat—"caloric"—as chemical elements), he discovered the composition of water and established that organic compounds contain carbon, hydrogen, and (in many cases) oxygen. (*See also* ACIDS AND BASES.) During the French Revolution, the extremist Jean-Paul Marat accused him of preventing the air in Paris from circulating—Lavoisier had supervised the construction of a city wall. His activities as a tax collector marked him as an opponent of the revolution, and he met his end at the guillotine. His widow married Count Rumford, who helped to establish the atomic theory of HEAT.

Photoelectric Effect

Photons striking a metal can cause electrons to be ejected

The fact that light striking a material which conducts electricity causes it to emit electrons plays a role in our everyday life. For example, some types of alarm systems work by transmitting a beam of light, visible or infrared, to a *photoelectric cell*—a device in which electrons emitted from a metal surface are "pushed" into a wire behind the metal. If something moves through the beam of light, the flow of electrons stops, that change is registered by electronic equipment, and the alarm is triggered.

The discovery of this so-called photoelectric effect in the late 19th century posed a serious problem, for nothing that was known about light or metals would explain it. It's not that the classical theory wouldn't allow light to kick electrons out of a metal—it would. Like a cork being moved slowly toward the shore, the electromagnetic waves of the incoming light could be expected, eventually, to float electrons free from the metal. The problem was that the photoelectric effect didn't seem to fit with this simple explanation. For one thing, the electrons appeared virtually instantaneously when the light was turned on. For another, it seemed that even very faint light beams could shake electrons loose, and that cranking up the intensity of the beam didn't seem to have much effect on the energy of the emitted electrons. This is all in contradiction to the classical picture of the interaction of light with electrons.

It was Albert Einstein who finally resolved the issue, and in the process gave a huge impetus to the fledgling science of QUANTUM MECHANICS. Max Planck had shown that BLACK-BODY RADIATION could be understood by assuming that atoms emitted light in discrete bundles of energy called *quanta*. He believed that this had something to do with the structure of atoms, and not with the nature of light. Einstein took Planck's idea seriously, and assumed that light itself comes in discrete bundles of energy—we now them *photons*. These photons sometimes behave as if they were particles (*see* COMPLEMENTARITY PRINCIPLE). In particular, when a photon hits an electron in an atom, the interaction is something like the collision of two billiard balls, with the electron shooting out immediately. Even a single photon can knock out an electron in this sort of collision. Furthermore, while increasing the intensity of the light may increase the number of photons (and hence the number of emitted electrons), it cannot increase the electron's energy.

Einstein's equation for the energy of photoelectrons is simple:

$$E = h\nu - \phi$$

where ν is the frequency of the incoming light, h is the PLANCK CONSTANT, and ϕ is the energy it takes to get the electron out of the metal once it has been shaken loose from its atom.

Photo-synthesis

Plants convert sunlight to stored chemical energy in two stages: the energy in sunlight is first trapped, then used to fix carbon in organic molecules

Green plants—what biologists call *autotrophs*—are the basis for all life on our planet, at the beginning of nearly all food chains. They convert the energy that falls on them in the form of sunlight into energy stored in carbohydrates (*see* MOLECULES OF LIFE), most importantly the six-carbon sugar known as glucose. This conversion process is known as photosynthesis. Other organisms then eat the plants to gain access to this stored energy, thereby creating the food chain that supports the global ecosystem.

Photosynthesis is also the process that supplies the oxygen in the air we breathe. The general reaction is

water + carbon dioxide + light → carbohydrate + oxygen

so that plants take in the carbon dioxide that is the result of respiration and give off oxygen as a waste product (*see* GLYCOLYSIS AND RESPIRATION). Photosynthesis also plays a crucial role in the planet's CARBON CYCLE.

It is rather surprising that, given the importance of the process, scientists were so long in starting to examine it. After the VAN HELMONT EXPERIMENT, performed in the 17th century, it wasn't until 1905 that the English plant physiologist Frederick Blackman (1866–1947) performed the experiments that defined its basic processes. He showed that, starting at low levels of illumination, the rate of photosynthesis increases if light levels are raised, but after a certain point further increases in light levels do not lead to increased activity. He found that raising the temperature has no effect at low levels of illumination, but increasing both light and temperature increases photosynthesis rates by much more than could be accounted for by the increase in light levels alone.

From these experiments, Blackman concluded that two processes are at work: one that depends strongly on light levels but not on temperature, and one that depends strongly on temperature but not on light levels. This insight underpins our modern understanding of photosynthesis. The two processes are sometimes called the "light" and "dark" reactions, respectively, but this is rather confusing since the "dark" reactions turn out to require the products of the "light" reactions, even though they do not themselves need the presence of light.

Photosynthesis begins when photons from the Sun strike specific kinds of pigment molecules, known as *chlorophyll,* in a leaf. The chlorophyll is contained within the leaf cells in the membranes of special structures known as *chloroplasts* (they are what give leaves a green color). The energy is captured in a two-step process, involving separate clusters of molecules known as *Photosystem I* and *Photosystem II.* The numbers refer to the order in which these processes were discovered, and it is one of

Van Niel's Hypothesis

The process of photosynthesis can be summarized by the chemical equation

$$CO_2 + H_2O + light \rightarrow$$
$$carbohydrate + O_2$$

The general thinking at the beginning of the 20th century was that the oxygen given off as a byproduct came from the breakdown of carbon dioxide. In the 1930s, however, Cornelis Bernardus Van Niel (1897–1986), then a graduate student at Stanford University in California, was doing experiments on purple sulfur bacteria (below), which require hydrogen sulfide (H_2S) for photosynthesis to occur, and which give off atomic sulfur as a byproduct. For these bacteria, the above equation becomes

$$CO_2 + H_2S + light \rightarrow$$
$$carbohydrate + 2S$$

The similarity between these two processes, suggested Van Niel, implies that it is the water which is the source of the oxygen in normal photosynthesis, not the carbon dioxide, and pointed to the fact that when the oxygen in water was replaced by sulfur in his bacteria, photosynthesis gave that sulfur back as a byproduct. Our current detailed understanding of photosynthesis bears this conjecture out, as the first step in the process (in the so-called Photosystem II reactions) is the breakup of a water molecule.

those amiable anomalies in the sciences that in a real leaf the reactions in Photosystem II occur before those of Photosystem I.

When a photon encounters the 250–400 molecules that make up Photosystem II, the energy is bounced around until it reaches the chlorophyll molecule. At this point two chemical reactions take place: The molecule gives up two electrons (which are taken by another molecule known as an electron acceptor) and a molecule of water is broken up. The electrons from the two hydrogen atoms in the water replace the electrons lost by the chlorophyll.

The high-energy electron is then passed, like a hot potato, down a chain of molecular carriers. In the process, some of its energy is used to make a molecule called adenosine triphosphate (ATP), one of the basic energy carriers of the cell (*see* MOLECULES OF LIFE). In the meantime, a slightly different chlorophyll molecule in Photosystem I has absorbed energy from a photon and given up an electron to another acceptor molecule. This electron is replaced in the chlorophyll by the electron being passed down the chain from Photosystem II. The energy of the electron from Photosystem I, together with the hydrogen ions that result from the earlier breakup of the water molecule, are used to produce another energy-carrying molecule, called NADPH.

The net effect of the light-trapping process is that the energy from two photons becomes stored in molecules that the cell can use to run reactions, and one oxygen molecule is created as a byproduct. (I should point out that there is a second, much less efficient process involving only Photosystem I that also contributes ATP molecules.) Once the energy of sunlight has been captured and stored, the next step is to use it to produce hydrocarbons. The main mechanism that plants use to do this was discovered by Melvin Calvin in a classic series of experiments in the 1940s. Calvin and his co-workers exposed algae to carbon dioxide containing radioactive carbon-14. By stopping photosynthesis at various stages, they eventually worked out the chemistry.

The so-called *Calvin cycle*, which converts the solar energy into carbohydrates, is like the Krebs cycle (*see* GLYCOLYSIS AND RESPIRATION) in that it consists of a series of chemical reactions that start when an input molecule combines with a "helper" molecule to initiate the series. Once the reactions have yielded the final product, they reproduce the "helper" molecule to start the cycle again. In the Calvin cycle this helper molecule is ribulose biphosphate (RuBP), a five-carbon sugar. A molecule of carbon dioxide hooks up to RuBP to start the cycle. Using energy from the ATP and NADH that has been trapped from sunlight, chemical reactions first link

An aerial view of rain forest in Central America. In using photosynthesis to get their energy, trees and other plants absorb carbon dioxide and give out oxygen.

carbon atoms together to make hydrocarbons, then go on to re-create the RuBP. Six circuits of the cycle incorporate six carbon atoms into the precursors of glucose and other carbohydrates. This cycle of chemical reactions, which will continue for as long as energy is supplied, is how the energy of sunlight is made available to living things.

In most plants, the Calvin cycle operates as outlined above, with carbon dioxide entering directly into the reactions by binding to RuBP. These are called C3 plants, because the carbon dioxide–RuBP complex breaks down into two smaller molecules that have three carbons each. In some plants (most notably maize, sugarcane, and many tropical grasses, including crabgrass), the cycle works differently. The problem is that carbon dioxide normally enters a leaf through openings in the surface called *stomata*. At high temperatures, the stomata close to prevent excessive loss of moisture. In C3 plants, this closing also cuts off the supply of carbon dioxide, which both slows down and changes the chemical reactions of photosynthesis. In maize, carbon dioxide is attached to a three-carbon molecule in the outer part of the leaf, then shunted to regions in the leaf's interior where further reactions release the carbon dioxide and begin the Calvin cycle. This process, while complex, allows photosynthesis to proceed even in very hot, dry weather. Plants that operate in this way are called C4 plants, because the carbon dioxide is initially transported in a four-carbon molecule. In general, temperate plants are C3, while C4 plants evolved in the tropics.

MELVIN CALVIN (1911–97) American biologist. He was born in St Paul, Minnesota, to Russian immigrant parents. He gained a BS degree in chemistry in 1931 from the Michigan College of Mining and Technology, and was awarded a Ph.D., also in chemistry, from the University of Minnesota in 1935. Two years later he joined the University of California at Berkeley, becoming professor there in 1948; the previous year he had been made director of the Bio-organic Division of the Lawrence Radiation Laboratory at Berkeley, where he used technological spinoffs from World War II, such as new techniques in chromatography, to study "dark" photosynthesis. Calvin received the 1961 Nobel Prize for Chemistry.

Planck Constant

The Planck constant defines the boundary between the Newtonian and the quantum worlds

Max Planck became involved in what turned out to be the origins of QUANTUM MECHANICS by thinking about how the newly discovered electromagnetic radiation interacted with atoms—what is called the problem of BLACK-BODY RADIATION. He realized that to explain the observed spectrum of radiation emitted by atoms, he had to assume that atoms could not absorb light at any possible energy or frequency, but only in discrete "packets" he called *quanta*. The energy E of one packet of light (now called a *photon*) was given by

$$E = h\nu$$

where ν was the frequency of the radiation and h was a new constant of nature, now called the Planck constant, that defines the energy scale in the subatomic world. Its numerical value is 6.63×10^{-34} joule seconds (in the SI system of units). Each atom could then emit a large number of related frequencies depending upon the arrangement of its electrons—a simplified version of the picture of radiation emission that would soon be provided by the BOHR ATOM.

Planck first published this result in 1900, but it is clear from his writings that at first he never accepted the reality of his quanta. When Einstein published his explanation of the PHOTOELECTRIC EFFECT five years later, people began to accept that Planck's result wasn't a fluke having to do with peculiarities of the structure of atoms, but a statement about the quantized nature of the subatomic world.

The constant appears repeatedly in the formulations of quantum mechanics. It defines, for example, the region in which HEISENBERG'S UNCERTAINTY PRINCIPLE becomes important. Essentially, it tells us that we need worry about quantum effects only for subatomic objects—that the uncertainties associated with, say, grains of sand are small enough to be ignored. In a sense, it defines the boundary between the Newtonian and the quantum worlds. Planck's constant, originally invoked to explain a specific experimental result and to remove a specific theoretical dilemma, is one of the most fundamental numbers in nature.

MAX KARL ERNST LUDWIG PLANCK (1858–1947) German physicist. He was born in Kiel into a prominent family—his father was a professor of law. Planck was a skilled pianist, and at one point in his youth he had to choose between music and science as a career. He was awarded his doctorate from the University of Munich in 1889 for a thesis on the second law of THERMODYNAMICS, and later that year he moved to the University of Berlin. In 1892 he was appointed professor of theoretical physics there, and held the chair until he retired in 1928. Planck was one of the founders of QUANTUM MECHANICS. Today, a string of German research establishments are named the Max Planck Institutes in his honor.

Plate Tectonics

The surface of the Earth consists of several large, interlocking plates which move slowly with respect to one another

When rocky planets form, they go through a period of heating for which the main source of energy is the infall of debris from space (*see* NEBULAR HYPOTHESIS). Nearly all of the kinetic energy of an impacting object is immediately converted into heat energy, as its speed of perhaps several tens of miles per second is suddenly reduced to zero at the moment of impact. For all the solar system's inner planets—Mercury, Venus, Earth, Mars—this heat was sufficient to cause them if not to melt, either completely or partially, then at least to become highly plastic and fluid. During this period, dense materials gravitated toward the center of each planet, forming a *core,* and less-dense materials rose toward the surface, much as a salad dressing will separate if it is left standing too long, to form a *crust.* This process, which is called *differentiation,* explains the interior structure of the Earth.

For the smallest inner planets, Mercury and Mars (and for the Earth's Moon), this heat was eventually carried to the surface and dissipated in space. The planets then solidified and (certainly in Mercury's case) have experienced little geological activity in the last few billion years. The subsequent history of the Earth was very different. As the largest of the inner planets it stored the greatest amount of heat in its interior. And the larger a planet is, the smaller is its surface area compared to its volume, and the less easy it is for it to radiate heat away. Consequently, the Earth cooled more slowly than did the other inner planets. (The same goes for Venus, just slightly smaller than the Earth.)

In addition, radioactive elements inside the Earth have been decaying since our planet began to form, adding to the store of heat in the interior. Consequently, you can think of the Earth as a spherical stove. Heat is continually being generated in the interior, being transported to the surface, and leaking away into space. The transport of heat causes the layer of the Earth known as the *mantle,* which extends from a few tens of miles to about 1800 miles (2900 km) beneath our feet, to move in response (*see* HEAT TRANSFER). Hot material from deep in the mantle rises, cools, and then sinks again to be replaced by new hot material—the classic example of a convection cell.

It would not be too wrong to picture the rock of the mantle as "boiling" like water in a pot—in both cases heat is being transferred by convection. Some Earth scientists believe that the rock may take several hundred million years to pass through a complete convection cycle—a very long time by human standards. We know that there are many materials which will deform slowly over time, although they seem completely solid and rigid as perceived on the human timescale. For example, there are panes of glass in medieval cathedrals in Europe which are thicker at the bottom than at the top because over many centuries the glass has flowed downward under the influence of gravity. If solid glass can do this in a few hundred years, it isn't too hard to imagine solid rock doing it in a few hundred *million* years.

On top of the convection cells in the Earth's mantle float the rocks that make up the solid surface of the Earth—the so-called *tectonic plates*. These plates are made of basalt, the commonest type of igneous rock. They are roughly 5 to 75 miles (10–120 km) thick and move around on the surface of the semi-molten mantle. The continents themselves, made up of relatively light rocks such as granite, form the topmost layers of the plates. In general, plate structures are thicker under continents and thinner under oceans. Over time, processes that operate inside the Earth move plates around, causing them to collide, to grind against one another, to be created, or to disappear. It is this never-ending, slow-motion dance of the plates that gives our planet its dynamic and constantly evolving surface.

It is important to appreciate that "plate" is not synonymous with "continent." The North American Plate, for example, extends from the middle of the Atlantic Ocean to the western coast of the North American continent. Part of the plate is covered with water, part with land. The Anatolian Plate, which underlies much of Turkey and the Middle East, is covered completely with land, while the Pacific Plate is covered completely by the Pacific Ocean. Thus, plate boundaries and the coastlines of continents do not necessarily coincide. Incidentally, the word "tectonic" comes from the Greek word *tekton* ("builder"), and has the same root as the word "architect," suggesting the process of building or putting together.

Plate tectonics is at its most visible where plates meet. There are three different types of plate boundaries:

A relief map of the Earth, including the ocean floors, showing the plates that make up its surface (there are several smaller plates too, but for simplicity they are not shown). Plates are separating at mid-ocean ridges or colliding at major mountain ranges.

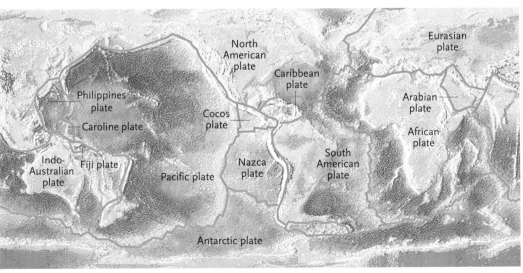

Diverging boundaries

In the middle of the Atlantic Ocean, hot magma formed deep in the mantle rises to the surface. There it breaks through the surface and flows sideways, slowly filling the gap between a separating pair of plates. Over time, this new rock solidifies and becomes part of the plates. As a result of this spreading of the sea floor, Europe and North America have been moving apart at the rate of several inches a year. (This motion has actually been measured by comparing arrival times of radio waves from distant quasars at radio telescopes located on the two continents.)

If a diverging boundary is under an ocean, the result is a mid-ocean ridge—a mountain chain formed by material piling up as it comes to the surface. The Mid-Atlantic Ridge, extending from Iceland to the Falklands, is the longest mountain chain on the surface of the Earth. A diverging boundary between two plates under a continent literally tears the continent apart. The Great Rift Valley, extending from Jordan down into east Africa, is an example of this process happening today.

Converging boundaries

If new plate material is being formed at diverging boundaries, then it must be destroyed someplace else—otherwise the Earth would be increasing in size. When two plates are pushed together, one of them slides under the other (a process known as *subduction*). The material in the subducted plate is carried down into the mantle. What happens at the surface above a subduction zone depends on whether the plate boundary is under a continent, at a continental margin, or in mid-ocean.

If a subduction zone is below oceanic crust, the result is a deep mid-ocean trench. The Marianas Trench by the Philippines, the deepest spot in the world's oceans, is an example of the process going on today. Material in the subducted plate, once it has been melted, often rises back up to the surface to form a series of volcanoes. The chains of volcanoes on the eastern rim of the Caribbean and on the western coast of the United States are both examples of this phenomenon.

If both the plates at a convergent boundary carry continents, the result is very different. Continental crust is made of light materials, and the two plates just float above the subduction zone. As one plate is subducted beneath the other, the two continents collide and crumple, creating a mid-continent mountain range. The Himalayas were formed by just such a process when the Indian subcontinent docked onto the Eurasian landmass about 50 million years ago. Similarly, the Alps were formed when Italy attached itself to Europe. The Ural Mountains, an old mountain chain, are the welding scar from the joining of the European and Asian landmasses.

If only one of the plates has a continent on it, the continent will be folded up and wrinkled as it floats over the subduction zone. The Andes mountains on the western coast of South America are an example. They are being formed as the South American Plate floats up over the subducted Nazca Plate under the Pacific Ocean.

Transform fault boundaries

It occasionally happens that two plates neither diverge nor subduct, but simply scrape past each other. The most famous example of this sort of plate boundary is the San Andreas Fault in California, where the Pacific Plate and the North American Plate slide by each other. In this situation the plates stick for a while, then let go, releasing much energy and causing severe earthquakes.

A final word about plate tectonics. Although it embodies a theory in which continents move around, it is not the same as the theory of *continental drift* which was proposed in the early 1920s. That theory was (correctly, I think) rejected by geologists because of a number of experimental and theoretical problems. The fact that our current theory incorporates one aspect of continental drift—the motion of continents—does not mean that scientists rejected plate tectonics early in the century only to adopt it later. The theory they adopted was quite different from the earlier one.

ALFRED LOTHAR WEGENER (1880–1930) German meteorologist and geologist. He was born in Berlin, where he obtained his doctorate, in astronomy, in 1905. Before World War I he lectured in astronomy and meteorology at the University of Marburg; after the war he became professor of meteorology and geophysics—a specially created chair—at the University of Graz, in Austria. Part of the evidence he produced to support his theory of continental drift consisted of measurements showing that Greenland and Europe were moving apart. It was in Greenland that Wegener died, during his fourth expedition there.

Predator–Prey Relationships

The relationships between predators and their prey go through cycles that illustrate neutral equilibrium

The cyclic variations in the numbers of a predator species and its prey.

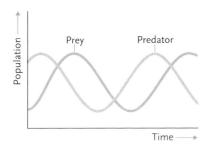

A simple mathematical model can sometimes give a good description of a complex biological system. The long-term relationship between predator and prey species in an ecosystem is an example. The basic mathematics of population growth for a single species (*see* EXPONENTIAL GROWTH) shows how limits on population density can be described by simple equations that yield a characteristic S-shaped curve. This curve represents a population growing exponentially when it is small, then leveling off as it approaches the limits of the ecosystem's ability to support it. A simple extension of this idea allows us to understand an ecosystem in which two species—predator and prey—interact.

So, if the number of herbivorous prey is H, and the number of carnivorous predators is C, then the probability that a predator will encounter a herbivore is proportional to the product HC. In other words, the more numerous is either species, the more likely are encounters between the two. In the absence of predators, the prey population will grow exponentially (at least initially), and in the absence of prey the predator population will shrink to zero, either through starvation or migration. Now, if dH is the change in the herbivore population in a time dt, and dC is the change in the carnivore population in the same time interval, then the two populations are described by the equations

$$dH/dt = rH - AHC \quad \text{and} \quad dC/dt = -qC + BHC$$

Here, r is the growth rate of the herbivore in the absence of predators, and q is the rate of decline of carnivores in the absence of herbivores. The constants A and B are the rates at which predator–prey encounters remove herbivores from the population and the rate at which these encounters allow predators to add to their population. The minus sign in the first equation indicates that the encounters lower the prey population, and the plus sign in the second indicates that the encounters increase the predator population. As you can see, any change in the number of herbivores affects the number of carnivores, and vice versa. The two populations have to be considered together.

The solution to these equations shows both populations going through cycles. If the herbivore population increases, the chance of a predator–prey encounter goes up, and so (after a suitable time delay) does the population of predators. But the increase in the predator population drives the herbivore population down (again, after a suitable delay), which lowers the number of predator offspring, which increases the number of herbivores, and so on. The two populations do a waltz through time, a change in one generating a change in the other.

Proteins

Proteins are strings of amino acids, and their most important function is to act as enzymes that govern the chemical reactions of living organisms

Living systems are, at bottom, chemical in nature. In every cell in your body, thousands of chemical reactions are taking place, and it is the sum of these reactions that makes you what you are. In this grand chemical scheme, molecules called proteins play a crucial role.

Let's begin our discussion of proteins by talking about how they are made. There are two ways that you can imagine building complex molecules: by using either a modular system, in which many large molecules can be assembled from a small number of building blocks, or a system in which each molecule is custom-made. Think of old and new methods of constructing buildings. Formerly, parts were usually made for one building alone, and did not appear in other buildings. This kind of structure, if it can be restored, is considered to be very beautiful and desirable today. The modern way to construct a building is to take ready-made parts, or modules, like bricks, windows, and doors, and put them together. In a system like this, the same parts, assembled in different ways, can give rise to widely varied structures. Similarly, in living systems, complexity is achieved through modularity. This is what you would expect from the theory of EVOLUTION, because it allows a steady progression in complexity as new modules are introduced.

The basic module from which proteins are built is called an amino acid. These are a class of molecules that can be thought of as having a common structure, but with variation in the detail. There is a backbone of atoms, with a positively charged hydrogen ion (H^+) on one side and a negatively charged combination of oxygen and hydrogen, called a hydroxyl group (OH^-), on the other. Branching off from this main chain is a side group, which differs from one amino acid to another. In living systems there are 21 different kinds of amino acids present.

A protein is made from amino acids in a procedure somewhat similar to stringing beads together to make a necklace. Two amino acids come together, and the H^+ from one side combines with the OH^- from the other to form a molecule of water, leaving the amino acids linked together. Thus, complex permutations are possible. The sequence of amino acids along the "necklace" is called the *primary structure* of the protein. Because each "bead" on the necklace can be any one of the 21 amino acids, there are a huge number of possible primary structures even for short proteins. For example, there are over 10 million *billion* different ways of constructing a protein that is only ten amino acids long!

Once the primary structure of a protein has been determined, the electrical interactions between various side groups of the amino acids, as well as between the amino acids and the surrounding water, cause the protein to assume a complicated three-dimensional shape. The most important proteins from our point of view are those that fold up into a complex spherical shape, because these are the proteins that govern chemical reactions in living things. (Other kinds of proteins, such as those in hair and other bodily structures, have different shapes.)

When complex molecules interact with each other, they form a CHEMICAL BOND between specific atomic sites, one in each molecule. It is not enough for two molecules to be capable of interacting, but they must also approach each other aligned in such a way that the atoms that are capable of forming chemical bonds can do so, like two spacecraft docking in orbit. Three-dimensional geometry therefore plays an extremely important role in the chemistry of living systems.

It is extremely unlikely that two complex molecules, left to themselves, will just happen to have the right orientation to allow an interaction to take place. What they need in order for that reaction to happen at an appreciable rate is the action of a type of molecule known as an *enzyme* (*see* CATALYSTS AND ENZYMES). An enzyme attracts two other molecules to itself and holds them in the right orientation so that the molecules can interact. Once the interaction has taken place, the enzyme molecule has done its job, and is free to repeat the task with a further pair of molecules.

Because of their complex shape, proteins make ideal enzymes. For each primary structure there is a specific shape for the protein, and therefore a specific chemical reaction for which that protein can serve as an enzyme. In living organisms the primary structure of proteins is coded for in DNA (*see* MOLECULAR BIOLOGY). Thus DNA controls an organism by determining what kinds of proteins can be made, and hence which chemical reactions can take place.

In principle, it should be possible to predict the shape of a protein from its primary structure, and therefore to predict the nature of the chemical reaction in which it will participate. In fact, this so-called *protein folding problem* is so complex that it cannot yet be solved, even with the best computers and computational techniques available. It remains one of the outstanding problems in the field of molecular biology.

Quantum Chromo-dynamics

The strong force, which holds quarks together in elementary particles, is itself mediated by the exchange of particles called color gluons

In the STANDARD MODEL, our current best theory of the ultimate nature of matter, QUARKS come together to make up all of the different ELEMENTARY PARTICLES, which in turn make up the nuclei of atoms. Quantum chromodynamics (QCD for short) is the theory that describes how the quarks interact with one another through the exchange of other particles called *gluons.*

In the standard Newtonian physics, a force is simply a push or a pull—something that causes a change in the motion of an object. But in modern quantum theories the force, now visualized as acting between particles, is given a somewhat different interpretation. It is seen as the result of the exchange of one kind of particle between two others.

An analogy may help here. Imagine two ice skaters approaching each other across a rink., As they approach, one skater throws a bucket of water at the other. The skater who does the throwing will recoil—and will therefore change direction—and the recipient will also recoil and change direction when the water hits. It is because of the exchange of the water that both skaters change their direction of motion. According to NEWTON'S LAWS OF MOTION, this means that a force has acted. In this example, it's not hard to see that the force comes from (a physicist would say "is mediated by" or "is generated by") the exchange of the water.

All modern theories that describe forces aspire to do so in terms of this sort of exchange (*see* THEORIES OF EVERYTHING). These ideas are called *gauge theories,* and are based on various symmetries and invariances in the system of particles and fields. The equations describing the system remain unchanged when an operation is applied to all the particles involved. For example, if the positive and negative electric charges within a system are swapped, the forces between the particles remain the same.

QCD gets its name from the first successful theory of this type, which is called quantum electrodynamics, or QED.* In the QED theory, the electromagnetic force between electrically charged particles is mediated by the exchange of photons (which you can think of as being something like bundles of light).

QCD operates in the same way, except that instead of electrical charge there is a different property that determines the way quarks interact with one another. It is called *color,* and it can have three different values (or "hues," perhaps). Physicists have chosen to call them *red, green,* and *blue,* but you shouldn't take this literally. Unfortunately, in the 1970s particle physicists went through a period of silliness and saddled the subject with names that, in retrospect, could have been better chosen—"charm" and "strangeness" are two other examples. In any case, saying that a quark has a red color charge has no more (and no less) meaning than saying that an electron has a negative electrical charge.

---

* The significance of the initials is not lost on physicists—they are also traditionally placed at the end of a rigorous mathematical proof.

The gauge theories for QED and QCD differ in one important respect—the type of symmetry involved. In QED, undertaking one operation followed by another gives the same final result as doing the same operations but in the reverse order. However, this is not the case in QCD, which makes it a good deal more complex than QED.

Color is not a property that can be attributed to particles composed of quarks. Particles such as the proton and neutron are made of three quarks (red, green, and blue) that, taken together, make neutral. The other kind of particles made from quarks are made from quark–antiquark pairs and thus are also white. One rule of QCD, then, is that the only color combinations allowed in nature are those where the colors of the quarks cancel each other out.

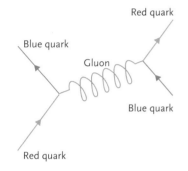

Red quark

Blue quark

Gluon

Blue quark

Red quark

Interactions between particles can be represented in a Feynman diagram, named after physicist Richard Feynman (1918–88). In this diagram, a red quark and a blue quark exchange a gluon and change color as a result.

The force between quarks is mediated by a family of eight particles called gluons (they "glue" the quarks together) that are exchanged between quarks just as the bucket of water is exchanged between skaters in our analogy. Unlike the photon in QED, which never carries electrical charge even though it acts between electrically charged particles, the gluons do carry the color charge and can change the identity of the quarks with which they interact. (It would be as though the bucket of water could change the gender of the skaters, turning men into women and vice versa!) For example, if the absorption of a gluon changes a blue quark into a red quark, then the gluon must have carried one unit of red charge and minus one unit of blue charge. So long as the total color charge stays the same, such interactions are allowed (in fact, they turn out to be necessary) in QCD.

QCD has been around since the early 1980s, and has actually passed a number of experimental tests in which its predictions of the outcome of various high-energy collisions between particles were compared to actual data. At the moment, it seems to be doing quite well. In fact, physicists routinely use the theory to design new experiments, assuming that it will give them realistic results. What more could you ask of any theory?

Quantum Mechanics

In the subatomic world, particles must be described by wave functions

The word "quantum" comes from the Latin for "so much," or "bundle." "Mechanics" is the old-fashioned word for the science of motion. Quantum mechanics, then, is the study of the motion of things which come in bundles (or, to use the modern term, which are *quantized*). "Quantum" was first used by the German physicist Max Planck (*see* PLANCK CONSTANT) to describe the interaction of light with atoms.

Quantum mechanics often defies our commonsense expectations of how things should behave. This is because our "common sense" comes from our everyday experience with common, macroscopic objects, and things behave differently on an atomic scale. HEISENBERG'S UNCERTAINTY PRINCIPLE reveals some of these differences very clearly. In the macroscopic world, when we measure the position of something—this book, say—there is no ambiguity in the result. Whether we use a ruler, radar, or an acoustic "tape measure," or take a photograph and measure its position on that, or use some other approach, the act of measuring does not affect the book's position (of course if you're clumsy with the ruler that might not be true!). There will be some uncertainty in the position we measure, determined by how accurate our measuring device is, but if we need a more accurate position then we simply use a more accurate device.

Now, if instead of this book it is an electron whose position we want to know, we find that we cannot ignore the interaction associated with the process of making the measurement. The force exerted on the book by the ruler or whatever is used to measure its position is negligible, but to determine where an electron is we have to bounce a photon, another electron, or some other particle off it, and then the force exerted on the electron is far from negligible. The very act of measuring the position of the electron therefore causes it to change its position, so there is an intrinsic uncertainty in the result—it has nothing at all to do with the quality of our measuring device. This, then, is the situation in the subatomic world. We cannot make a measurement without an interaction, and we cannot have interactions that do not affect the thing being measured.

A precise statement can be made about the effects of the interactions:

$$\text{uncertainty in position} \times \text{uncertainty in velocity} > h/m$$

or, in fully mathematical form,

$$\Delta x \times \Delta v > h/m$$

where Δx and Δv are respectively the uncertainties in position and velocity of the particle being measured. The symbol h on the right sides of these equations is the PLANCK CONSTANT, and m is the particle's mass.

So, it is not just the position of the electron that has an intrinsic uncertainty when it is measured, but also other properties of the electron, such as its velocity. The precision with which the value of one such pair of linked quantities can be found (another linked pair is the electron's energy and the instant of time at which the energy is measured) directly affects the precision that is possible for the other quantity. If we measure the position of an electron very accurately, then our measurement of its velocity *at the same instant* will be very ambiguous, and vice versa. In fact, if we were able to measure the position of the electron exactly, we would have absolutely no idea of its velocity, and an electron whose velocity we could measure exactly could be anywhere in the entire universe! Actual measurements fall between these two extremes. For example, if we can determine the position of an electron to within 10^{-6} meter, we can at best only find its velocity to within 650 meters per second.

Because of the uncertainty principle, the way in which the state of some entity is described in the quantum world has to be different from the way objects are described in our everyday Newtonian world. Instead of using position and velocity, as you would for a billiard ball, for example, quantum objects are described by something called a *wave function*. The height of the "wave" at a point represents the probability that the particle would be found there if a measurement was made. The progression of this wave is described by SCHRÖDINGER'S EQUATION, which tells us how the state of a quantum system changes with time.

The picture of quantum events painted by Schrödinger's equation is that electrons and other particles are similar to tidal waves moving across the surface of the ocean. Over time, the peak of the wave (corresponding to the place where an electron, say, is most likely to be found) moves from place to place in the manner prescribed by the equation. What we think of as a particle, then, actually displays some of the characteristics of a wave in the quantum world.

The reconciliation of the wave and particle aspects of quantum objects (*see* DE BROGLIE RELATION) came when physicists accepted that quantum objects are neither particles nor waves, but some other entity that has properties similar to both but which has no analog in the Newtonian world. Although this resolution of quantum paradoxes still causes puzzlement (*see*, e.g., BELL'S THEOREM), the theories that quantum mechanics has thrown up remain our best understanding of the subatomic world.

Quantum Tunneling

*It is sometimes possible
for a quantum particle
to penetrate a barrier
that cannot be crossed
by a classical particle*

Imagine a ball rolling around in a bowl-shaped hollow in the ground. At any moment, the ball's energy is split between the kinetic energy associated with its motion and the gravitational potential energy associated with how far up the wall of the hollow it is (*see* the first law of THERMO-DYNAMICS). As the ball rolls up toward the lip of the bowl, one of two things can happen. If the total energy of the ball is greater than the potential energy it would have at the top of the hollow, then the ball will be able to roll out of the hollow. But if the total energy of the ball is less than the potential energy at the top of the hollow, then the ball will slow down as it rolls up the side, stop at the level where the potential energy *is* equal to the total energy of the ball, and then roll back down. In the latter case, the only way to get the ball out of the hollow is to give it more energy—by kicking it, for example. There is no way consistent with NEWTON'S LAWS OF MOTION that the ball can escape if it doesn't have enough energy to get it over the top.

Imagine now that the hollow has a raised rim, rather like a crater on the Moon. If the ball can get over the top of the rim, it will keep on rolling. What is important to bear in mind is that, in the Newtonian world inhabited by the ball and the hollow, the fact that the ball will keep on rolling once it is over the rim is irrelevant unless it has enough energy to get it over the rim in the first place. Unless it does, it will never escape from the hollow, no matter how low the ground on the other side is.

In the world of QUANTUM MECHANICS, though, things are different. Imagine a quantum particle in a counterpart of that hollow. This isn't a physical hollow, but some situation in which the particle needs an energy boost to escape—what physicists call a *potential well*. There is also an energy equivalent of the rim around the hollow—a *potential barrier*. Now, if the energy level outside the potential well is less than the energy possessed by the particle, then there is a mechanism by which the particle can penetrate the barrier, even if it can't climb over the top in the Newtonian sense. This mechanism is called quantum tunneling.

Here's how it works: In quantum mechanics, the particle is described by a wave function, which is related to the likelihood that the particle will be found at a specific position. When a quantum particle meets a potential barrier, SCHRÖDINGER'S EQUATION tells us that the wave function does not simply vanish inside the barrier, but drops off very quickly—exponentially, in fact. Potential barriers in the quantum world are fuzzy. They do inhibit the motion of particles, but they are not the solid, impenetrable boundaries of the Newtonian world.

If the barrier is thin enough, or if the total energy of the particle is close enough to the energy at the top of the barrier, the wave function, though dropping rapidly, can actually extend all the way through the barrier. There is then a possibility that the particle will be located on the other side of the barrier—something which could never happen in the Newtonian world. If the barrier is similar to the lunar crater, then once the particle has penetrated to the other side, it is free to move down the slope and away from the hollow within which it had been trapped.

Quantum tunneling can be regarded as the particle "leaking" through the barrier, then taking off once it gets to the other side. There are many examples of this process in nature, and in modern technology as well. For example, a common type of RADIOACTIVE DECAY is when a large nucleus emits an alpha particle, which consists of two protons and two neutrons. One way to picture this process is to imagine that an alpha particle is held inside by nuclear forces, much as the ball is held in the hollow in our example. Even if the alpha particle doesn't have enough energy to tear itself loose from the nucleus, it can still tunnel its way out through the barrier that confines it. When we see that particle outside, we know that the tunneling process is complete.

Another important example of tunneling in nature occurs in the nuclear fusion process that powers stars (*see* STELLAR EVOLUTION). One step in the process is the coming together of two deuterium nuclei (each consisting of a proton and a neutron) to make a nucleus of helium-3 (two protons and a neutron). Because of COULOMB'S LAW, there is a strong electrical repulsion—a barrier—between the two deuterium nuclei. In a Newtonian world, these nuclei could not get close enough together to undergo fusion. In the core of a star, however, the temperature is so high that the nuclei are almost at the top of the barrier, so tunneling proceeds quickly and the star shines.

A picture of atoms taken with a scanning tunneling microscope, which utilizes the phenomenon of quantum tunneling. The gold atoms (shown here as yellow, red, and brown) form a layer three atoms thick on a graphite substrate.

Tunneling also plays a crucial role in the scanning tunneling microscope (STM). In this instrument, a sharp metal tip is brought near a surface. There is a barrier that acts to keep electrons in the metal from flowing into the surface. As the tip of the microscope is moved across the surface, it moves past the atoms within the surface. The barrier is lower in the vicinity of an atom than when the tip is between atoms. Thus, in the neighborhood of an atom, there will be a relatively large current flowing through the point as electrons tunnel out, while between atoms the current will drop. Thus, by monitoring the current through the point, we can create a "picture" of individual atoms in a surface. In a sense, then, the STM provides us with the ultimate proof of the existence of atoms.

Quarks and the Eightfold Way

The fundamental constituents of the particles in the atomic nucleus are quarks

For the past two centuries, scientists interested in the structure of the universe have searched for the basic building-block of matter—the ultimate constituent from which everything else is built. The ATOMIC THEORY explained the plethora of chemical compounds by postulating that there were a small number of elementary entities called atoms, and that it was the combinations of these that gave the chemical world its diversity. Thus, complexity at one level gave way to simplicity at a lower level.

But the simple atomic picture soon ran into trouble. For one thing, many kinds of atoms were found, and for another, unexplained regularities among the atoms began to appear, which became clearer as the PERIODIC TABLE of the elements began to be assembled. Once again the world started to look complex.

In the early decades of the 20th century atoms were found to be not elementary at all, but were made of things more elementary still—the protons and neutrons that made up the atomic nucleus, and the electrons that surrounded the nucleus. Once again, complexity at one level was replaced by simplicity at the next level down. But this apparent simplicity itself began to break down as more and more ELEMENTARY PARTICLES were discovered. The most difficult of these to understand were the so-called *hadrons*—particles that appeared to have a fleeting existence within the nucleus.

Furthermore, unexplained regularities—in what was the particle physicists' equivalent of the periodic table—were found among hadrons. Using clues from a branch of mathematics called *group theory*, physicists were able to show that when the hadrons were put together in groups of eight, arranged around the points of a hexagon with two at the center, the particles that lay along lines passing through the center of the hexagon had properties in common, just as elements in the same column in the periodic table have similar chemical properties. The arrangements were given the name of the *eightfold way* (a reference to a concept from Buddhist theology). In the early 1960s, theorists saw that these regularities could be understood if the elementary particles weren't elementary at all, but were made up of entities more elementary still.

These entities were given the name of *quarks,* a word taken from a line in James Joyce's famously perplexing novel *Finnegans Wake*. These new denizens of the subatomic world were strange beasts. For one

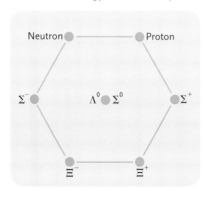

A diagrammatic arrangement of eight particles in a hexagon. The particles at the bottom are the Xi particles, which come in two varieties, with positive and negative charges. Across the center are the three Sigma particles, with positive, zero, and negative charges. The eighth particle is the Lambda zero, indistinguishable from the Sigma zero yet distinct— each a kind of doppelganger of the other.

thing, they had fractional electrical charges: ⅓ and ⅔ of the charge on the electron or proton (*see* the table). For another, as the theory developed it became clear that quarks can never be seen as free particles, but have an existence that can only be inferred from the properties of the particles made from them. You can get some sense of this notion, called *quark confinement,* by imagining that you have an elastic string, each end which represents a quark. If you add energy to the system by stretching the string, you may be able to break it. If you do, however, you don't get a free end, but two shorter lengths of string, each of which has two ends. In the same way, trying to pull a quark out of an elementary particle just creates more "ends" and more particles, but never a free quark.

Today, our theories tell us there should be six quarks, and indeed particles containing all six have been found in the laboratory. The commonest quarks are the p and n quarks (named in analogy with the proton and neutron). The next doublet are the s and c quarks (for *strange* and *charmed*), while the last doublet, most recently discovered, are the b and t (for *bottom* and *top*). Each of these quark "flavors" comes in three *colors.* This doesn't mean that they actually have a color in the conventional sense—"color" in this sense is a just a label for a property of particles like electrical charge (*see* QUANTUM CHROMODYNAMICS).

In the STANDARD MODEL, our best theory of the elementary nature of the universe, the succession of deeper levels of reality ends with quarks—there is nothing more elementary. Some physicists have speculated that we might go on "peeling the onion," but that's just speculation. My own guess is that the standard model is right and that, in this piece of science at least, we've come to the end of the line.

The six types of quark, and their charge as a fraction of the charge on the electron. In addition, each quark has its own ANTIPARTICLE.

Quark		Charge
u *or* p	(up *or* proton)	$+\frac{2}{3}$
d *or* n	(down *or* neutron)	$-\frac{1}{3}$
c	(charmed)	$+\frac{2}{3}$
s	(strange)	$-\frac{1}{3}$
b	(bottom)	$+\frac{2}{3}$
t	(top)	$-\frac{1}{3}$

MURRAY GELL-MANN (1929–) American physicist. He was born in New York City to Austrian immigrant parents. Gell-Mann's book *The Eightfold Way* (1964), cowritten with Yuval Ne'eman, made sense of the burgeoning numbers of subatomic particles in the same way that Dimitri Mendeleyev had found patterns among the elements. This work, and Gell-Mann's theory of quarks and sub-atomic interactions, won him the 1969 Nobel Prize for Physics. Gell-Man later turned to the study of COMPLEX ADAPTIVE SYSTEMS.

Radioactive Decay

The number of nuclei in a sample of radioactive material that decay in a given time period is proportional to the number of nuclei present

Most atomic nuclei are unstable. Sooner or later, they will spontaneously disintegrate, leaving behind pieces of themselves as decay products. (*Parent nuclei* are said to change into *daughter nuclei*.) Every familiar element—iron, oxygen, calcium, and so on—has at least one stable isotope. (Different isotopes of an element have the same number of protons in their nuclei, but different numbers of neutrons.) The fact that they are familiar implies that they stay around for a long time, that they last a long time without decaying. But every element has a number of unstable isotopes—nuclei which can be created by nuclear processes and which then proceed to decay over time.

There are three common ways in which a radioactive nucleus can decay, known by the first three letters of the Greek alphabet. In *alpha decay,* a nucleus of helium is emitted. This consists of two protons and two neutrons, and is often called an alpha particle. Since the loss of the helium nucleus removes two protons from the parent nucleus, it changes into the nucleus of an element two places nearer the beginning of the PERIODIC TABLE. In *beta decay,* an electron is lost from the nucleus, and the element moves one position toward the end of the periodic table (effectively, a neutron has turned into a proton and an electron). *Gamma decay* is the emission of a high-energy photon (a gamma ray). The nucleus thus loses energy, but the element stays the same.

The fact that a particular isotope is unstable, however, doesn't mean that if you have a collection of its nuclei they will all decay at once. Watching radioactive nuclei decay is a lot like watching popcorn pop. The kernels don't pop all at once, but sporadically here and there until they have all had their turn. Radioactive nuclei are the same—if you watch a sample, the nuclei decay one by one, just like popcorn kernels popping. The law cited above simply says that the number of nuclei decaying in any given time interval is proportional to the number of nuclei still there—to the number of "unpopped kernels." In equation form, if N is the number of undecayed nuclei at a given time, and dN is the number that decay in a time interval dt, then

$$dN = \lambda N \, dt$$

where λ is known as the *decay constant* and is determined by measurement. From this equation, it follows that the number of undecayed nuclei left after a given time t is given by

$$N = N_0 \, e^{-\lambda t}$$

where N_0 is the number of nuclei that were present at the beginning.

The decay constant determines how quickly the nuclei decay, but it's more usual to measure the rate of decay in terms of a quantity called the *half-life.* This is the amount of time it takes for half of the original sample of nuclei to undergo decay. The half-lives of isotopes can (and do) range from microseconds to billions of years: Some radioactive nuclei stick around for eons, while others are gone in a flash. (Note that after one half-life, half of the original complement will be present; after two half-lives, one-quarter; after three half-lives, one-eighth; and so on.)

There are many ways in which radioactive elements can be created. For example, the Earth's atmosphere is constantly being bombarded by cosmic rays—high-energy particles from space (*see* ELEMENTARY PAR- TICLES). When these particles hit the nuclei of atoms in the Earth's atmosphere, they create fragments that include radioactive isotopes of some elements. The isotope carbon-14, with 6 protons and 8 neutrons (*see* RADIOMETRIC DATING), for example, is constantly being made in this way, from nitrogen-14 in the atmosphere.

More commonly, however, radioactive elements are created in one stage of a *decay sequence* (or *decay chain*). This is a string of events in which the original parent nuclei decay into daughter nuclei that are themselves radioactive and decay into radioactive "granddaughters," and so on until the sequence finally ends in a stable isotope. For example, the isotope uranium-238 (92 protons, 146 neutrons) has a half-life of about 4.5 billion years. This is the same as the age of the Earth, so the entire his- tory of our planet has been only one half-life of this isotope (which means that half of the uranium-238 that was here when the Earth formed is still around). Uranium-238 decays into thorium-234 (90 protons, 144 neutrons), which has a half-life of 24 days. The thorium-234 decays into palladium-234 (91 protons, 143 neutrons), which has a half-life of about 6 hours … and so on. This process goes through about a dozen decays until it reaches the stable isotope lead-206.

Several points can be made about decay sequences. First, even if we start with a pure sample of a single radioactive isotope, we will soon have many other elements mixed in as the inexorable process of nuclear decay goes on. Second, decay sequences show that radioactivity has always been a natural phenomenon, and is not a recent human addition. The ura- nium-238 that is initiating decay sequences as you read these words has been around since the Earth began and has nothing to do with the evolu- tion of human beings, much less with the introduction of nuclear energy.

Finally, the inevitability of the decay process can lead to problems (and opportunities) for human beings. The decay sequence initiated by uranium-238, for example, produces a colorless, odorless gas called radon-222. This is a noble gas that doesn't form CHEMICAL BONDS. It is chemically inert, so that it just bubbles up out of the ground. Normally, it has no effect on us—it just dissipates into the air until it too decays. If it enters an unventilated basement, however, it can accumulate and produce decay products that, if inhaled, can constitute a health hazard. This is the so-called indoor radon problem.

On the other hand, radioactivity can be useful. If, for example, a sample of radioactive phosphorus is injected into the veins of someone with a broken bone, it will migrate (as phosphorus does normally) to active growth surfaces on bones. There it decays, sending its (harmless) decay products outside the body, where they can be detected and used to assemble a picture of the bone surface. Physicians can use that image to guide them in setting difficult breaks. There are hundreds of uses of these *radioactive tracers* in medicine and industry. They all work the same way: The chemistry of the atom—in essence, the behavior of the electrons—takes the atom somewhere, then the nucleus, operating according to its own timetable, decays, sending out decay products that can be detected.

The radioactivity of a sample of a radioactive substance falls away over time. In fact, it's the opposite of EXPONENTIAL GROWTH.

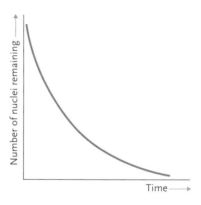

Radiometric Dating

*It is possible to measure
the age of an object if
it contains the products
of radioactive decay*

Scientists often want to know the age of an object. An archeologist might want to know when a particular piece of pottery was made, for example, or a paleontologist might want to date a fossil. The standard technique for doing this sort of thing makes use of radioactive isotopes in the object being dated (*see* RADIOACTIVE DECAY). It can be applied to objects that contain an isotope whose half-life is known, because it is then possible to determine how much of that isotope was in an object when it was made (or in an organism when it died). Comparing this quantity with the amount that hasn't yet undergone radioactive decay reveals how many half-lives of the isotope have passed since the object was made, and hence gives the age of the object. For example, if there is half as much of the isotope now as there was when the object was made, then the object is one half-life old.

The hard part is to figure out how much of a given isotope was present when an object was made (the half-lives of isotopes are quite easy to measure). The first practical radiometric dating technique, developed in the middle of the 20th century at the University of Chicago, used the isotope carbon-14, and is a good example of how this procedure works. Carbon-14 is an isotope of carbon that has 6 protons and 8 neutrons in its nucleus. It is a slightly heavier version of normal carbon (carbon-12, with just six neutrons), and participates in chemical reactions in just the same way that carbon-12 does. Right now, carbon-14 is being created high in the atmosphere by cosmic rays colliding with nitrogen nuclei. It pervades the atmosphere, and is taken up by plants and incorporated into the structures of animals that feed on those plants. Roughly speaking, in living things about one carbon atom in a million is carbon-14.

Carbon-14 is unstable, with a half-life of about 5700 years. For as long as an organism is alive, any carbon-14 that decays in it is replaced by carbon-14 from the environment. When an organism dies, however, it stops cycling carbon (*see* CARBON CYCLE) and the carbon-14 begins its inexorable decay. After 5700 years, a piece of wood will have only half the amount of carbon-14 it had when the tree of which it was a part died. If that tree was cut down to make a piece of furniture, then this carbon-14 date gives a pretty good estimate of when the furniture was made. Most dating of organic materials—papyrus scrolls, partially burned wood from ancient campfires, leather items—is done in this way.

In the case of carbon-14, the amount of isotope present at the start can be determined directly. This is because there is a species of tree, the bristlecone pine, that lives for thousands of years. By comparing the patterns of tree rings in living trees with those in dead ones, scientists can construct a continuous record of tree rings going back some 10,000 years (this is called *dendrochronology*). Counting back along the rings allows

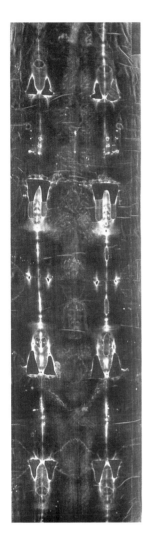

The Turin Shroud, which was believed by some to have been wrapped around the body of Christ and to bear his image. Radiometric dating carried out in 1988 shows the cloth to have been made no earlier than 1260.

you to determine exactly when a given ring was added to a tree, and measuring the amount of carbon-14 left in that ring tells us how much carbon-14 there was in the environment when it formed.

Carbon-14 dating (also known as *radiocarbon dating,* or simply *carbon dating*) is widely used for objects that are up to tens of thousands of years old—and that includes most archeological artifacts. The practical limit of modern techniques is about 50,000 years. In objects older than that, there just isn't enough carbon-14 left in them. To find the age of older objects, such as rocks and meteorites, other "clocks" have to be found.

Potassium–argon dating is a radiometric dating method that can be applied to very old rocks. Potassium is a common atom found in minerals, and the isotope potassium-40 decays into argon-40 (a noble gas—one that's extremely unreactive) with a half-life of 1.25 billion years. (The argon remains trapped in the structure of the rock because its atoms are too large to squeeze out of the crystal structure.) The amount of potassium in a given mineral is fixed by the mineral structure, and the fraction of potassium-40 is fixed. To apply potassium–argon dating, a suitably prepared sample of mineral is ground up and the amount of argon-40 present is measured. Each argon-40 nucleus comes from the decay of a potassium-40 nucleus, so the number of decays (and hence the number of half-lives) that have elapsed since the mineral formed can be determined. Similar techniques based on the decay of uranium-238 to lead-206 (half-life 4.5 billion years) and rubidium-87 to strontium-87 (half-life 49 billion years) are also widely used. It is the use of methods such as these to date meteorites and Moon rocks that has allowed planetary scientists to estimate the age of the solar system.

Two points should be made about the radiometric dating of rocks. First, if a rock melts, the radiometric clock is reset—for example, accumulated argon-40 will escape from a mineral if it is melted. What radiometric dating actually measures is the time that has elapsed since the rock last solidified from a molten state. (The same material may have undergone several episodes of melting and freezing since it was first formed.) The second point is that it cannot be used to date sedimentary rocks, which are made from bits and pieces of an assortment of rocks (*see* ROCK CYCLE). The technique could tell you when each grain crystallized, but not when the grains were assembled into the rock. Since fossils occur only in sedimentary rocks, this means that the dating of fossils requires some care. Typically, a fossil-bearing stratum of undatable sedimentary rocks will lie between layers of lava or volcanic ash, which can be dated. The date of the fossils, then, lies between the dates of the strata above and below (*see* SUPERPOSITION).

Rayleigh Criterion

Two sources of light can be distinguished when viewed through an aperture if the diffraction maximum of one falls on the diffraction minimum of the other

Lord Rayleigh was one of the many "gentleman scientists" of Victorian Britain. Known for his discovery of ARGON, he made contributions to many areas of physics. It was while studying the scattering of light that he arrived at the criterion that now bears his name.

You are driving down a long, straight road at night. Another car is coming in the opposite direction. At first, its headlights appear as a vague, undifferentiated blur. As it approaches, you begin to see that there are two headlights. You are, of course, observing that car through the optical instrument we call the human eye, and the light reaches your retina by passing through the iris. Now, how far away will the car be when this optical apparatus first allows you to distinguish the two sources of light?

According to the classical theory of DIFFRACTION, light from any distant object, when passed through a circular aperture, will produce an image consisting of a series of light and dark rings surrounding a bright central spot—what is called a *diffraction pattern*. The laws of optics tell us that the original point source of light will be smeared out when that light is viewed through an optical instrument. If two light sources are close to each other, their smeared images will overlap. What Rayleigh showed was that, as long as the central bright spot of one image is at least as far away from the central bright spot of the other as the first dark ring of its own diffraction pattern, you can resolve the two. If the linear separation between the two light sources is d, and they are a distance D away from you, and the wavelength of the light is λ, and the diameter of the aperture is A, then Rayleigh's condition will be met if

$$d/D > 1.22\lambda/A$$

In other words, if the light sources are separated by at least d, then someone looking through the aperture will be able to distinguish them —otherwise, they will blur together. The ratio d/D is the angle (measured in radians; multiply the number by 57.3 to convert radians to degrees) between the two light sources. The Rayleigh criterion, then, sets the limit of the angular resolving power of any optical instrument, be it telescope, camera, or human eye. (The numerical factor of 1.22 comes from the details of the mathematics, and requires the size of the aperture and the wavelength of the light to be measured in the same units.)

Self-portrait of John Strutt, later Lord Rayleigh, made in 1870.

The criterion gives a resolution of about 25 seconds of arc, less than a hundredth of a degree, for the human eye. In practice, even someone with good eyesight can resolve only between 3 and 5 minutes of arc—poorer by a factor of about ten. (The reason for this lies mostly in the structure of the retina.) Thus a pair of car headlights should be theoretically seen, on a dead straight road, as twin light sources at a distance of about 6 miles (10 kilometers). In practice, though, they will be resolved only at a distance of around ⅔ of a mile (1 kilometer). A driver probably won't see them as two lights until they are much closer—but that is because he or she is (or should be) concentrating on the road.

JOHN WILLIAM STRUTT, THIRD BARON RAYLEIGH (1842–1919) English physicist. Rayleigh was born in Witham, Essex, and inherited his title upon the death of his father in 1873. His long and productive scientific career saw him conducting experiments at his family estate and, eventually, gain a professorship at Cambridge. His work spanned a wide variety of fields: his book *Theory of Sound,* written in 1872, is still to be found in acoustics laboratories, while his 1904 Nobel prize was for the discovery of argon, an achievement in chemistry. In between, he did important work on the interaction of light with matter. He was, for example, the first to prove that blue light is scattered by atoms more readily than red—the key to understanding why the sky is blue (*see* DISPERSION).

Reflection, law of

When light is reflected from a surface, the angle of incidence equals the angle of reflection

Imagine shining the beam of a flashlight on a flat reflecting surface—a mirror, perhaps, or a polished piece of metal. You know that the beam will be reflected, and that the reflected beam will leave the surface traveling in a specific direction. In this situation, the angle between a line drawn perpendicular to the surface and the incoming beam of light is called the *angle of incidence,* while the angle between that perpendicular and the outgoing beam is called the *angle of reflection.* The law of reflection says that these two angles must be equal. This accords with our intuition. A beam coming in at a low angle, barely grazing the surface, will go out in similar fashion—low, and close to the surface. A beam coming in at a steep angle, close to the vertical, will be reflected at the same steep angle, as the law requires.

The law of reflection, like all laws of nature, is derived from observation and experiment. It can also be derived more formally from FERMAT'S PRINCIPLE (although that doesn't invalidate its grounding in experiment).

The key point about the law is that it defines all angles with respect to the perpendicular to the surface at any point. For a flat surface such as a mirror, this isn't particularly important because the perpendicular to the surface always points in the same direction. A parallel beam of light, as emitted by a flashlight, can be thought of as a bundle of parallel rays of light. Adjacent rays in an incoming beam striking a flat surface will therefore be reflected through the same angle, and in the outgoing beam the rays will retain their positions relative to one another. This is how a flat mirror forms a faithful image.

For a rough surface, however, the direction of the perpendicular varies from one spot to the next. This means that adjacent rays in a light beam will be reflected through different angles, so that the image in the incident beam is lost. This is why rough surfaces can reflect light but do not act as mirrors.

Not all mirrors are flat. Curved mirrors are useful because they change the image in some way. The main mirror in a reflecting telescope is concave so that light from a distant object can be brought to a focus. Rear view mirrors in cars are often convex to give a wider view of the road behind, and bent and twisted mirrors at funfairs show you amusing distorted images of yourself.

It is not only light that obeys the law of reflection. All electromagnetic waves—radio, microwaves, X-rays, and so on—behave in the same way. This is part of the reason that the shape of the receiver in a radio telescope or TV satellite dish has much the same shape as the main mirror of a large optical telescope—they're both designed to use reflection to focus radiation that falls on them.

Relativity

The laws of nature are the same in all frames of reference

They say that Albert Einstein had his momentous insight while riding on a streetcar in Bern, Switzerland. He is supposed to have looked at a clock-face on one of the city's towers and realized that if his streetcar were to accelerate to the speed of light, it would appear to him that the clock had stopped—that time had ceased to pass. This led him to one of the central insights of relativity—the notion that different observers will see events differently and will disagree even on such fundamentals as time and distance.

In technical language, what Einstein saw that day was the fact that the description of any physical event depends on, or is "relative" to, the *frame of reference* of the observer (*see* CORIOLIS FORCE). If someone in a moving streetcar drops her sunglasses, for example, she will see the glasses fall straight down, but you, standing at the side of the road, will see them fall in an arc because the car moves while the glasses are falling. Each of you has a different frame of reference.

But while descriptions of events may vary from one frame of reference to another, some things do not. If instead of asking for a description of the fall of the glasses you had asked for the law of nature that governed their fall, both the stationary and the moving observer would have given the same answer. The law of COMPOUND MOTION would be valid both on the ground and in the car. In other words, whereas descriptions of events are relative to the observer, the laws of nature are not. In technical language, the laws of nature are said to be *invariant*. This is the *principle of relativity*.

Like all hypotheses, the principle of relativity has to be tested against nature. From this principle, Einstein produced two rather different (though related) theories. The *special theory of relativity* proceeds from the proposition that the laws of nature are the same in all frames of reference in a state of constant motion (i.e., unaccelerated frames of reference), while the *general theory of relativity* extends that proposition to all frames of reference, including those that are accelerating. The special theory was published in 1905, while the general theory, technically more difficult, was not completed until 1916.

Special relativity

Most of the counterintuitive outcomes of travelling close to the speed of light are predictions which come out of the special theory. The most famous is the slowing down of moving clocks—an effect called *time dilation*. A clock that is moving with respect to a particular observer will appear to be ticking slower than an identical clock in the observer's own frame of reference.

As well as time dilation, the lengths along the direction of motion of objects moving at close to light speed are reduced when seen by an external observer. This effect, known as *Lorentz–Fitzgerald contraction,* was first derived by Hendrick Lorentz (1853–1928) and George Fitzgerald (1851–1901) to explain why the MICHELSON–MORLEY EXPERIMENT had failed to detect the Earth's changing motion through space. Their equations were later incorporated into special relativity by Einstein. The masses of objects also increase as their velocities approach the speed of light. At 260,000 km per second (87% of the speed of light), the mass of an object seen by an outside observer will have doubled.

Since Einstein's time, all of these predictions, as counterintuitive as they seem, have been amply confirmed by direct experiment. In one dramatic experiment, scientists at the University of Michigan strapped an atomic clock to a seat in a commercial airliner that regularly flew around the world. They found that when the clock returned to its home base it had run ever so slightly more slowly than an identical clock that had been left on the ground. Giant machines called accelerators have been the main tool of particle physicists for the past half century. In these machines, beams of subatomic particles such as protons and electrons are accelerated to almost the speed of light, then made to strike targets. In these machines, the particles' increase in mass have to be taken into account—if they weren't, the experiments carried out in the accelerators wouldn't work as intended. In this sense, then, special relativity has moved from being a new theory to being an engineering tool, just like NEWTON'S LAWS OF MOTION.

Speaking of Newton's laws, I should point out that special relativity, though it appears to be profoundly non-Newtonian, in fact reproduces

Albert Einstein at a blackboard with some of the equations of special relativity. The revolutionary theories of relativity and QUANTUM MECHANICS *provided the foundations for 20th-century physics.*

all of normal Newtonian science when it is applied to bodies moving at speeds much less than that of light. Thus, it does not invalidate Newtonian physics so much as incorporate and extend it. (This point is discussed more fully in the Introduction.)

The principle of relativity also helps explain something that often puzzles people—the importance of the speed of light, as opposed to any other speed, in the theory. The appearance of the speed of light comes about because it is built into the laws of nature (*see* MAXWELL'S EQUATIONS). Thus, according to the principle of relativity, the speed of light in a vacuum must be the same in every frame of reference. This is counterintuitive because it implies that light from a moving flashlight, however fast it is moving, moves at the same speed, *c*, as light from a stationary flashlight. But that's how it is.

Because of its special role in the laws of nature, the speed of light plays a central role in both versions of relativity.

General relativity

General relativity applies to all frames of reference, and is mathematically much more complicated than special relativity (which explains the hiatus between the publication of the two theories). It incorporates special relativity (and hence Newton's laws) as a special case, but it goes further. Its most important result is a new theory of gravitation.

In general relativity, there are four dimensions: the three conventional spatial dimensions, plus time. The four must be considered together, so we are no longer dealing with the distance between two objects, as we would with three dimensions, but with the interval between events, which incorporates both linear distance and separation in time. Space and time are now a four-dimensional *spacetime*. In spacetime, observers moving relative to one another will disagree about sequences of events. A stationary observer might see two events at different physical locations as occurring simultaneously, while a second, moving observer sees them occurring at different times. Fortunately for our sanity, such differences between observers do not extend to our being able to see effects preceding causes.

NEWTON'S LAW OF GRAVITATION tells us that there is an attractive force between any two objects in the universe. In this way of looking at things, the Earth orbits the Sun because there is an attractive force between them. General relativity gives us an alternate way of looking at the Earth–Sun system. In this theory, the gravity is the consequence of a large mass like the Sun warping the fabric of spacetime around it. Imagine a flexible rubber sheet stretched taut, onto which is dropped a bowling ball. The sheet will deform, creating a depression around the ball.

In general relativity, the Earth rolls around the depression in spacetime the way a marble would roll around the depression in the sheet. What we perceive as gravity is an effect of the altered geometry of spacetime rather than a force in the Newtonian sense. At the moment, general relativity is our best theory of gravitation.

An artist's conception of the warping of spacetime by the presence of matter, one of the main features of general relativity. The rather small warping caused by the Sun is shown on the left, while the enormous warping caused by a black hole is shown to the right. The mass in the center is a neutron star.

General relativity turns out to be difficult to test, because it differs from Newtonian gravitation only in very small ways under normal laboratory conditions. However, it has been experimentally verified in a few crucial tests and seems to explain everything we can see in the cosmos. It explains, for example, small deviations in the orbit of the planet Mercury that cannot be explained by Newton's laws, as well as the bending of electromagnetic radiation from stars when the line of sight passes close to the edge of the Sun.

It is only in situations where there are very strong gravitational forces, that the theory of relativity predicts results that differ significantly from those predicted by Newton's laws. This means that it becomes important when considering either extremely accurate measurements of very massive objects, or BLACK HOLES, to which none of our normal intuitions apply. In the meantime, devising new tests of general relativity remains an important task for the experimental physics community.

Reproductive Strategies

There are two extremes of reproductive strategy—having many offspring with little subsequent investment, and having few offspring with a large subsequent investment

Every organism has to expend energy and resources to reproduce and, according to the theory of EVOLUTION, those strategies that produce the most successful offspring—the ones that survive to have offspring themselves—will be the ones that are adopted. There are, in fact, two extreme strategies that are seen in nature, which are called the *r strategy* and the *K strategy*, each associated with a different type of environment.

In habitats in which conditions are constant (or at least seasonably predictable), with very few fluctuations in the environment, populations tend to be more or less constant in size. Success then depends primarily on competition among adults, so the most successful strategy is to concentrate resources in a few offspring and make sure they get into the game. Leaving young to fend for themselves just doesn't work—they can't compete with adults. This strategy, exemplified most strongly among primates, who spend a great deal of time and effort raising their young, is called the K strategy. (The name arises because the population size is said to be near the environment's *carrying capacity*, which is represented by the symbol K.)

Diametrically opposed to the K strategy is one that works well in environments that are unpredictable, or suffer from intermittent periods of unfavorable circumstances, such as large storms or floods. In this kind of situation there are periodic massive extinctions, followed by periods in which populations can grow rapidly, without competition. In these periods of growth, the main competition is between young organisms and the best strategy is to have large numbers of offspring ready to move into the newly vacant niche. This strategy works even during periodic disasters, since fatalities then tend not to depend on competitive advantage, but on more or less random events. Thus, there is no advantage to be gained from nurturing offspring, since their survival doesn't really depend on how much effort the parents put into raising them. In this so-called r strategy, then, organisms produce many young, only a few of which can be expected to survive under normal circumstances, but which can take rapid advantage of the decimation of populations of other organisms that follows the disruption of the environment. Dandelions, which routinely put hundreds of seeds into the air and move in rapidly on disturbed soil areas, are an example of an organism that follows the r strategy. (This label originates in the symbol for reproduction rate, r.) In an unstable environment, the population begins at a low level and begins by following EXPONENTIAL GROWTH. The rate of reproduction is of greater significance to the r strategist than the environment's carrying capacity.

Reynolds Number

A dimensionless number derived from velocity, length, and viscosity determines whether fluid flow is smooth or turbulent

Osborne Reynolds was, in a sense, the last of the old-time mechanical Newtonians. Later in life, he developed a model of the luminiferous ether (*see* MICHELSON–MORLEY EXPERIMENT) in which mechanical ether particles rubbed against one another like marbles in a bag. He felt that "of mechanical progress there is not end … so that what for the time being may appear to be a … limit will turn out to be but a bend in the road."

To understand the importance of his most lasting discovery, you need to know a little about *dimensionless numbers*. Imagine that you have to measure a room. You could take a ruler and measure the distance from one wall to the one opposite. It might, for example, be 15 feet across if your ruler were calibrated in feet, but it might be a little under 5 meters if the ruler were metric. The actual number that represents the size of the room, in other words, can change if you change the system of units you are using, even though the room itself doesn't change at all.

However, the room has some properties that will have a fixed value whatever units of measurement we use. For instance, the shape of a rectangular room can be expressed as its aspect ratio—its length divided by its width. If the room is 20 feet long and 10 feet wide, then its aspect ratio is 2. We get the same result measuring in meters—6.096 m divided by 3.048 m is still 2. Aspect ratio is thus a dimensionless number.

Now we can get down to looking at fluid flow. Different fluids vary in their ability to flow through pipes or over surfaces. A thick gooey substance such as honey is said to have a higher *viscosity* than a thin runny material like gasoline. The viscosity is measured by a number given the symbol η, and called the *coefficient of viscosity*. Gooey substances like honey have a higher η value than do runny ones like petrol.

Reynolds found a dimensionless number associated with the flow of a viscous fluid. Although he first came to this number as the result of an exhaustive series of experimental studies, it can also be derived from NEWTON'S LAWS OF MOTION as applied to a moving fluid. The number, now known as Reynolds number and denoted by *Re*, is given by

$$Re = vL\rho/\eta$$

where ρ is the liquid's density, v is a typical fluid velocity, and L is some characteristic length, such as the diameter of a pipe through which the fluid flows. (Don't be misled by the two letters in *Re*—it's just one symbol, not R times e.)

Now, the units (or dimensions) of these quantities are:

—for η, newton-seconds per square metre, or Ns/m^2. Converting the newton to its basic units of kilogram-meters per second squared ($kg\ m/s^2$) we get $kg/m\ s$ for the units of η

— for ρ, kilograms per cubic metre, or kg/m^3

— for v, metres per second, or m/s

— for L, metres, or m

Thus the units of the Reynolds number are

$$(m/s) \times (m) \times (kg/m^3) \div (kg/m\ s)$$

which simplifies to $(m^3\ kg\ s) \div (m^3\ kg\ s)$. All the units therefore cancel out, showing that the Reynolds number is indeed dimensionless.

What Reynolds found was that when this number exceeds about 2000–3000, the flow of the fluid is completely turbulent, while when it is below a few hundred the flow is smooth. Between these values the flow is sometimes smooth, sometimes turbulent.

We can think of the Reynolds number as being something derived from experiment, nothing more. We can, however, give it an interpretation in terms of Newton's laws. A moving fluid has its own momentum, something theorists call an "inertial force." Essentially, a fluid, once set in motion, tends to remain in motion. In a viscous fluid there is a frictional force which acts to smooth things out, slow things down. The Reynolds number is the ratio between these two forces—inertial divided by viscous. High Reynolds numbers correspond to situations in which the viscous force is comparatively small, so that the fluid has little ability to smooth out turbulent eddies and ripples. Low Reynolds numbers, on the other hand, correspond to situations of high friction and smoothed flow.

The Reynolds number is useful because the behavior of a fluid in two different situations will be the same, provided the Reynolds numbers are the same in each case. If you make a scale model of an airplane and place it in a wind tunnel, for example, and adjust the airflow so that the Reynolds number for the model is the same as it would be for a full-sized airplane in flight, then you can be sure that what you see in your wind tunnel is what will happen in the air. (Now that computers have become so powerful, aeronautical engineers no longer feel the need to use wind tunnels to test their designs—they do it by computer simulation. The Boeing 747 was the first aircraft to be designed in this way. Oddly enough, racing yachts and skyscrapers are still tested in the old-fashioned way.)

OSBORNE REYNOLDS (1842–1912) Irish engineer and physicist. He was born in Belfast into a long line of Anglican clergymen. After a short apprenticeship in an engineering firm, he studied at Cambridge. Despite his comparative youth, he was given a newly created professorship in civil engineering at Owens College (now the University of Manchester), a position he retained for 37 years. Reynolds helped to advance hydrodynamics and hydraulics, introducing theories of lubrication and turbulence, and improving the design of centrifugal pumps. For a study of estuarine flow, he built a scale model of the mouth of the River Mersey.

Robotics, the three laws of

A robot may not injure a human being or, through inaction, allow a human being to come to harm

A robot must obey the orders given it by human beings, except where such orders would conflict with the First Law

A robot must protect its own existence as long as such protection does not conflict with the First or Second Law

A poster for the 1956 movie Forbidden Planet, which featured Robby the Robot. Robots existed in science fiction long before they were a reality in laboratories and factories.

* In these stories the laws are said to be from the 56th edition of the *Handbook of Robotics*, published in 2058. Asimov credited his editor, John W. Campbell, with their exact formulation.

These aren't "laws" in the sense in which the term is used elsewhere in this book. In fact, they're complete fictions, created in the fertile mind of the late Isaac Asimov for his classic science-fiction series of stories about robots.* In those stories, every robot was built with these three laws implanted in its "positronic brain," and the interpretation of the laws was frequently woven into the story line.

I have spent a great deal of time over the past few years talking to people engaged in research on the brain, on artificial intelligence, and on the nature of consciousness. I have been amazed at how often these purely fictitious laws have come up in those conversations. I believe that this is because, even though no one has any idea of how to build a robot that will obey Asimov's laws, we all agree that these are the kinds of laws we would want to build into intelligent machines.

I expect, therefore, that as we get better and better at building machines that mimic human action and thought, we'll hear a lot more about Asimov's three laws, fiction or not.

ISAAC ASIMOV (1920–92) Russian-American science-fiction author and science popularizer. He was born in Petrovichi, Russia, and his family emigrated to America when he was three. He studied chemistry at Columbia University, taking his MA in 1941 and, after war work, his Ph.D. in 1948. The next year he became a professor of biochemistry at Boston University. For some time he pursued a dual career as an academic and as a science-fiction writer, but the latter came to dominate. Asimov's best-known science-fiction works are the *Foundation* series of future-history novels about a Galactic empire. He was a prodigious writer, with nearly 500 books on a huge range of subjects to his credit.

Rock Cycle

Rocks at the surface of the Earth are cycled through different forms

Early in its life the Earth went through a period when its surface was molten. As the planet cooled, this melt crystallized to form the first rocks. This type of rock is called *igneous*, meaning "fire-formed." Igneous rocks are still forming today—as volcanic lava, for example, or in magma upwellings around mid-ocean ridges (*see* PLATE TECTONICS). Because they were the first rocks to form on our planet, we can start our story of the rock cycle with them.

After the first rocks formed, the temperature of the Earth continued to drop until it fell below the boiling point of water. Once the first raindrop had fallen, the igneous rocks began to weather. The first grain became detached from a rock and washed down to become the first grain on the first beach. As time went on, more and more sediment washed into the newly formed ocean, eventually building up a layer several miles deep on the sea floor. At the bottom of this accumulation, water flowed through grains of sand, leaving behind sticky residues that formed a kind of cement between the grains. This, along with the intense pressure of the layers above, turned the loose accumulation of grains into solid rock— a rock known as sandstone. The process is similar to pouring glue over a pile of sand and then placing a weight on it.

Once life appeared on the planet, another common form of sediment began to form. Plankton in the ocean extracted calcium from the water to incorporate into hard, microscopic shells. When the plankton died, their shells rained down on the ocean floor like snow. Over the millennia this material was also transformed into stone—in this case, limestone.

Rocks formed from the process of sedimentation are called, not surprisingly, *sedimentary rocks*. There are various types of sedimentary rocks, depending on the kind of sediment—sand leads to sandstone, calcium compounds to limestone, mud to shale. Right now, large rivers like the Amazon, the Nile, and the Mississippi are pouring tons of mud out into their deltas, mud that may someday be transformed into shales.

Sedimentary rocks are often easy to recognize. Because they form on ocean and lake bottoms they come in layers, and looking at them is something like looking at the pages of a book end on. You can often see such rocks in road cuts as you pass through hilly or mountainous country. The fact that you often see sedimentary rocks in high mountains, far from any ocean, is one of the most vivid demonstrations that the surface of the Earth is in a constant state of flux (*see* PLATE TECTONICS).

Once sedimentary rocks form, many things can happen to them. They can weather away, contributing grains to some future generation of sedimentary deposits. Or they can be buried by tectonic activity and subjected to pressure and heat deep beneath the surface, in which case the crystal structure of the minerals that make up the rock can be modified, changing its nature. For example, limestone subjected to these sorts of conditions is converted into marble, and shale can become slate. Rocks which have been transformed in this way are called *metamorphic*.

A single grain weathered from an igneous rock, then, can make its way in and out of the Earth's crust in many ways. It can form into a sedimentary rock which can, in turn, weather again to form another generation of the same type of rock. Alternately, it can metamorphose into another kind of rock. Ultimately any of these rocks can be carried below the surface by one process or another, such as continental collisions and subduction (*see* PLATE TECTONICS). In this case, they may be melted and their atoms could come to the surface as igneous rock to start the whole cycle again.

The life cycle of rock: melting, eruption, erosion, sedimentation, compression, and melting again.

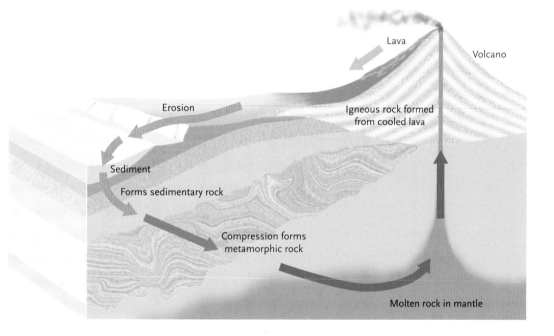

Rutherford Experiment

The atom consists of a tightly packed, positively charged nucleus with negatively charged electrons around it

The experiment that revealed the existence of the atomic nucleus.

Ernest Rutherford was a rarity in science—a man who did his most important work *after* he had received his Nobel prize. It was in 1911 that he carried out the experiment which not only gave physics a key insight into the structure of the atom, but which has also become a paradigm of elegance and depth.

Using naturally radioactive materials, Rutherford constructed a device that produced a beam of particles. It consisted of a lead box, with a small slit on one wall, into which a lump of radioactive material was placed. The radiation (in this case alpha particles, each consisting of two protons and two neutrons) from the material was emitted in all directions, but unless it went through the slit, it was absorbed by the lead. Further sets of lead plates with small openings were arranged so that all alpha particles except those moving in a specific direction were eliminated. The result was a tightly focused stream of alpha particles, and a target consisting of a thin gold foil was positioned so that it would be struck by the beam. After interacting with the foil, individual alpha particles in the beam would go on and register their presence by striking a plate, located beyond the foil, which was coated with a material that would give off light when a particle hit. In this way, the experimenter could tell what had happened in the collision.

Experiments of this type had been done before. The idea was to observe what happened when an alpha particle interacted with an atom, and from this information to deduce something about the nature of the atom. In the first decade of the 20th century, the atom was known to contain negatively charged electrons. The prevailing picture was of a thin, tenuous web of positively charged material in which the electrons were embedded like raisins in a bun—in fact, the model was known as the "raisin bun atom." From the results of experiments similar to Rutherford's, scientists had deduced the general features of the atom—its size, for example.

Rutherford noticed that in these previous experiments, no one had looked to see whether any alpha particles had been deflected through large angles. There was nothing dense or heavy enough in the raisin bun atom to deflect the particles significantly, so no one had thought it necessary to look for any such events. Rutherford got one of his students to set up the experiment to look for large-angle scattering of

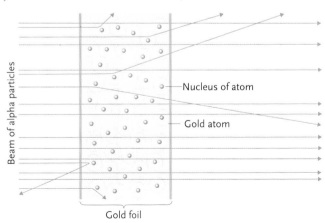

Beam of alpha particles

Nucleus of atom

Gold atom

Gold foil

alpha particles, more or less as an exercise to rule out an improbable result. As a detector, the student used a screen coated with a substance (zinc sulfide) that emits a flash of light when hit by an alpha particle. The screen could be rotated around the gold foil. Imagine his surprise, then, when the student began finding alpha particles going into the gold foil and being turned completely around—in effect, being scattered through an angle of 180°.

In the prevailing model of the atom, there was no way to explain this result: there was nothing in the raisin bun capable of turning the ponderous alpha particle around. Rutherford was forced to conclude that most of the mass of the atom was concentrated in an incredibly dense structure at the center. The rest of it was far more tenuous than had previously been imagined. From the behavior of the scattered particles, it was clear that this structure, which Rutherford called the *nucleus,* also housed all of the atom's positive charge.

In later years, Rutherford would often use a favorite analogy to explain his result. A customs official in a South American country has been warned that rebels are going to smuggle in a shipment of rifles hidden inside a bale of cotton. Confronted with a warehouse full of bales, how is the official going to find the rifles? Rutherford's solution was for the official to walk through the warehouse firing his pistol into the bales. When he found a bale where the bullets bounced back out, he knew he had his rifles. In the same way, when Rutherford saw the alpha particles coming right back out of the gold foil, he knew he had discovered a concentrated structure inside the atom.

The picture that sprang from Rutherford's work is familiar to us today. The atom consists of a tightly packed, positively charged nucleus surrounded by electrons. Although the structure of the atom was put on a firmer scientific basis by later scientists, beginning with the BOHR ATOM, it all started with a small piece of radioactive material and a thin gold foil.

ERNEST RUTHERFORD (LATER FIRST BARON RUTHERFORD OF NELSON) (1871–1937) New Zealand physicist. He was born in Nelson, the son of a farmer-craftsman, and won a scholarship to Cambridge University in England. Moving to McGill University in Canada, he and Frederick Soddy (1877–1966) established the basics of RADIOACTIVITY, for which he received the 1908 Nobel Prize for Physics. He then moved to Manchester University, where he and

Hans Geiger (1882–1945) invented the Geiger counter, and he instigated the experiment which established the existence of the atomic nucleus. During World War I, Rutherford worked on sonar systems for detecting submarines. In 1919 he was made Cavendish Professor of Physics at Cambridge, and the same year announced his discovery that the nucleus could be split by bombarding it with energetic particles.

Rydberg Constant

The wavelengths of radiation emitted by a given type of atom depend on the differences between the inverse squares of integers known as quantum numbers

In the 19th century, scientists had learned that different species of atoms gave off different colors of light—light of different wavelengths (*see* KIRCHHOFF–BUNSEN DISCOVERY). We can appreciate this if we look at streetlights. The ones used on major highways often have a yellowish cast. Because they contain sodium vapor, and sodium is an atom that radiates primarily at a couple of wavelengths corresponding to yellow light.

As the science of SPECTROSCOPY developed, it was found that the atoms of each chemical element emitted a characteristic "fingerprint" of colors, which allowed the atom to be identified even in distant stars. In 1855 the Swiss mathematics teacher Johann Balmer (1825–98) took the first step toward a systematic understanding of the wavelengths of visible light emitted by hydrogen by finding a formula that predicted the values of the wavelengths (hydrogen has a simpler and more regular spectrum than those of other atoms, and so was especially suited for the purpose). Later, Johannes Rydberg extended Balmer's work with a formula that summarized the emission of radiation by the hydrogen atom in all regions of the ELECTROMAGNETIC SPECTRUM. This equation, which now bears Rydberg's name, says that the wavelength λ of light emitted by the hydrogen atom will be given by

$$\frac{1}{\lambda} = R\left(\frac{1}{n_1^2} - \frac{1}{n_2^2}\right)$$

where R is the Rydberg constant, and n_1 and n_2 are integers (whole numbers). These integers are the *principal quantum numbers* for the atom. n_1 takes the value 2 for all the hydrogen lines in the visible spectrum, while $n_2 = 3$ gives the red line, $n_2 = 4$ gives the blue-green line, $n_2 = 5$ gives the first blue line, and so on. $n_1 = 1$ generates a series of lines in the ultraviolet, while $n_1 = 3, 4, 5, \ldots$ produces several series of lines in the infrared. The value of R was initially determined by experiment.

When Rydberg first wrote down this formula, it was strictly an empirical relation—something that matched the data. With the development of the BOHR ATOM, it became clear why this formula works. When Bohr worked out the energy of an electron in the nth orbit from the nucleus, he found that it was simply proportional to $-1/n^2$.

JOHANNES ROBERT RYDBERG (1854–1919) Swedish physicist. He was born in Lund and educated at the university there, obtaining a doctorate in mathematics in 1879. He remained at Lund for his working life, and was appointed professor of physics in 1901. His studies of atomic spectra was motivated by his desire to refine the PERIODIC TABLE of the elements, and his great insight was to link the patterns in the table with atomic structure.

Schrödinger's Equation

The wave–particle duality of quantum particles can be summarized in an equation

According to physics folklore, it happened like this: In 1926 a theoretical physicist named Erwin Schrödinger was giving a seminar at the University of Zurich. He was talking about strange new ideas which were floating around, that objects in the subatomic world often seemed to behave like waves rather than like particles. A senior faculty member is supposed to have stood up and said, "See here, Schrödinger, this is all nonsense. We all know that if you want to talk about waves, you have to use the wave equation." Schrödinger took this advice to heart, and went on to develop the science of quantum mechanics.

A word of explanation: in our everyday world, there are two ways of moving energy around. It can be carried by matter moving from one place to another (a speeding locomotive, for example, or the wind), in which case we say that particles are involved. Alternately, it can be carried by a wave (which is how most TV broadcasts reach your home). In our world, everything that carries energy has to be one or the other—a particle or a wave. And if it's a wave, it has to be described by a particular type of equation called, appropriately enough, the *wave equation*. Every wave —water waves on the ocean, seismic waves in rocks, radio waves from distant galaxies—can be described in this way. The point of the remark quoted above is that if we are going to describe things in the subatomic world in terms of waves of probability (*see* QUANTUM MECHANICS), then those waves, too, should be described by the wave equation.

Schrödinger adapted the notion of probability waves to the classical idea of a wave equation, and came up with the equation that now bears his name. Just as an ordinary wave equation describes the progression of a ripple across a lake, Schrödinger's equation describes how a probability wave associated with a particle moves through space. The peaks of this wave, where the probability is highest, denote where the particle is most likely to be found. Although Schrödinger's equation involves higher mathematics, it is so crucial to our understanding of modern physics that I give it here, in its simplest version (called the "one-dimensional time-independent equation"). The Greek letter ψ (psi) represents the probability mentioned above. Don't worry if you don't follow the math—just accept that the equation shows that the probability varies in a wave-like fashion.

$$\frac{d^2\psi}{dx^2} + \frac{8\pi^2 m}{h^2}(E - U)\psi = 0$$

where x is distance, h is the PLANCK CONSTANT, and m, E, and U are respectively the mass, total energy, and potential energy of the particle.

The picture of quantum events that Schrödinger's equation gives us, then, is that electrons and other subatomic particles are similar to tidal waves moving across the surface of the ocean. Over time, the peak of the wave (corresponding to the place where the electron is most likely to be found) moves from place to place in the manner prescribed by the equation. What we think of as a particle, then, actually displays some of the characteristics of a wave in the quantum world.

When Schrödinger first published his result, it led to a minor tempest-in-a-teapot among theoretical physicists. His contemporary Werner Heisenberg (known for HEISENBERG'S UNCERTAINTY PRINCIPLE) had published another, more mathematically complex version of quantum mechanics called "matrix mechanics," and people were worried that there might be two mutually contradictory ways of looking at the subatomic world. This conflict was quickly resolved when it was shown that each of the two theories could be derived from the other, so that they could be thought of as two ways of getting to the same answer. Today, because it is simpler and easier to teach, the Schrödinger version (which is sometimes called "wave mechanics") is used almost universally.

Erwin Schrödinger, one of the pioneers of quantum mechanics. His wave equation has been adopted as the standard for thinking about the science of the subatomic world. The stamp on the opposite page is one of several issued down the years by Austria to commemorate its native physicists.

But accepting that something like an electron can behave as a wave causes problems. As we have seen, in the everyday world everything is either a particle or a wave. A baseball is a particle, sound is a wave, and that's it. In the quantum world, apparently, things aren't so clear cut. In fact, it quickly became apparent from experiments that everything in the quantum world shares this property of being different from the objects we're used to. Light, which we normally think of as a wave, can sometimes act like a particle (when it does, the particle is called a photon), and particles such as electrons and protons can act as waves (*see* COMPLEMENTARITY PRINCIPLE).

This problem is usually referred to as *wave–particle duality,* and appears to be a feature of the subatomic world (see BELL'S THEOREM). It tells us that in the microscopic realm our everyday intuition about what form matter takes and how it behaves simply doesn't apply. The fact that we use an equation designed to describe waves to discuss the motion of things we picture as particles is ample proof of this. As pointed out in the Introduction, there is no essential contradiction here. There is, after all, no reason why what we see in the familiar macroscopic world should be copied in the microscopic. Nevertheless, it remains a troubling aspect of quantum mechanics for many people, and you could say that the trouble all started with Erwin Schrödinger.

ERWIN SCHRÖDINGER (1887–1961)

Austrian physicist. He was born in Vienna, the son of a wealthy manufacturer with intellectual interests, and received most of his early education at home. Schrödinger didn't attend lectures in theoretical physics until his second year at the University of Vienna, but went on to take his doctorate in that subject. During World War I he served as an artillery officer, but still found time to study the papers of Albert Einstein.

After accepting and leaving posts at several universities, Schrödinger settled in Zurich. It was there that he developed the wave mechanics theory which is still at the heart of modern quantum mechanics. In 1927 he accepted the chair of theoretical physics at the University of Berlin, just vacated by Max Planck. He was an unbending opponent of the Nazis, and left Germany for Oxford in 1933, the year he was awarded the Nobel Prize for Physics.

He became homesick, however, and returned to Graz, Austria, in 1936, just before that country's Anschluss with Germany. Dismissed from his post without notice, Schrödinger returned to Oxford with little more than what he could carry. An extraordinary series of events then began to unfold. Eamon de Valera, the prime minister of Ireland, had been a professor of mathematics at Oxford. Anxious to bring Schrödinger to Ireland, de Valera founded an Institute for Advanced Studies in Dublin. While the institute was being set up, Schrödinger accepted a guest lectureship in Ghent in Belgium. When war broke out in 1939 he found himself an enemy alien, and again de Valera came to his rescue with a letter of safe conduct that brought Schrödinger to Ireland. The Austrian remained in Dublin until 1956, when he returned to a chair in Vienna that had been created especially for him.

In 1944 Schrödinger published a book called *What Is Life?* It inspired a generation of scientists, with its vision of a physics untainted by military applications and its prediction of a GENETIC CODE hidden in the molecules of life.

Shock Waves

When an object moves faster than the waves it generates in a material, a shock wave is set up behind it

When an object moves through a medium, it will create waves in that medium. An airplane moving through the atmosphere, for example, will jostle the molecules in the air as it passes. Once the plane has passed, the sound wave caused by that jostling will move radially outward, according to the laws that govern the movement of waves in that particular medium. Thus, each point in the path of the airplane can be thought of as the source of a circular wave.

At subsonic speeds the waves spread out like ripples in a pond, and we hear the normal sound of an airplane. When the airplane moves faster than sound, each ripple originates from a point further along the airplane's track than the radius of the ripple preceding it. The ripples therefore do not follow one another sequentially, and at a particular angle to the airplane's direction of flight they reinforce one another. This happens all around the airplane, and the result is that the sound is reinforced over the surface of a cone with the airplane at its apex. The reinforced sound then expands outward as a cone of ever-increasing size. The sound over the surface of the cone is more intense than normal, and is called a shock wave. Shock waves cause sharp and abrupt changes to materials as they pass through them, and can cause damage. If you happen to be within the cone, you experience the shock wave as a sharp bang—a *sonic boom*. This is not a true explosion, just the result of hearing the noise which was emitted by the airplane over a relatively lengthy interval compressed into a much shorter period of time.

The angle of the cone is defined by the equation

$$\cos \theta = v_s / v$$

where v and v_s are the speeds, respectively, of the airplane and of sound. The ratio v_s / v is called the *Mach number*. An airplane moving at the speed of sound is said to be traveling at Mach 1, for example.

Sound waves are not the only type of wave to cause shock waves. When a particle moves through a medium at a speed greater than the speed of light in that medium, it produces a shock wave of light known as CHERENKOV RADIATION. The detection of this radiation allows physicists to detect the presence of a particle as well as to determine the speed at which that particle is traveling.

ERNST MACH (1838–1916) Austrian physicist. He was born in Turas (now Turany, Czech Republic) and educated by his father, who emphasized practical skills as well as academic learning. Mach got his Ph.D. at Vienna in 1860, and returned there as professor in the history of science in 1895. He was most influential as a philosopher of science, but he also worked in psychology, investigated shock waves, and formulated Mach's principle—that an object's inertia comes from its interaction with the rest of the universe.

Snell's Law

*The angle through which
a light beam is bent when
it crosses the boundary
between two media depends
on the refractive indexes
of those media*

*Willebrord Snell discovered
the simple law that governs
the path of a refracted ray.
Total internal reflection (see
the next page) happens when
the refracted ray strikes a
surface at more than a critical
angle to the perpendicular.*

The theory of RELATIVITY has taught us that nothing can move faster than the speed of light, but there is a subtle point about this dictum that is often missed. When theorists talk about the speed of light, denoted by the letter c, they actually mean the speed of light in a vacuum. When light moves through a transparent material such as glass or water, its speed is actually much lower than c because it is continually interacting with the atoms of the material.

What happens to wavefronts of light when they cross a boundary between different media is the subject of Snell's law, which is named for Willebrord Snell. An important example of such a boundary crossing is the passage of light from air to glass or vice versa, which occurs in almost all optical equipment, be it a sophisticated piece of laboratory equipment or a humble pair of spectacles. Imagine a line of hikers walking diagonally across a square field which is bisected by a marshy strip parallel to two of the sides. The hikers can walk more quickly over the field than through the marsh, so when the leading hikers reach the boundary with the marsh, they will slow down while their companions behind them maintain their speed. As the rest of the hikers reach the boundary, they will also slow down, until they are all moving slowly in the marsh. The net effect is that the line of hikers winds up moving in a different direction in the marsh from the direction it had in the field. In the same way, a beam of light entering a medium with a different velocity of light will change direction. Like the hikers, the beam will shift its direction of travel toward a line perpendicular to the surface if the speed in the second medium is slower than it is in the first. If the light moves faster in the second medium, as it would when moving from glass into air, it will be deflected away from that perpendicular.

The ratio of the speed of light in a vacuum to the speed of light in a

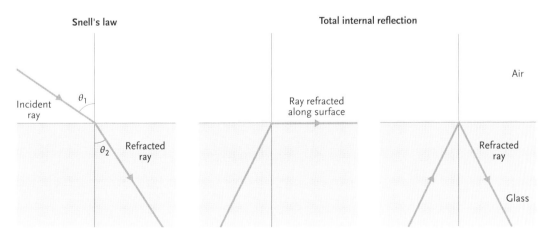

Snell's law

Incident ray

θ_1

θ_2 Refracted ray

Total internal reflection

Ray refracted along surface

Air

Refracted ray

Glass

Total Internal Reflection

Imagine a beam of light inside a block of glass approaching one of the surfaces. As the light passes through the surface and into the air, it is refracted away from the perpendicular because air has a lower refractive index (about 1) than glass (about 1.5). By Snell's law, if the beam approaches the surface at an angle to the perpendicular of (say) 30°, then it will leave the glass at an angle of about 49°. As the approach angle increases, the beam will exit at greater and greater angles, until with an approach angle of about 42° (known as the *critical angle*), the exit beam will emerge at 90° to the perpendicular, and thus just skim over the surface of the glass.

What if the approach angle increases further? Since increasing the exit angle over 90°, would cause the beam to be refracted back into the glass, what happens is that the beam stays inside the glass, but is *reflected* from the boundary instead of being refracted. This process is called total internal reflection. The critical angle is defined by the equation

$$\sin \theta > n_2 / n_1$$

When θ is larger than the critical angle, the light beam coming up through the glass can no longer penetrate into the air, but is reflected back into the glass.

You can observe total internal reflection for yourself the next time you enjoy a candlelit dinner. Hold your wineglass up and look up through the wine. When the wine is high above your head, you will be able to see through the surface. As you lower the glass you will reach a point where the surface of the wine suddenly becomes opaque. What is happening is that the light from the candle is undergoing total internal reflection at this angle, blotting out any other light coming through the surface.

But total internal reflection is more than a curiosity—it's the basis for a major modern technology, *fiber optics*. If light enters one end of a thin glass fiber at a large enough angle, then it will move down that fiber, undergoing internal reflection whenever it encounters the fiber's surface, and will emerge from the other end of the fiber undiminished. If a group of fibers are bound together into a bundle, then the pattern of light and dark spots that enters at one end will emerge at the other. In this way, a picture can be transmitted through the fibers. This is the principle behind many modern medical techniques (such as arthroscopic surgery) in which a thin bundle of optical fibers is passed into the body through an orifice or a small incision so that the surgeon can see what is happening as he or she operates.

Total internal reflection is also routinely used for high-speed transmission of data through telephone lines. Because the wave being sent is light rather than an electrical pulse, fiber-optic cables can carry much more information in a given time than can ordinary copper wires. In fact, most telephone systems in industrialized countries now use optical fibers routinely.

By inserting a tube with a fiber-optic cable into this patient's mouth, doctors can examine her stomach without the need for surgery.

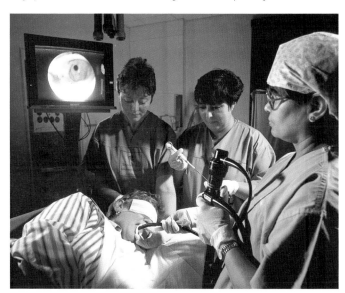

material is called the *refractive index* of that material. The refractive index of glass, for example, is about 1.5 (depending on the type of glass), which makes the speed of light about a third slower in glass than in a vacuum. In general, different materials have different refractive indexes.

Snell's law quantifies the change in a light beam's direction. If θ_1 and θ_2 are the angles which the light beams in the first and second media make with the perpendicular to the surface between the media, and n_1 and n_2 are the refractive indexes of the two media, then

$$n_1 \sin \theta_1 = n_2 \sin \theta_2$$

The point about this is that if you know the refractive indexes of the two materials and the direction of the incoming light, you can predict the amount of bending that the light will experience when it crosses the boundary.

Have you ever stood at the side of a swimming pool and noticed that your friend's legs seem very short when he's standing in the water? The light beam that comes to your eye from your friend's foot is bent when it leaves the water and enters the air. Consequently, it comes to your eye at a flatter angle than it would have done had the water not been there. Your eye traces back along the direction of the beam and tells your brain that your friend's feet are higher up than they actually are.

WILLEBRORD VAN ROIJEN SNELL (1580–1626) Dutch mathematician and physicist. He was born in Leiden, the son of the professor of mathematics at the university there. He studied mathematics and law at various European universities and traveled widely, making the acquaintance of many scientists, including Johannes Kepler. He succeeded to his father's university position in 1613. Snell took up the new science of geodesy (measuring the Earth), and he saw the importance of surveying by triangulation. In 1621, after much experimental work in optics, he discovered the law of refraction now named after him. Snell did not publish his result, which lay hidden for years until it was included in a book by René Descartes.

Social Darwinism

Social inequalities result from the operation of Darwinian natural selection

In the 19th century, the introduction of Darwin's theory of EVOLUTION led to revolutions in thought in many fields. The British philosopher and sociologist Herbert Spencer (1820–1903) was one of the most enthusiastic converts to Darwinian ideas. In fact, it was he who coined the phrase "survival of the fittest." He was the first to attempt to apply evolutionary principles to the analysis of social organization.

This trend, in the hands of those with less sophistication than Spenser, led to the doctrine known as social Darwinism. Its proponents argued that, just as nature progresses through unbridled competition and survival of the fittest, so does human society proceed. The unbridled competition of late 19th-century capitalism and the great social inequity it engendered were thus perceived as the "natural" state of society—a powerful argument against things like social reform and labor unions.

Actually, social Darwinism rested on a fundamental misunderstanding of the principle of natural selection. The "fittest" (or at least the genes of the "fittest") may indeed survive in nature, but *fitness* in this sense is defined in only one way. The winner in the Darwinian game—the fittest individual—is the one who gets most of his or her genes into the next generation. From the Darwinian point of view it makes no difference whatsoever how much money or power an individual has accumulated. All that matters is how many children carry that individual's genes.

When I talk to my students about social Darwinism, I always use the railroad magnate Leland Stanford (1824–93) as an example. He started life as a grocery clerk, but eventually rose to a position of enormous prominence in America. It was he who drove the "golden spike" at Promontory Point, Utah, to complete the first trans-American railroad line across North America. Stanford founded the great university that now bears his family name in memory of his only son, who died as a teenager (in fact, if you look closely at the crest, you will see that it's called Leland Stanford *Junior* University).

By the standards of society, Leland Stanford was enormously successful, a man who changed the face of the world he lived in. By Darwinian fitness criteria, however, he was a complete loser, since he transmitted not one gene to another generation. This example alone, of course, is enough to show the basic scientific falsehood of the doctrine of social Darwinism. Other arguments (e.g., that "is" and "ought" are not the same) can also be made against it, but seem to me to be unnecessary.

Truth in advertising: I am a proud graduate of Stanford University, having completed my graduate studies in the Physics Department there, and remain enormously grateful to Leland Stanford for using his vast wealth to create such an important institution.

Species–Area Law

The number of species that can be supported in a given ecosystem decreases as the area of the ecosystem decreases

It is reasonable to ask whether smaller ecosystems will be able to support fewer species than can larger ones. On the one hand, you might expect that the larger the area, the more different ecological niches there should be available for exploitation. On the other hand, it's hard to see why dividing something like a meadow in half should lower the biological diversity. It may lower the total number of organisms, but why should it also alter the kinds of organisms to be found?

This is, at bottom, an observational question. In fact, ecologists have carried out many studies in pursuit of the answer—in one, for example, the available area of a small island off the coast of Florida was changed by means of a chain saw! Data on island and mainland ecosystems all point to the same conclusion. If A is the area of an ecosystem and S is the number of species, then the relation between the two is given by the equation

$$S = KA^n$$

where n is a number between 0.1 and 0.3, and K is a constant that represents the number of species present on an ecosystem of unit area. This equation is called the species–area law.

The first thing you have to note about the law is that, although the number of species does depend on the area of the ecosystem, it doesn't depend on it very strongly. Even with a value of n of 0.3, for example, doubling the size of an ecosystem only increases the number of species by around 23%.

Nevertheless, the law does have important implications for conservation efforts aimed at preserving biodiversity. It tells us, for example, that when we reduce the size of an ecosystem we will get not just a smaller version of the original, but a version with rather fewer species. Ten small patches of wilderness, in other words, may be able to support only half the species that would be found on a single patch of the same total area.

Conversely, increasing the size of a given ecosystem will not generate a proportionate expansion in biodiversity, so that resources are often better directed toward establishing nature reserves in entirely new ecosystems rather than toward expanding existing ones.

Spectroscopy

Atoms can be identified by the light they emit or absorb

According to the simple model of atomic structure provided by the BOHR ATOM, we can picture the electrons as occupying certain, well-specified ("allowed") orbits around the nucleus. But they can actually move directly from one orbit to another in what is called a *quantum jump*. If the electron jumps to a lower orbit, it has to lose energy in the process, so it emits a photon of a specific energy and specific wavelength. We perceive these photons as specific colors—the blue of copper in a fire, for example (*see* FLAME TEST). Similarly, if an electron jumps to a higher orbit, it requires an input of energy. Usually, the electron absorbs a photon in order to move up (*see* KIRCHHOFF–BUNSEN DISCOVERY).

This interaction between light and atoms is the basis of an important branch of science known as spectroscopy. Because the nuclei of different chemical elements contain different numbers of protons, they have different allowed orbits (or, in today's more sophisticated picture of the atom, *orbitals*, where the electron only has a probability of being somewhere, rather than a definite position). This means that the energy difference between allowed orbitals in atoms of different chemical elements will be different and so will emit light at different wavelengths. Thus sodium emits light primarily at two close wavelengths in the yellow part of the spectrum (hence the yellow of sodium streetlights), whereas mercury emits at several wavelengths mostly in the blue region (hence the bluish cast of mercury lights).

The simple fact that you can tell what element's atoms are emitting light by looking at the wavelengths of the light makes possible the science of spectroscopy. Each chemical element has a unique array of wavelengths in the light it emits. Furthermore, the pattern of permitted wavelengths changes to another unique set if the atom becomes ionized. Thus, if we detect a set of frequencies in the light emitted by a particular object (e.g., an unknown material heated in the flame of a Bunsen burner), we can be sure that a particular chemical element is present in the object.

This is the basis of what is called *emission spectroscopy*. By comparing the intensities of the light from various elements, we can deduce the relative abundances of the emitting atoms. In this way, we can find the chemical composition of an object without ever having to bring a sample of it into our laboratories.

Absorption spectroscopy works in a similar way. A group of atoms illuminated by light that is a mixture of all frequencies will absorb those wavelengths needed to cause electrons to jump to higher orbitals. If we look at the light coming through the atoms, we will see dark lines corresponding to the wavelengths removed by the absorption process. Absorption spectroscopy is widely used in astronomy to determine the chemical elements present in stars, planets, nebulae, and galaxies.

Spontaneous Generation

Living things arise spontaneously from inorganic materials

Larvae like those of the common housefly—maggots—were once thought to be generated spontaneously from dead matter. This idea was disproved in the 17th century, but it took Louis Pasteur to bury it for good.

From the earliest times, people believed that living things arise from simpler materials. Leave a pile of grain out in the rain, for example, and it will soon generate mice; leave meat out, and it will soon be crawling with maggots. In the 17th century, Francesco Redi put meat out in various containers—some open, some sealed, some covered by gauze—and showed that maggots never appeared in meat that was protected from flies.

Although the Redi experiment demolished the notion that complex organisms can arise spontaneously, the discovery of microorganisms in the 19th century led to a rebirth of the notion of spontaneous genera-

tion. Even decomposing materials shielded from flies seemed to produce organisms which were visible under a microscope. By 1860, the controversy over spontaneous generation had become so heated that the French Academy offered a prize to anyone who could resolve the issue. The French scientist Louis Pasteur (*see* GERM THEORY OF DISEASE) performed a series of careful experiments that eventually settled the issue, and he received the Academy's prize in 1864.

Pasteur took flasks with long, narrow, curved necks and filled them with liquid. The liquid was boiled to kill any microorganisms present, and the glass neck acted as a trap for fungal spores and other microorganisms that might contaminate the liquids. He showed that microorganisms appeared only in flasks whose necks were subsequently broken—only if the liquid was exposed to airborne organisms.

It's a historical irony that in the 1870s a new controversy arose, centered on the supposed spontaneous generation of yeast molds in the process of wine fermentation. Once again, Pasteur showed by careful experiments, in which he removed material from the inside of a grape under sterile conditions and isolated it from the air, that the yeast spores were carried by the air, and did not arise spontaneously from the material in the grape.

Today, the result of the long debate over spontaneous generation is summarized in the biologist's slogan "Life comes from life."

FRANCESCO REDI (1626–97) Italian physician biologist, linguist, and poet, born in Arezzo. Educated in philosophy and medicine at the University of Pisa, he returned to Arezzo as head physician at the Tuscan court and supervisor of the ducal pharmacy. He studied the effects of snake venom, proving that viper venom is harmless if swallowed, and became an expert on insects and parasites.

Standard Model

All matter is composed of quarks and leptons, and mediators

The standard model is the name given to what at the moment is the best theory we have of the ultimate nature of the universe. It takes statements about how matter is constructed from its most basic constituents and combines them with statements about the nature of the forces that act between those constituents. (*See also* QUARKS AND THE EIGHTFOLD WAY, THEORIES OF EVERYTHING, and ELEMENTARY PARTICLES.)

On the structural side of things, the elementary particles that make up the nucleus of the atom are built from particles more elementary still. These truly basic building blocks are called *quarks,* and have electrical charges $1/3$ and $2/3$ times those on the electron and proton. The lowest-mass quarks are called the up (u) and down (d) quarks. (Sometimes they are called p and n quarks because in some respects they resemble the proton and neutron.) Up and down quarks have charges of $2/3$ and $-1/3$, respectively. The proton itself is made from two u quarks and one d quark, while the neutron is made from two ds and a u. (Add the charges up yourself to verify that they work out correctly.) Two other pairs of quarks make their appearance in more exotic particles. They are called, respectively, the strange and charmed quarks, and the bottom and top quarks. All of these quarks have been detected in particles in laboratory experiments.

The other set of basic building blocks are the *leptons.* The most familiar of the leptons is the *electron,* which exists in the atom but does not take part in the nuclear reactions, only in reactions between atoms. There are two other particles called, respectively, the *mu* and the *tau,* which have properties like the electron except that they are more massive. Each of these leptons is paired with a partner called a *neutrino*—respectively, the electron, mu, and tau neutrinos. These particles are massless (or almost massless) and chargeless.

The quarks thus come in three families of two each, as do the leptons. This symmetry has not escaped theoretical physicists, but no one has yet come up with a convincing explanation of why it exists. In any case, the quarks and leptons are the basic building blocks from which all matter in the universe is built.

To understand the other side of the coin—the basic forces that act between quarks and leptons—we have to understand how scientists now interpret the idea of force. An analogy may help here. Suppose two punts are passing each other on the River Cam, in Cambridge. One of the punters, feeling generous, throws a bottle of champagne to the other punter. The effort of throwing the champagne pushes the first punt off course, and when the second punter catches the bottle, the momentum of the bottle also moves the second punt. Thus the exchange of the champagne has caused both boats to change direction. According to NEWTON'S

Lepton		Charge	Mass
e^-	(electron)	−1	1
ν_e	(electron neutrino)	+1	14
μ^-	(mu)	−1	207
ν_μ	(mu neutrino)	+1	0.5
τ^-	(tau)	−1	3491
ν_τ	(tau neutrino)	+1	61

The six leptons, with their charges and mass relative to those of the electron. Leptons and quarks (see the table at QUARKS AND THE EIGHTFOLD WAY) are believed to be the fundamental particles from which all other subatomic particles are made.

LAWS OF MOTION, this means that a force has acted. In this example, it's not hard to see that the force is caused by (a physicist would say "mediated by" or "generated by") the exchange of the champagne.

In just the same way, we say that the forces between particles are generated by the exchange of particles. In fact, we say that the fundamental forces between particles are different from each other because the particles that mediate them are different. There are four such forces—the *strong force* (that holds quarks together in particles), the *electromagnetic force,* the *weak force* (responsible for some radioactive decays), and the *gravitational force*. The strong force is generated by the exchange of massless mediators called *gluons*. This exchange is described by the QUANTUM CHROMODYNAMICS theory. The electromagnetic force is mediated by the exchange of bundles of electromagnetic radiation called *photons*. The weak force, on the other hand, is mediated by the exchange of particles called *vector bosons,* which have masses 80 to 90 times that of the proton, and were first observed in laboratories in the early 1980s. Finally, the force of gravity is thought to be generated by the exchange of massless mediators called *gravitons,* but these have not yet been detected in the laboratory.

Force	Mediator
Strong	g (gluon)
Weak	W^+, W^-, Z^0 (gauge bosons)
Electromagnetic	γ (photon)
Gravitational	graviton

The four fundamental forces of nature, together with their mediators—the particles that enable the force to act. The weak force is mediated by three different particles. Gluons and gravitons have been proposed by theorists, but have not yet been discovered.

In the standard model, the first three of these fundamental forces are not regarded as distinct, but as different aspects of a single unified force. For an analogy, we can return to the punts on the Cam, and suppose that another pair of punts are passing each other, but this time only an ice cream is thrown across. This action still causes the boats to swerve. It might then appear to an outsider that there were two forces operating between the boats, one due to the exchange of a solid (the ice cream), and the other to the exchange of a liquid (the champagne—we'll ignore the fact that the champagne has to be contained in a bottle since for most of us it's the liquid that's of interest). Imagine now that the whole process is repeated on one of those rare hot summer days in Cambridge, when the ice cream will have melted. The increase in temperature shows us that in fact there is only one force in operation, the force mediated by the exchange of a liquid. The only reason that it looked as if there were two forces was that the temperature was low. Raise the temperature, and the true underlying unity of the forces is made evident.

For the forces operating within the universe a process analogous to the melting of the ice cream occurs. At high enough temperatures (or, equivalently, at high enough energies for particles colliding with one another), the forces will become indistinguishable. The first pair of forces to merge—or *unify,* to use the accepted term—in this way are the weak nuclear force and the electromagnetic force. They merge into what is called the *electroweak force* at collision energies attainable within current particle accelerators, so the process can be observed directly in the laboratory. In the EARLY UNIVERSE, energies were high enough for the weak nuclear and electromagnetic forces to be combined for the first 10^{-10} seconds after the big bang. After this time, or below temperatures of 10^{14} K, all four forces appear to be distinct. Above that temperature, there are only three fundamental forces: the strong force, the unified electroweak force, and gravity.

For the electroweak force and the strong nuclear force to unify, the temperature needs to exceed 10^{27} K. Such temperatures (or the equivalent collisional energy) cannot currently be achieved in the laboratory. The world's highest-energy accelerator, the Large Hadron Collider now being built astride the border between France and Switzerland, will be able to reach only 0.000000001% of the energy needed to unify the electroweak and strong nuclear forces. So it is likely to be a while before experimental proof of the unification is obtained. For the first 10^{-35} seconds of the life of the universe, the temperatures were higher than 10^{27} K, and for that brief interval there were just two forces in operation; the *electronuclear force* and gravity. Theories that describe this unification are called *grand unified theories,* or GUTs. These theories cannot be checked directly, but they do make a number of predictions about lower-energy processes that can be checked. At the moment, the GUTs seem to do a good job of making accurate predictions.

The standard model, then, is a theory of the universe in which matter is composed of quarks and leptons, and the strong, electromagnetic, and weak forces are described by the grand unified theories. The model must be incomplete, for it does not include the force of gravity. Presumably at some time in the future it will be replaced by a more complete theory (a THEORY OF EVERYTHING), but for the moment it's the best we've got.

States of Matter

Matter can exist in solid, liquid, and gaseous form, as well as in the form of a plasma

Matter is composed of atoms, and the bulk form of the matter depends on how those atoms are arranged and how they interact with one another. In general, you can think of matter as coming in three forms—solid, liquid, and gas.

A *gas* will expand to fit both the size and the shape of any container into which it is put. If you could look at gases at the atomic level, you would see the atoms zipping around, colliding with one another and with the walls of the container, but not interacting to any great degree. If you expand or shrink the container, the gas molecules spread themselves evenly through the new volume. The KINETIC THEORY of gases explains how the atomic properties of a gas can be related to its macroscopic properties such as temperature and pressure

Unlike a gas, a *liquid* keeps a constant volume, though, it too will assume the shape of its container below its surface. The easiest way to picture a liquid in terms of atoms is to imagine the atoms in contact with one another, but free to roll around, rather like marbles in a jar. If you pour a liquid into a container, the atoms will roll around and assume its shape, but they will not expand to fill any new volume.

A *solid* has its own shape and will not adjust to the shape of a container. To think of a solid at the atomic level, imagine the atoms locked together by CHEMICAL BONDS, so that their relationship to one another is fixed. This locking together can be in a rigid, lattice arrangement (as in crystals) or in a more disordered structure, as in plastics, in which molecules are arranged like the intermingling strands of spaghetti in a bowl.

The states of matter described above are the traditional three. There is, however, another state in which matter can exist, one which scientists generally think of as a fourth state of matter. This is the *plasma*. In a plasma, electrons have been torn from their atoms, but remain in the material. Thus, a plasma is electrically neutral—there are as many positive charges as negative—but the positive and negative charges are free to move independently. We can have a plasma in which only a few electrons are removed from their atoms (as in a fluorescent light bulb), or in which all the electrons are torn loose (as in the interior of the Sun).

At extremely low temperatures, the velocities of atoms become very small and so are known to a high degree of accuracy. According to HEISENBERG'S UNCERTAINTY PRINCIPLE, their positions become very uncertain. When the uncertainty in the position of an atom becomes as large as the group of atoms to which it belongs, that group starts to behave as a single entity. Such groups of atoms are called *Bose–Einstein condensates,* and can be regarded as a fifth state of matter.

Steady-State Theory

The universe is expanding, but matter is being created continuously in the space between galaxies, so the universe had no beginning and will have no end

After the discovery of HUBBLE'S LAW, most astronomers accepted the notion of the BIG BANG, a theory according to which the universe began at a specific point in the past. During the 1940s, however, a group of astrophysicists led by Fred Hoyle offered an alternate theory.

The basic idea of this theory was that as galaxies moved away from one another in the Hubble expansion, new matter was created in the widening spaces between them. This newly created matter would eventually form itself into galaxies which would in turn move away, vacating space where new matter would be created. In this way, the observed expansion was made consistent with a "steady-state" universe which preserved its overall density, implying no single point of origin (as did the big bang theory), but at the cost of having to postulate a new process for the creation of matter.

The steady-state theory enjoyed a kind of parallel existence among a minority of astronomers until well into the 1960s. Its main advantage was philosophical. It was argued that it was in keeping with the COPERNICAN PRINCIPLE of mediocrity in that it did not single out a specific moment of time as being special.

Evidence against the theory soon began to accumulate. For one thing, precise experiments failed to turn up matter creation in the laboratory. More importantly, new discoveries in cosmology such as the cosmic microwave background (*see* BIG BANG) showed that there were many aspects of the universe that could be understood in the big bang scenario, but not in the steady state. For example, as telescopes became powerful enough to probe deep into the universe, and therefore further back in time, it became clear that all the most distant galaxies were young, unevolved systems. This is just what would be expected in a big bang universe, but it was not consistent with the steady-state picture. Eventually, most of the steady-state advocates, overwhelmed by the counter-evidence, just gave up.

One lasting legacy of this episode, however, is the term "big bang" itself. It was originally coined by Hoyle as a way of making fun of his rivals, and he must have been very surprised indeed to have them adopt it enthusiastically.

FRED HOYLE (1915–2001) English cosmologist and astrophysicist. He was born in Bingley, Yorkshire, and graduated from Cambridge University in 1936, where he became Plumian Professor of Astronomy in 1958. Hoyle's greatest achievement was to establish how chemical elements are manufactured inside stars (*see* STELLAR EVOLUTION). He was not afraid to promote what others regarded as maverick ideas, such as the theory that bacteria brought to Earth on interstellar dust particles have been responsible for the outbreak of hitherto unknown diseases.

Stefan– Boltzmann Law

The amount of energy radiated by a black body is proportional to the fourth power of the temperature

When objects are heated, they radiate energy. When we describe something as "red hot," it's because it is hot enough to radiate visible light, mostly in the red region of the spectrum. At the atomic level, this radiation arises through the interaction of atoms and photons, in just the same way that BLACK-BODY RADIATION does. The law which specifies the energy of this radiation is named after the Austrian physicists Josef Stefan and Ludwig Boltzmann (*see* BOLTZMANN CONSTANT). Stefan modified existing theories of heat radiation to match new data, while Boltzmann worked out the theoretical justification for the law.

To understand how the law works, imagine light being emitted by an atom inside the Sun. The light will be absorbed by another atom, reemitted, then absorbed by other atoms and reemitted again until equilibrium is reached. In this equilibrium, the light and the atoms will come to a state in which whenever light of a particular frequency is absorbed by an atom in one place, light of the same frequency will be emitted by another atom somewhere else. Thus, the amount of light at each frequency will stay the same.

The temperature inside the Sun decreases from the center to the surface. As radiation moves outwards, it will thus have a higher temperature (as specified by the Stefan–Boltzmann law) than its surroundings. However, as it is absorbed and reemitted, the reemission occurs at the temperature of the gas (i.e., more photons will be emitted but at lower energies, so that the total energy involved remains constant). Thus by the time light works its way to the surface, it will have the frequency distribution characteristic of the surface temperature (5800 K) instead of the temperature at the center (15,000,000 K).

Energy that comes to the Sun's surface (or to the surface of any hot object) is radiated away, and the Stefan–Boltzmann law tells us how much energy actually leaves it. In equation form, the law says that

$$E = \sigma T^4$$

where σ is the *Stefan–Boltzmann constant* and the temperature T is measured in degrees kelvin. The law says that as the temperature of the object goes up, so too does the energy it radiates, but by much more. Double the temperature, and the amount of radiated energy will increase by a factor of 16.

According to this law, every object above a temperature of absolute zero is giving off energy. Why, then, don't they all cool down? Why, for example, can you yourself radiate the energy characteristic of an object at your own body temperature (a little over 300 K) all the time without having your temperature fall?

There are actually two parts to the answer to this question. In the first place, the food you eat supplies energy which your metabolic processes turn into the heat needed to maintain your body temperature. The heat you lose according to the Stefan–Boltzmann law, then, is replaced by energy derived from the food you eat. When an animal dies, its body cools off as this source of replacement energy disappears.

More important, though, is the fact that the law applies to every object above a temperature of absolute zero. Thus, while you are radiating energy away into your surroundings, those surroundings—walls and furniture, for example—are radiating energy back toward you. If the surroundings are at a lower temperature than you are (the usual situation), then radiation from them will make up for part (but not all) of the energy you radiate away. If, however, the temperature of the environment is near or above body temperature, you won't be able to get rid of all the heat generated by your metabolism through radiation. In this case, a second mechanism will come into play. You will sweat, and the excess heat will be used to evaporate water through your skin.

An athlete sits under a spray of water. The heat to evaporate the water from his skin comes from his body, which cools in the process. Without the water spray he would have to rely on Stefan–Boltzmann radiation to lower his temperature.

The Stefan–Boltzmann law as stated above applies to a perfect black body—an object capable of absorbing all of the radiation that falls on it. For real materials, the T^4 law generally still applies, but the Stefan–Boltzmann constant will need to be replaced with a different constant to take account of the fact that the object in question isn't a perfect black body. Such constants usually need to be found experimentally.

JOSEF STEFAN (1835–93) Austrian experimental physicist. He was born in Klagenfurt and educated at the University of Vienna, where he served as professor of higher mathematics and physics from 1863, and director of the university's Institute for Experimental Physics from 1866. Stefan's investigations covered a wide range of subjects, including electromagnetic induction, diffusion, and the kinetic theory of gases. But his reputation comes from his work on heat transfer by radiation. He established the experimental foundation for the Stefan–Boltzmann law by measuring the heat lost from a platinum wire; the theory was provided by his student Ludwig Boltzmann. Stefan used the law to make the first good estimate of the Sun's surface temperature, which he estimated at 6000°C.

Stellar Evolution

The life cycles of stars depend on their stellar mass: Low-mass stars end up as white dwarfs, while high-mass ones become supernovae

Although stars appear to be eternal on any human time scale, they are born and they die like everything else. According to the generally accepted NEBULAR HYPOTHESIS, a star begins its life in the collapse of an interstellar cloud of dust and gas. As the cloud collapses forming a *protostar*, the temperature at the center rises, which makes the particles at the center move faster and faster. Eventually, two protons can approach each other with sufficient speed to overcome their mutual electrical repulsion (*see* COULOMB'S LAW) and begin to undergo a nuclear fusion reaction (*see* NUCLEAR FUSION AND FISSION).

The net result of the fusion reaction is that four protons come together to form a single helium nucleus (two protons and two neutrons) and some miscellaneous particles. The particles in the final state have less mass than the four original protons, which means that the reaction has liberated energy (*see* RELATIVITY). The central core of the nascent star heats up and energy begins to flow outward through the star. At the same time, the pressure at the center of the star begins to increase (*see* IDEAL GAS LAW). Thus, by consuming—or "burning"—hydrogen in a fusion reaction, the star generates a pressure which will counteract the inward pull of gravity. Stars in a hydrogen-burning phase are said to be "on the main sequence" (*see* HERTZSPRUNG–RUSSELL DIAGRAM). The transformation of one type of element into another inside stars is known as *nucleosynthesis*.

The Sun, for example, has been burning hydrogen for about 5 billion years, and has enough hydrogen reserves in its core to keep on doing so for another 5½ billion years or so. Stars much more massive than the Sun

The Crab Nebula is an expanding cloud of gas, the remains of a supernova. Asian and Native American astronomers recorded the AD 1054 supernova that gave rise to the nebula.

contain much more hydrogen, but also have to contend with a much stronger gravitational force, so they have to burn their fuel faster to stop themselves from collapsing. In fact, the more massive a star is, the shorter is its lifetime, and very massive stars literally burn themselves out in a few tens of millions of years. Very small stars, on the other hand, have the potential to live out uneventful lives for hundreds of billions of years. Within this vast range of lifetimes, the Sun is pretty average.

Sooner or later, though, every star has to run out of usable hydrogen. What happens next, like the star's lifetime, depends on its mass. The Sun (and stars up to eight times as massive) enter a relatively simple end game. As the hydrogen in the central core begins to give out, the force of gravity, which has been waiting in the wings since the star was born, now comes into its own and causes the star to contract. This has two effects: It raises the temperature in the region just outside the core to a level where the hydrogen there can now undergo fusion reactions; and it also raises the temperature in the core to a level where helium nuclei, the "ashes" of the previous nuclear fire, can now become fuel for a new kind of fusion reaction in which three helium nuclei fuse together to make a carbon nucleus. This kind of process, by which the ashes of one nuclear fire become fuel for the next, is one of the key features in the life cycle of stars.

The secondary burning in the region around the core generates sufficient energy to make the outer envelope of the star swell up. The Sun's outer envelope, for example, will extend past the current orbit of Venus during this stage of its life. At this point, the star is emitting something like the same amount of energy as before, but through a much larger surface, so the outer surface is very cool. This kind of star is called a *red giant*.

For stars like the Sun, once the secondary burning is over, gravitational collapse sets in again. Temperatures never rise to a level where further nuclear reactions can initiate. The star will continue to collapse until a new force arises to counteract gravity. That force is the *degeneracy pressure* of electrons (*see* CHANDRASEKHAR LIMIT). Electrons have always been present in the star, but have so far played no role in its evolution— they have been moving around freely while the nuclei have participated in reactions. But now the star is so compressed that the electrons no longer have "elbow room," and begin to exert a force that permanently counteracts the star's gravity. The star is now a *white dwarf*. It will remain in space, radiating away its heat until it becomes a cinder.

For stars much more massive than the Sun, there is a far more spectacular ending. Once they have gone through the stage of helium burning, their mass is enough to raise the temperature at the core (as well as in outer envelopes) to fuse carbon, then silicon, then magnesium, and then a succession of other atomic nuclei. As each new reaction begins, the previous one continues in a new shell outside the core. In fact, all the elements up to iron are created in a star of this type. Iron, however, cannot act as a fuel for any kind of nuclear reaction. Energy has to be supplied to split iron in fission reactions, and energy has to be supplied to add particles to iron in fusion reactions. So an iron core builds up over time—a core that cannot serve as fuel for further nuclear reactions.

Eventually the pressure and temperature in the core get high enough for electrons to begin to interact with protons in the iron nuclei to make neutrons. In a very short space of time—a matter of seconds, some theorists suggest—the electrons disappear, all the iron in the core turns into neutrons, and the whole thing goes into free fall. The outer envelope of the star, with its support pulled out from under it, collapses inward. The core rebounds and collides with the infalling outer envelope, and the star literally explodes as a *supernova*. For short while the energy released in a supernova event can exceed the energy generated by all of the stars in a galaxy. In the resulting nuclear maelstrom, all of the chemical elements up to uranium are manufactured and shot into space.

For stars up to about thirty times the mass of the Sun, the core collapse will go on until the neutrons begin to demand *their* elbow room and exert their own degeneracy pressure. This will happen, typically, when a star is about 10 miles (15 km) across. The result is a rapidly rotating neutron star that sends out pulses of electromagnetic radiation, and is therefore referred to as a *pulsar*. For masses greater than this value, nothing can stop the collapse and the supernova results in the creation of a BLACK HOLE.

Stem Cells

Some cells in the fetus and the adult body retain the ability to give rise to many different types of specialized cells

The human body begins as a single cell (*see* CELL THEORY)—a single fertilized egg, or *zygote*. The DNA in that cell will eventually be copied into every cell in the adult body. But the DNA in cells changes as an individual matures. At the start, all of the genes in the DNA of a zygote are "switched on"—in the language of genetics, they can be *expressed*—in other words, they can all work. As an individual ages, however, the cells start to specialize, which requires various genes to be switched off so they can no longer be expressed. For example, every cell in your body contains the genes for producing insulin within its DNA, but insulin is produced only in the pancreas. In all of the other cells in your body (e.g., your skin, or nerve cells in your brain) the gene for insulin is switched off.

The same is true for every cell in your body—human development, carried forward by molecular processes we don't yet understand, produces specialized cells by turning off all but a few genes in each cell's DNA, and the specialization depends on which parts of the DNA are turned off. Once this switching process has happened, the specialized cells are locked in forever—muscle cells produce only muscle cells, skin cells only skin cells, and so on. This fact of development has enormous consequences for human welfare. Muscle cells destroyed by a heart attack cannot be replaced by other cells; dopamine-generating cells in the brain destroyed by Parkinson's disease cannot be replaced; nerve cells in the spinal cord, once cut, cannot be repaired. The inability of the body to replace cells because of their specialization lies behind a great deal of human misery.

Once the zygote starts to divide to produce an embryo, the cells temporarily retain their ability to develop into any kind of tissue. Cells like this, capable of developing into any cell in the body, are called *embryonic stem cells*. In the late 1990s, scientists learned how to take these cells and keep them in culture indefinitely. This has opened up amazing possibilities, because it means that we may soon be able to create new cells in the laboratory, and possibly entire new organs.

Scientists could use stem-cell techniques together with CLONING, to extract the DNA from the cell of an adult, place it in a human ovum, and then produce embryonic stem cells with that adult's DNA. It should therefore be possible to grow organs to replace those that have failed without having to worry about the body rejecting the implanted tissue.

More recently, it has been found that some cells in the adult body seem to possess at least some of the capability of creating stem cells that one finds in the embryo. If this turns out to be possible, it will remove one of the political objections to the use of embryonic stem cells—the objection that it is necessary to destroy a human embryo to obtain them.

Stern–Gerlach Experiment

It is possible to show experimentally that the spins of atoms and particles are quantized

Many subatomic particles have their own ANGULAR MOMENTUM, which gives them a quantity called *spin*. Because of its spin, a charged particle such as an electron or proton gives rise to a kind of electrical current, so it may also behave like a tiny magnet. Thus, we expect both atoms and the particles from which they are made to possess magnetic fields associated with orbital motion or intrinsic spin.

It might seem reasonable to expect the north and south poles of these atomic and subatomic magnets to orient themselves in any direction they please. The rules of QUANTUM MECHANICS, however, say differently. Like everything else in the quantum world, these directions are quantized—they can point only in certain specified directions relative to an external magnetic field. Otto Stern and Walther Gerlach carried out the first experiment to verify this prediction experimentally, in 1921.

At the heart of their experiment was a powerful magnet with its north and south poles close together, arranged such that the magnetic field between them was not uniform, but varied along a line from one pole to the other. In such a field, a tiny atomic or subatomic magnet will feel a force and will be deflected either up or down, depending on which way its north pole is pointing.

If the atom were governed by ordinary Newtonian laws (i.e., unaffected by quantum mechanics), and the poles were oriented every which way, you would expect atoms directed between the poles of the magnet to be deflected through all possible angles. No matter where we aimed our detectors on the "downstream" side of the magnet, we would expect to find atoms or particles going by. If the quantum mechanical predictions are right, however, the tiny magnets will point only in certain specified directions, and the beam of atoms or particles will be deflected only in certain directions. We would then see them coming through on the downstream side only at certain angles.

In the original experiment, conducted with atoms, some atoms were deflected only in two directions (corresponding to the atomic magnets oriented "up" and "down" only). Later, Stern showed that electrons and protons going through his apparatus behaved in the same way. The laws of quantum mechanics were confirmed one more time.

OTTO STERN (1888–1969) German-American physicist. He was born in Sohrau (now Zory, in Poland) and obtained his doctorate in 1912 from the University of Beaslau (now Wrocław). Walther Gerlach (1889–1979) joined Stern briefly when both were at Frankfurt in 1921. As professor of physical chemistry at the University of Hamburg, Stern studied the magnetic properties of the fundamental particles, for which he would receive the 1943 Nobel Prize for Physics. Stern left Germany for the USA in 1933, and spent the rest of his career at the Carnegie Institute of Technology, Pittsburgh.

String Theories

The ultimate particles of matter can be thought of as microscopic strings vibrating in a multi-dimensional world

Fasten your safety belts, because I'm about to describe one of the strangest theories now making the rounds among scientists whose quest is the ultimate nature of the universe. It is a theory so weird that it might even be right!

String theories are the current leading candidates for a THEORY OF EVERYTHING. This is the holy grail of particle physicists and theoretical cosmologists—a theory which encapsulates in a few equations all of the information about the interactions of the most fundamental constituents of the universe. String theory has been combined with a concept called *supersymmetry* to produce *superstring theory*, and that may prove to be our closest approach so far to unifying the four fundamental interactions (forces). Supersymmetry is itself built on the modern notion of forces being associated with the exchange of one kind of particle between others (*see* STANDARD MODEL). Think of the particles undergoing interactions as the bricks of the universe and the particles being exchanged as the mortar.

In the standard model the bricks are quarks and the mortar is provided by the mediators (called gauge bosons) that are exchanged between them. The idea of supersymmetry is that for every brick (quark or lepton) there is a much heavier but as yet undiscovered particle corresponding to mortar (mediator), and that for every mediator (mortar) there are as yet undiscovered particles corresponding to the quarks or leptons (bricks). None of the so-called supersymmetric partners has ever been seen in the laboratory, but they have been given names nonetheless—the supersymmetric partner of the electron, for example, is the *selectron*, the supersymmetric partner of the quark is the *squark,* and so on. The existence of these as yet unseen particles is one of the firm predictions of the theories.

The picture of the universe given to us by these theories is, in some ways at least, easy to visualize. On distance scales of 10^{-35} m, a scale fully 20 orders of magnitude smaller than a particle such as the proton or the quark, the structure of matter is very different from what we're used to. At these tiny distances (and incredibly high energies), matter appears to be a series of oscillating, stringlike fields. As with a guitar string, there are many ways, or modes, in which the string can vibrate: there is the main note, the fundamental, and the overtones, the higher harmonics. Each mode then corresponds to a different energy. By the principle of RELATIVITY, energy and mass are equivalent, so higher-energy vibrations correspond to higher-mass particles.

But while vibrating strings may be easy to visualize, the fact that the theory requires these vibrations to take place in 10 or 11 dimensions is not. The world we're used to has four dimensions: up–down, right–left,

If string theorists are right, elementary particles consist of minute strings, bundled up in a multiplicity of dimensions.

How Many Dimensions?

We humans have been happily using three dimensions to describe our surroundings since prehistoric times (the saber-toothed tiger is 60 feet ahead, 10 feet to the left and 15 feet up, on that rock!). Relativity has accustomed most of us to the concept that time is in some manner a fourth dimension (the sabre-toothed tiger is there *now!*). In the 20th century, though, theoretical physicists began to postulate the existence of 10, 11, even 26 spatial dimensions. Of course, they then have to explain why we do not perceive all those extra dimensions, however many there might be. This is where compactification comes in.

Imagine a garden hose. If you are close to the hose, you can see that it is a three-dimensional object. But from a distance, it appears just to have the one dimension of length: its thickness has become too small to perceive. The thickness of the hose has become compactified— curled up on too small a scale to be detected.

This, the theorists suggest, is what happens to the extra dimensions needed to explain the properties of the subatomic world—they are compactified down to a scale of 10^{-35} m and so are too small to be detected by any of our present-day techniques. Of course, this could all be wrong. It is only speculation —informed speculation perhaps, but speculation nonetheless.

front–back, and past–future. The world of strings has to have many more dimensions than this (*see* sidebar). Theorists get around the problem of there having to be more dimensions than we can observe by saying that the extra dimensions are tucked away (the technical term is *compactified*) and can't be seen at normal energies.

Recently, the concept of strings has been extended to multidimensional membranes, or *branes,* which are essentially strings that have been expanded into sheets. As one wag put it, the theorists' picture of the world has gone from a plate of spaghetti to a plate of lasagna.

So these are some of the characteristics of the best current candidate for a THEORY OF EVERYTHING. At the moment, though, there are some problems. For one thing, working out the mathematics of these theories turns out to be excruciatingly difficult. Almost all of the work over the last 20 years has been devoted to learning how to do calculations and proving that different versions of the theory are equivalent to one another. More importantly, the simple fact is that string theorists have yet to produce a single prediction that can actually be tested in the laboratory. Until they do, I'm afraid that their work, while a fascinating and esoteric exercise in logic, will remain outside the mainstream of science.

Superposition, law of

In undisturbed sedimentary rocks, the lower rocks are older than the higher ones

One of the most dramatic displays of sedimentary rock (*see* ROCK CYCLE) is to be found in the Grand Canyon in Arizona, where layer upon layer of brightly colored rocks sit atop one another, between them spanning millions of years of geological history. Sedimentary rocks form from layers of mud and other sediments deposited on ocean or lake bottoms in horizontal layers. Naturally enough, newer sediments are deposited on top of older ones. As we look at deeper and deeper layers of sediment in the Grand Canyon and elsewhere, we are seeing older and older rocks—in effect, we are moving back through time.

This law of superposition was the first tool that paleontologists used to unravel the history of life on our planet. It may seem obvious enough to us now, but when it was first proposed, in the 17th century, the idea that the Earth should have a long geological past in which it had appreciably changed was quite revolutionary. Later, at the beginning of the 19th century, came another important realization: If lower sediments are

The upper stratum of clay in this rock formation in Norfolk, England, was laid down after the lower strata of sandstone. This nicely illustrates the principle of superposition, according to which, in general, the lower a layer is found, the earlier it formed.

older than those higher up, then life forms found in lower sediments must have appeared before those in higher ones. This is the *law of faunal succession*. (This was a time, remember, when the idea of EVOLUTION was held by most people to be unorthodox, even heretical.) The story of a particular species of plant or animal, then, begins when the earliest layer of sediment containing its remains was laid down, and its extinction is marked by the highest layer to contain its remains. With this and the law of superposition, paleontologists could begin to establish the relative ages of rock strata from the fossil remains they contained.

If a succession of sediments is unbroken and contains good enough fossil specimens, it should be possible to trace the transformation of one type of organism into another—to monitor, in other words, the process of natural selection by evolution. In practice, fossils are not usually sufficiently well preserved to allow this to be done (although there are a few sites around the world where conditions allow it). Instead, one sees a fossil record that is patchy and incomplete, and paleontologists have to resort to theoretical reasoning to relate what they find to what was actually there in the past. For example, it is not unusual to have a record in which an organism is found in a few lower strata, then not found in a few strata, and then found again still higher up. Obviously, what has happened is that the organism was present throughout the entire range, but failed to be preserved as a fossil in the intermediate strata. (This phenomenon has been called the "Lazarus effect," after the Biblical story in which a man is raised from the dead.)

A few words of caution: First, there is nothing in the law of superposition by which we can tell how long ago sediments were formed—only the relative ages of different strata. For absolute ages, we must use RADIO-METRIC DATING. Second, the word "undisturbed" in the statement of the principle is extremely important. Because the surface of the Earth is constantly changing (*see* PLATE TECTONICS), sedimentary rocks, once formed, can be folded over, so that in the region of the fold older rocks overlie younger ones.

Surface Tension

Molecules at the surface of a liquid experience a force that keeps them from moving outward

The molecules in a liquid experience forces that tend to attract them to other molecules—indeed, it is this force that keeps the liquid together. For a molecule in the bulk of the liquid, these forces are exerted in all directions by neighboring molecules. But molecules at the surface have no neighbors above them, so all of the attractive forces are directed inward. The net result is a force that tends to pull the surface of the liquid inward. This force is called surface tension. We usually think of it as being exerted along the surface, pulling the fluid together and creating the effect of a flimsy skin.

One consequence is that you need energy to expand a liquid surface, since you have to do work against the force of surface tension. This means that, left to itself, a liquid will assume the shape that has the smallest possible surface. This shape, of course, is a sphere, which explains why a falling raindrop assumes an almost spherical shape (I say "almost" because the shape of the raindrop is distorted slightly by air resistance.) It also explains why water on the body of a newly waxed car gathers into small beads.

Surface tension effects are exploited commercially in the manufacture of approximately spherical metal objects such as lead shot. Drops of molten metal dropped from a height form themselves into spheres under the influence of surface tension as they fall, and harden in time to survive the impact when they land.

There are many everyday examples of surface tension in action. When wind blows across the surface of a lake or the ocean, for example, it creates small ripples. These ripples are waves in which the upward force of pressure in the liquid is countered by the downward force of surface

This pond skater uses surface tension to support itself as it skims along the surface of the water. Because the weight of the insect is less than the upward force exerted by the surface tension of the water, it doesn't sink.

tension. The two forces alternate, like a spring alternately stretching and compressing, creating the ripples.

Whether a liquid will "bead up" or spread out on a solid surface (a property called "wetting") depends on the relation between the intermolecular forces behind surface tension and the forces of attraction between the liquid and the solid. Liquid water, for example, is held together by hydrogen bonds (*see* CHEMICAL BONDS) between the molecules. Water will wet a glass surface because there are many oxygen atoms in glass, and water molecules can temporarily bond with them as easily as with other water molecules. If the surface is greasy, however, these hydrogen bonds cannot form, and the water molecules will be more strongly attracted to one another than to the surface. Consequently, the water beads up under the influence of surface tension.

Engineers often add substances known as *surfactants* to water to prevent it from beading—they are added to detergents used in dishwashers, for example. When they find themselves in the surface of a liquid, surfactant molecules act to weaken the forces of surface tension. Water to which a surfactant has been added will spread out evenly over a surface, and will not form drops that leave spots when they dry. (For a discussion of the action of detergents, *see* "LIKE DISSOLVES LIKE.")

Symbiosis

*Symbiosis can take
the form of mutualism,
commensalism, or
parasitism*

Symbiosis (from the Latin for "living together") is the close association of organisms of different species. It can take several different forms, depending on the nature of the relationship between the two species and whether that relationship leads to benefit or harm. If the relationship benefits both parties, it is called *mutualism*. If it benefits one party and neither benefits nor harms the other, it is called *commensalism*. If it harms one party while benefiting the other, it is called *parasitism.*

There are many examples in nature of symbiotic relationships that benefit both partners. For example, an important link in the NITROGEN CYCLE depends on a symbiotic relationship between plants such as legumes and soil bacteria known as *Rhizobium.* These bacteria live in the roots of the plants and have the ability to "fix" nitrogen—that is, to break the strong bonds that bind nitrogen atoms into molecules in the atmosphere so that the nitrogen can be incorporated in molecules, such as ammonia, that the plant can use. In this case the mutual benefit is obvious—the roots provide a home for the bacteria, the bacteria supply an essential nutrient for the plant.

There are also many examples of symbiotic relationships that benefit one party and neither harm not help the other. The human gut, for example, is host to many bacteria whose presence does not harm us. In the same way, there are plants called bromeliads (the pineapple is an example) that live on the branches of trees, but get all of their nutrients from the air. These plants use the tree for support, but do not take nutrients from it.

Parasitism is also common. Mistletoe is a parasite that feeds on the trees onto which it fastens, draining nutrients from the host without providing any compensatory benefit. Every bacterium or virus that causes a disease can also be thought of as a parasite, as can organisms like tapeworms. A good deal of modern medicine and public health resources are devoted to keeping these sorts of parasites away from human beings.

A particularly interesting example of mutualism can be seen in the evolution of modern complex cells. Today, living cells come in two types. There are *prokaryotes* ("before the nucleus"), primitive cells in which the DNA is found loose inside the cell body, and *eukaryotes* ("true nucleus"), advanced cells in which the DNA is stored in a special structure called the nucleus. (For a discussion of the role of DNA in living systems, *see* MOLECULAR BIOLOGY.) All multicelled organisms (including human beings) are constructed from eukaryotic cells.

Strange though it may seem, there is actually a fossil record of single-celled organisms that goes back at least 3.5 billion years. Although cells have no hard parts that can fossilize in the conventional sense (*see* EVOLUTION), they can become trapped between layers of mud and sediment in river beds or ocean floors. When the mud is transformed into rock (*see*

ROCK CYCLE), an impression of the cell remains, like the impression of a leaf. These microscopic impressions can be examined, and a record can be drawn up of life on Earth before the development of skeletons. This record tells us that around a billion years ago, cells underwent a major change. It was at this point that eukaryotic cells started to appear.

These cells, in addition to their nuclei, contain many separate interior structures call *organelles*. One class of these, the so-called mitichondria, are where energy is generated in the cell—think of them as the cellular power plants. Both the nucleus and the mitochondria have a double cell membrane, and the mitochondria contain their own DNA as well. These facts have led to the theory that eukaryotic cells originated in an act of symbiosis. One cell absorbed another, and the two found that they could get along better together than either could on its own. This is called the *endosymbiosis* theory.

In this theory, the double cell membrane is easy to understand. The inner layer comes from the membrane of the original cell that was absorbed, while the outer layer comes from the part of the membrane of the absorbing cell that wrapped itself around the newcomer. Mitochondrial DNA is also easy to understand—it's what's left of the original DNA complement of the newcomer. So, many (perhaps all) of the organelles in eukaryotic cells started their existence as separate organisms, but joined forces about a billion years ago to create a new kind of cell. Our entire bodies, then, represent the result of one of the oldest partnerships in nature.

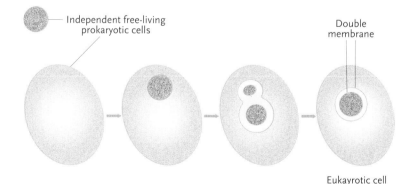

Independent free-living prokaryotic cells

Double membrane

Eukayrotic cell

The first eukaryotic cells may have formed when large prokaryotic cells engulfed smaller ones. This would account for the double membrane around the nucleus of eukaryotic cells.

Terminal Velocity

An object falling through a fluid will reach a constant velocity when the force of gravity is balanced by the force of fluid resistance

A group of skydivers spreadeagle themselves to minimize their terminal velocity.

When an object is dropped, NEWTON'S LAWS OF MOTION tell us that it will accelerate because an unbalanced force—the force of gravity—acts on it. Once an object starts moving through the air (or any other fluid), however, another force enters the picture. To move downward, the object has to push aside molecules in the air, and the motion of these molecules gives rise to a force that opposes the motion of the object. We call this kind of force *air resistance,* or *viscous drag.* The faster the object falls, the greater this opposing force becomes. When the viscous drag becomes equal to the force of gravity, the two cancel and there is no more acceleration. From that point on, the object falls at a constant velocity. This final maximum speed is called the terminal velocity of the object.

The magnitude of the terminal velocity, then, depends on how air moves around the falling body. In the simplest case, air can simply slip around the body without creating any turbulence—this is termed *laminar flow*. In laminar flow, the viscous force increases in proportion to the velocity of the object. The airflow around a falling raindrop smaller than a fine grain of sawdust, for example, will be laminar. The terminal velocity will then be rather small—about 3 miles per hour (5 km/h), the speed of a gentle stroll. This is why fine rain often appears to "hang" in the air. A similar terminal velocity is exploited in the MILLIKAN OIL-DROP EXPERIMENT.

For larger objects, other mechanisms begin to dominate drag. Should that raindrop grow even to one-tenth of a millimeter across, a phenomenon called *vortex shedding* appears. You have probably seen vortex shedding: When a car goes down a tree-lined street, and the trees have shed their leaves, the leaves on the ground don't just blow away from the car, but move in small circles. These circles, which trace the flow of the air in the wake of the moving car, are called *von Kármán vortices* after Theodore von Kármán (1881–1963), a Hungarian-born engineer who became one of the founders of modern aeronautical engineering at the California Institute of Technology. The existence of these vortices is normally the main cause of drag, and the wind resistance that slows down a car or an airplane is generally associated with them. If you've ever been passed by a large truck on the highway and felt your car being pushed back and forth in its wake, then you've experienced von Kármán vortices firsthand.

When large objects begin to fall through the air, they begin shedding vortices immediately, and soon reach a terminal velocity. Skydivers, for example, reach a terminal velocity of about 120 miles per hour (190 km/h) if they spreadeagle themselves to maximize their air resistance, but a velocity of 150 mph (240 km/h) if they assume a head-down position like a high-board diver about to enter the water.

Theories of Everything

All forces in nature are manifestations of a single unified force

There are four fundamental forces in nature, and everything that happens does so because of the action of one or more of them. These forces, in decreasing order of strength, are:

— the *strong force*, which holds the nucleus of the atom together and binds quarks into the elementary particles,
— the *electromagnetic force* between electric charges and between magnets,
— the *weak force,* responsible for certain types of radioactive decay, and
— the *gravitational force.*

In standard Newtonian physics, a force is simply a push or a pull—something that causes a change in the motion of an object. In modern quantum theories, however, the force (now visualized as acting between particles) is given a somewhat different interpretation. It is regarded as the result of the exchange of one kind of particle (which is said to *mediate* the force) between two others. In this way, the exchange of a photon between two electrons generates the electromagnetic force, and the exchange of other particles generates the other three fundamental forces. The modern view is that what distinguishes the four forces is the different particles that mediate them. (For more details, *see* STANDARD MODEL.)

Furthermore, the properties of the force depend on the properties of the particles being exchanged. NEWTON'S LAW OF GRAVITATION and COULOMB'S LAW have the same form, for example, because both are mediated by massless particles. The weak force operates only over very short distances (essentially inside the nucleus) because the gauge bosons are so heavy. The strong force is also short-range, but for a different reason, having to do with the confinement of quarks (*see* STANDARD MODEL).

The optimistic label "theory of everything" is used for any theory in which all of these forces would be seen to be simply different aspects of a single underlying force. In this case, the ultimate picture of the universe would be extremely simple. Matter would consist of quarks and leptons (*see* STANDARD MODEL) and there would be only one force acting between the particles. The equations describing this situation would undoubtedly be simple enough to write on the back of an envelope, and would, in some sense, contain within it the entire workings of the universe. It would be, as Nobel prizewinning American physicist Steven Weinberg has pointed out, a deep theory from which all arrows of inference radiate outward, with no logical constructs needed from something deeper still. As the word "would" in every sentence in this paragraph indicates, no theory of everything yet exists. What we can do is outline the process by which we can hope to arrive at one.

The route from four forces to a single force is called, appropriately enough, *unification*. To see how it works, imagine two pairs of skaters. The air temperature is below the freezing point of water. One pair of skaters exchanges a bucket of alcohol (which is still liquid) and the other pair exchange a bucket of water (which is in fact frozen into ice). It might seem as though two separate forces were operating—one corresponding to the exchange of a liquid, and the other to the exchange of a solid. But if the temperature were to climb above 0°C, the water would melt and we would see that there was really only one force operating—one associated with the exchange of a liquid. It only seemed as though there were two forces because the temperature was very low. Raise the temperature, and the true underlying unity of the forces becomes evident.

In the same way, unification theories operate on the assumption that when the energy of particle interactions is high enough—when collisions are violent enough—the "ice will melt" and the forces will be seen to be the same. The theories predict that this will happen not all at once for all four forces, but in a stepwise manner as the energy of particle interactions is increased.

The lowest energy at which one of these unifications can occur is pretty big, but just within the capabilities of present-day particle accelerators. Particle energies were very high during the early stages of the BIG BANG (*see also* EARLY UNIVERSE). For the first 10^{-10} seconds they were high enough for the weak nuclear and electromagnetic forces to be unified. Only after this time (or below the corresponding energy) do all four forces appear to have become distinct. Before then (or above that energy) there were only three fundamental forces—the strong force, the newly unified *electroweak force,* and gravity.

The next transition is at energies well beyond those attainable in terrestrial laboratories—energies not seen since the universe was 10^{-35} seconds old. At these energies, the strong force unifies with the electroweak force. Theories that describe this unification are called *grand unified theories* (known a little offputtingly as GUTs). These theories cannot be tested directly at high energies, but they do make a number of predictions about lower-energy processes that can be checked. At the moment, the GUTs seem to do a good job of making accurate predictions. The GUTs, however, mark the boundary of our reliable theories about the universe. Beyond the GUTs lie the theories of everything (known somewhat prosaically as TOEs)—gleams in the eyes of theorists. In TOEs, gravity is unified with the strong–electroweak force, and the universe becomes as simple as it can be.

Thermal Expansion, law of

The change in the dimensions of a material when it is heated is proportional to the change in the temperature

When materials are heated, they tend to expand. This is easy to understand in terms of the atomic theory of HEAT, because raising an object's temperature makes its atoms move faster. In solid materials the vibrations are more violent, and each atom needs a little more "elbow room" to move around. The result is an increase in the dimensions of the object. Similarly, liquids usually expand when their temperature is increased, because of the higher velocities of the particles (*see* BOYLE'S LAW, CHARLES'S LAW, and IDEAL GAS LAW for the expansion of gases.)

The basic law that governs thermal expansion says that if an object of length L has its temperature changed by an amount ΔT, then the change in the dimension, ΔL, is given by the equation

$$\Delta L = \alpha L \, \Delta T$$

where α is called the *coefficient of linear expansion*. There are similar formulae for increases in area and volume. In the simplest case, where this coefficient depends on neither the temperature nor the direction of expansion, the material will expand uniformly in all directions, and the change of length of each side will be given by the above equation.

Thermal expansion is a fact of life for engineers. For example, a steel bridge across a river in a continental city may well be exposed to temperatures ranging from below 0°C to over 40°C. The length of such a bridge may well vary by several feet between winter and summer, and if the structure is not to buckle, allowance has to be made for this fact. The usual way to do this is to build *expansion joints* into the bridge: Meshing sets of steel teeth set into the span across its length come together in the summer and pull apart in the winter. A long bridge will have several expansion joints.

Not all materials, especially not all solids, expand uniformly in all directions. And not all materials expand the same way at different temperatures—perhaps the most interesting example of this is water. Cool water, and it will shrink according to the above formula until it reaches a temperature of about 4°C. Below this temperature, water actually expands when cooled and shrinks when it is heated. (In the language of the above equation, we say that α becomes negative below 4°C.) This is why the water at the bottom of the Earth's oceans does not freeze, no matter how cold it gets, since water colder than 4°C will be less dense and will float upward.

Incidentally, the fact that water expands when it freezes is another (though unrelated) anomaly—an anomaly that is responsible for the presence of life on our planet. If this didn't happen, the ice would sink to the bottom of lakes and oceans and they would freeze all the way through, killing any living organisms in them.

THERMAL EXPANSION

Thermo-dynamics, first law of

Heat is a form of energy, and energy is conserved

The hunters shown in this 14th-century French tapestry are using the elastic energy stored in their bows to propel the arrows.

To a physicist, work is done whenever a force moves something over a distance. To raise this book, for example, you have to exert an upward force to overcome the downward pull of gravity over the distance through which you raise it, and you are doing work in the process. Similarly, energy is the ability to do work. Any object that can exert a force over a distance thus possesses energy. From the scientific point of view, there are three important things about energy: it comes in many types, the types of energy can be converted into one another, and energy is always conserved.

Energy of motion
A moving object can exert a force over a distance, and, in fact, will do so when it hits something. Think of an arrow on its way toward a target. When it strikes, it exerts a force that moves the fibers in the target apart. Consequently, a moving object is capable of doing work and, by definition, possesses energy. Energy of motion of this type is called *kinetic energy*. According to the atomic theory of HEAT, what we call heat arises from the motion of molecules, so heat can be thought of as a special kind of kinetic energy.

Energy of position
If you raise this book, it will then be capable to doing work just by virtue of its new position in the Earth's gravitational field. To see why, imagine dropping the book. During its fall it will speed up, which means it will acquire kinetic energy. When it hits the table or floor, it will exert a force over a small distance as the target is slightly distorted by the impact. We call the type of energy the book has when it is higher up *potential energy*—while it is raised, the book is doing no work, but it has the potential to do work if it is dropped. To be precise, we call the book's energy *gravitational potential energy* because it has energy only by virtue of its being in a gravitational field. If you lifted the book in a spaceship

---

THERMODYNAMICS, FIRST LAW OF

396

located in interstellar space, where there was no gravitational field, it wouldn't fall at all, so it would have no gravitational potential energy.[*] A stretched rubber band or a bent bow both have elastic potential energy, because either could do work if it were released.

An electrically charged particle located in an electrical field will have *electrical potential energy*. We see this in the atom (*see* ATOMIC THEORY), where an electron's energy depends on how far it is from the positively charged nucleus. A special kind of electrical potential energy is involved when atoms combine chemically. The electrons in each atom start off with a certain electrical potential energy because of their respective positions in their atoms. After the atoms have combined, those electrons will have different energies, associated with their new positions. In general, the total energies before and after will be different. We speak of the energy associated with this change of electron configuration as *chemical potential energy*.

There are many other kinds of potential energy, associated with magnetic and electrical fields, with various special properties of materials, and so on. All forms of potential energy appear in systems where work can be done, but is not necessarily being done at present.

Energy of mass

When Einstein developed the theory of RELATIVITY, he discovered a form of energy in nature that no one had suspected. It turns out that mass can be turned into energy, and vice versa, as governed by the equation $E = mc^2$, where c is the speed of light (3×10^8 meters per second). What this equation says is that a small amount of mass will produce a huge amount of energy—as happens when the fission of uranium nuclei is used to run nuclear reactors. It also says that large amounts of energy can create small amounts of mass, as happens in particle accelerators around the world. These machines speed protons up to nearly the speed of light, then smash them into targets. In the collisions some of the kinetic energy of the protons is converted into the mass of new particles.

Convertibility and conservation of energy

The different forms of energy are interchangeable—energy can flow from any form to any other. When an archer fires an arrow, for example, the elastic potential energy of the bow is converted into the kinetic energy of the arrow, and when the arrow strikes the target its kinetic energy winds up as heat. Except for heat, any form of energy can be completely converted into any other (the conversion of heat into work is limited by the second law of THERMODYNAMICS).

This conversion of one type of energy to another is not completely

[*] This might seem counter-intuitive. The explanation is this: In outer space, where there is no significant gravitational field, the potential energy must clearly be zero. Since potential energy is released as an object moves from outer space down towards the surface of a planet or star, its value must be *negative*. A one-kilogram book at the surface of the Earth would have a potential energy of about -6×10^7 joules. If the book were then lifted to a height of 1000 km, its potential energy would increase to -5×10^7 joules.

random because energy is a conserved quantity. This means that the total amount of energy in an isolated system cannot change, even though the form it takes can vary. Think of a fixed amount of money spread among many different types of bank accounts—a savings account, an account tied to stocks and bonds, and so on. You can do many things with this money. You can put it all in one account, you can spread it around evenly, you can put different amounts in each account. But no matter what you do, the total amount of money added up over all the accounts stays the same. (In our analogy we are ignoring any interest payments or fluctuations in the stock market values.) In the same way, even as it changes form, energy is neither created nor destroyed. This is a statement of the *conservation of energy,* and is stated explicitly as follows: In an isolated system, the total energy remains constant.

Thermo-dynamics, second law of

Heat does not flow spontaneously from cold to hot objects

It is impossible for an engine to convert heat to work with 100% efficiency

The entropy of a closed system cannot decrease

There is a kind of directionality in nature that isn't captured by most of the laws in this book. It's easy enough to break eggs to make an omelet, yet virtually impossible to turn an omelet back into eggs. Perfume from an uncorked bottle pervades a room, but never gathers itself up and returns into the bottle. The reason for this one-way operation of the universe has to do with the second law of thermodynamics, easily the most difficult and most misunderstood of the concepts of classical physics.

For one thing, there are the above three separate statements of the law, each made by a different scientist at a different time. While it may appear that the three of them have nothing in common with one another, in point of fact they are all logically equivalent. Given one of the statements, in other words, you can show mathematically that the other two both follow.

Let's start with the first statement, which is due to the German physicist Rudolf Clausius, who rediscovered the CLAUSIUS–CLAPEYRON EQUATION. Here is a simple illustration of it: Take an ice cube out of your refrigerator and put it in the sink. As time goes by, the ice cube will melt—heat will have flowed from a warm body (the air) into a cold one (the ice cube). From the point of view of energy conservation, there's no reason why the heat has to flow in this direction only—energy would still be conserved if the ice cube got colder and the air got warmer. The fact that this doesn't happen is an example of the directionality mentioned earlier.

We can easily see why things happen this way by looking at how the ice cube and the air interact on the molecular scale. We know from the KINETIC THEORY that temperature is related to the speed of molecules in an object—the faster they move, the higher the object's temperature. This must means that the molecules in the air are moving faster than the molecules in the ice cube. When a molecule in the air collides with a water molecule at the surface of the cube, our experience tells us that, on average, the fast molecule will slow down and the slow molecule will speed up. The molecules in the ice will thus move faster and faster or, equivalently, the temperature of the cube will increase. This is what we mean when we say that heat flows from the air to the ice cube. Seen this way, the first statement of the second law seems to follow quite naturally from the behavior of molecules.

Work is done whenever a force moves something by some distance, and the various forms of energy measure a system's ability to do work. Heat, a measure of the kinetic energy of molecules, is a form of energy, so it can be converted to work as well. But again, there is a directionality. You can convert work into heat with 100% efficiency—you do it every time you depress the brake pedal on your car, initiating a process that ultimately converts all of the kinetic energy of the car into heat in the brake pads. The second statement of the second law says that you can't go the other way. If you try to convert heat into work, you won't be able to extract all of the energy—some will be lost to the environment as waste heat.

There is a simple way to visualize how the second statement works. Imagine a cylinder, like the ones in your car's engine, filled with an explosive mixture of gas and in which a closely fitting piston is free to move. If the mixture is ignited, it creates a great deal of heat which we can convert to work by allowing the gas to expand against the piston, so that the piston moves. In an ideal world, we could do this with 100% efficiency— that is, all of the energy in the heated gas could wind up as work done by the piston.

To harness this work by making the cylinder part of a practical engine, we have to return the piston to its original position so it can go through the cycle again. Suppose we just push the piston back to where it started from (in your car, the piston actually goes down, then up, then down again, and finally back up to complete the cycle). The gas in the cylinder will be compressed and heated during the return stroke, so to get back to the actual starting situation we have to remove that heat and dump it into the environment. In a complete cycle, then, some of the work we get from the first stroke has to be converted back to heat and thrown away. The net effect is that some of the work done by the piston on the way down has to be used to return the engine to its original position.

This formulation of the second law is incorporated in CARNOT'S PRINCIPLE, as stated by the French military engineer Sadi Carnot. It has profound implications for the generation of electrical power, a major undertaking in today's society. It turns out that if, for example, we burn coal to produce electricity, fully two-thirds of the energy in the coal is lost to the environment. This is why the most striking feature of modern generating plants are huge cooling towers—this is where the waste heat is being returned to the environment. This low efficiency has nothing to do with bad design: Modern engineers, can achieve efficiencies only a few percentage points below the Carnot efficiency. Anyone who claims they can produce energy at a higher efficiency (e.g., by building a PERPETUAL-MOTION machine) would be claiming to outwit the second law of thermodynamics. They might as well say they can make an ice cube sitting in the sink get colder.

The third statement of the second law, usually credited to the Austrian physicist Ludwig Boltzmann (*see* BOLTZMANN CONSTANT), is probably the best known. *Entropy* is a measure of the disorder of a system—the higher the entropy, the greater the disorder. Boltzmann developed a straightforward mathematical way of defining the amount of order in a system. We can see how it works by applying it to water. Water in the

These cooling towers at a power plant in England cool the steam produced to power the plant's turbines. The heat is dispersed into the atmosphere, in accordance with the second law of thermodynamics.

liquid state is a relatively disordered system because the molecules are free to move around and assume many different configurations. Ice, on the other hand, is a relatively ordered system because the molecules are locked into a lattice structure. Boltzmann's statement of the second law says that it is possible for a system to go from ice to water (i.e., from low to high entropy) but not for it to go the other way. Thus it, too, reflects directionality in nature.

It's important to realize that this statement of the second law does not say that entropy can never be decreased. You can, after all, make ice cubes from water in your refrigerator. The point is that the statement applies to *closed systems*—those in which energy from the outside doesn't play a role. A working refrigerator isn't isolated, because it is plugged into the wall and, ultimately, connected to the power plant that generates the electricity that runs it. The closed system in this case is the refrigerator plus the power plant, and as long as the increase in disorder at the power plant is greater than the increase in order in the ice cubes, there is no violation of the second law.

Which leads, I suppose, to yet another statement of the second law: *A refrigerator won't work unless you plug it in.*

Thermo-dynamics, third law of

It is impossible, in a finite number of steps, to bring an object to a temperature of absolute zero

Absolute zero is one of those concepts that has a sexy name and a deceptively simple definition. In the days before QUANTUM MECHANICS, the definition really was simple. KINETIC THEORY provided a connection between the motion of atoms and the value of the temperature that was easy to visualize—the slower the atoms moved, the lower the temperature. In this picture, it is easy to see that there is a lowest possible temperature, which is reached when all the atoms in a material stop moving. This temperature is about 273°C degrees below zero (–456°F), and this is what is called absolute zero.

This temperature doesn't change when we bring in the ideas of quantum mechanics, but the way we think about the atoms does. If the atoms in a material just stopped dead, we would then be able to establish both their location and velocity. That would violate HEISENBERG'S UNCERTAINTY PRINCIPLE. We are therefore forced to think of each atom as being smeared out over a small volume, if we are considering it as a wave, or vibrating a little, if we are considering it as a particle). Instead of saying that the atom isn't moving, we say that it is in a state from which no further energy can be extracted (the energy that is left is called the *zero-point energy*). The bottom line, however, is the same—there is a lowest possible temperature, and it is –273°C.

Actually, the existence of zero-point energy illustrates an interesting point about the quantum world. As the temperature drops, the wave nature of matter (*see* SCHRÖDINGER'S EQUATION) becomes more evident and more important, and quantum effects begin to dominate the familiar "billiard ball" atom.

Incidentally, 273°C is the only temperature that is built into the laws of nature. It is utilized to define the most commonly used temperature scale in the sciences, the *Kelvin scale*. The zero point on this scale is taken to be absolute zero, and the size of the degree kelvin is taken to be the same as on the Celsius scale. Thus, absolute zero is at 0 K, water freezes at 273 K, and room temperature is about 300 K.

The third law says simply that absolute zero is unattainable—like the speed of light, it can be approached arbitrarily closely but never actually reached. In essence, the closer a system gets to absolute

The science of the very cold is called cryogenics. Here a technician wearing insulating clothing is retrieving cancer cells, stored at very low temperatures, from a cryogenic container.

Zero-Point Energy

A billiard ball rolling across a table will eventually lose its energy to friction and come to a stop, according to the first law of THERMODYNAMICS. A quantum particle like an electron in an atom, however, cannot do the same thing because of the UNCERTAINTY PRINCIPLE. This principle states that it is impossible to know the position and velocity of a particle with perfect accuracy at any given time. Were the electron to come to a stop, we would know its velocity (zero) and its position as well. Thus, quantum particles, unlike their classical analogues, must always be in a kind of fuzzy motion, never exactly in one place or moving with exactly one speed. And this, in turn, means that the quantum particle must always have some energy.

This zero-point energy associated with quantum mechanics is a little unusual, because it cannot be extracted or changed like other forms of energy. It is basically the lowest energy a particle can have and still not violate the laws of quantum mechanics. Calculating the zero-point energy of a quantum system—the innermost electron orbit in the BOHR ATOM, for example—often gives a reasonable approximation to the value for the orbit obtained by more precise and thorough calculations.

zero, the harder you have to work to extract energy from the system and lower its temperature farther. Actually, scientists have come remarkably close to absolute zero in the laboratory. Temperatures in the nanokelvin range (i.e., temperatures only a billionth of a degree above absolute zero) are now produced routinely.

There are many ways of lowering an object's temperature. A liquid can be evaporated from its surface, taking away heat in the process—this is how sweating works in humans. Alternately, a gas can be expanded—this is why a spray can cools when you use it. Using methods such as these, scientists can get to temperatures a few degrees above absolute zero. To go to really low temperatures, though, you have to suspend small numbers of atoms in electric and magnetic fields for long periods of time. Thus held, the atoms can be manipulated with precisely controlled laser beams, first getting them to give up some of their energy by interacting with the light, then by using the beams to expand the collection of atoms as if they were coming from a spray can. This is how to get down to nanokelvin temperatures. But no matter how clever we get, the third law tells us that we can never cross that final barrier separating us from absolute zero.

A physicist with a sense of humor once restated the three laws of thermodynamics this way:

First Law of Thermodynamics: You can't win
Second Law of Thermodynamics: You can't break even
Third Law of Thermodynamics: You can't even get in the game

Titius–Bode Law

The distances to some of the planets follow a simple arithmetical formula

There's something about numerology that seems to fascinate people. As a scientist who deals with the public, I routinely get mail from people who have discovered the secrets of the universe in the sequence of numbers in π or the masses of the elementary particles. Their logic seems to be that if you can find a numerical sequence that explains something in nature, there has to be a deep meaning to it. There aren't too many incorrect "laws" in this book, but this is one of them. (Actually, there was nothing wrong with the way the law was originally derived and tested, but eventually it failed—as we shall see.)

In 1766, Johann Titius, a German scientist, announced that he had noticed that the orbits of the planets follow a simple pattern. He started with the sequence 0, 3, 6, 12, …, formed by doubling each number (after the first). He then added 4 to each number and divided through by 10. The final sequence was a pretty good approximation to the distances of the planets from the Sun in astronomical units, at least for those planets that were known at the time. (The astronomical unit, or AU, is the distance between the Earth and the Sun.)

Planet	Titius–Bode law prediction	Actual number
Mercury	0.4	0.39
Venus	0.7	0.72
Earth	1.0	1.00 (by definition)
Mars	1.6	1.52
"missing planet"	2.8	—
Jupiter	5.2	5.2
Saturn	10	9.5

The fit looks impressive, especially when you consider that the prediction for the distance to Uranus, which wasn't discovered until 1781, was 19.6, compared with 19.2 AU for the measured value. The discovery of Uranus renewed interest in the "law," and in the mysterious gap at 2.8. Surely there had to be a planet orbiting between Mars and Jupiter—but it would have to be small to have escaped detection.

In the year 1800, 24 astronomers banded together to mount a search, calling themselves the Celestial Police, but it was another astronomer, Giuseppe Piazzi (1746–1826), who found the culprit, on New Year's Day, 1801. Its distance from the Sun was 2.77 AU. This 580-mile (933-km) wide object was clearly too small to be a major planet. Soon more small bodies were discovered between the orbits of Mars and Jupiter, and now we know of many thousands. They are known as minor planets, or asteroids, and are believed to represent material that never accreted into a larger planet (*see* NEBULAR HYPOTHESIS). I believe that it is this gap in the sequence that gave rise to the mythology of an exploding planet between Mars and Jupiter. Remember Krypton, the birthplace of Superman? Old ideas die hard!

The German astronomer Johann Bode was so impressed with Titius' work that he featured these results in his popular astronomy textbook, published in 1772. His role in publicizing the law is what gives it its hyphenated title. You sometimes see it referred to, quite unfairly, simply as Bode's law.

Confronted by a sequence of numbers like this, how is one supposed to react? I always start by remembering some advice from my old statistics prof. He would always talk about "the golf ball on the fairway." His argument went like this: If you calculate in advance the probability that your golf ball will land on a given blade of grass, the result will be almost zero. But, of course, the ball has to land somewhere, so it does no good trying to understand why the golf ball lands on a particular blade of grass, because if it hadn't landed there, it would have landed somewhere else.

Johann Bode, whose "law" turned out to be nothing more than a numerical coincidence.

As far as the Titius–Bode law is concerned, the six numbers representing the distances of the planets from the Sun are like the golf ball. Think of the blades of grass as all the possible numerical schemes that could generate those numbers. In point of fact, there are a lot more such schemes than there are blades of grass on a fairway. Some scheme is going to come close to the numbers, just as the golf ball is going to land on some blade of grass. That it was Titius' formula and not someone else's is a matter of chance, not science.

Actually, it isn't necessary to go through statistical arguments to disprove the Titius–Bode law. As it happens, the discoveries of Neptune and Pluto have already done that. Neptune is at a "wrong" distance from the Sun (the prediction is 38.8, the actual value 30.1 AU), and Pluto's orbit is so eccentric that the exercise becomes meaningless.

Does this make whole thing an exercise in pseudoscience? I don't think so. There was nothing wrong with Titius (and Bode) looking for regularities in the solar system—scientists do this sort of thing all the time. The problem is that no one after them tried to go beyond the numbers to find a physical reason why the planets should be orbiting where they do. In the absence of any work of that kind, the "law" remains mere numerology—and incorrect numerology at that.

JOHANN ELERT BODE (1747–1826) German astronomer and mathematical, born in Hamburg. Self-taught in astronomy, he was already publishing astronomical treatises in his teens. In 1772 he was hired by the Berlin Academy to work on its annual almanac; he was to edit the *Astronomisches Jahrbuch* until the year of his death, turning it into a prestigious and profitable publication. He proposed the name Uranus for the new planet discovered by William Herschel in 1781. In 1786 Bode was made director of the Academy's astronomical observatory. He compiled several important star atlases, including the *Uranographia* (1801), perhaps the most beautiful star atlas ever drawn.

JOHANN DANIEL TITIUS (1729–96) German astronomer, physicist, and biologist. He was born in Konitz (now Chojnice, in Poland) and educated at the University of Leipzig, where he received a degree in 1752. Four years later he moved to the University of Wittenberg, where he stayed for the rest of his life, holding chairs in mathematics and physics. Titius was inspired to develop the "law" while he was translating a book by Charles Bonnet, a French naturalist. Bonnet was commenting on the orderly hand of God at work in the solar system; Titius helped him out by introducing some numbers. Bode didn't credit Titius' work on the "law" until 1784.

Triune Brain

The brain evolved by accumulating evolutionary layers—reptilian, mammalian, and human

In the mid-20th century, an oddly appealing notion of the brain took hold. Over time, it was argued, the human brain had evolved by accreting layers, like the rings of a tree. At the very base of the brain were the cerebellum and brain stem, responsible for fundamental functions such as balance and maintaining internal body functions. This was supposed to be the "reptilian" part of the brain, a legacy from our distant forebears on the evolutionary tree. Overlaying this was the middle brain—the seat of hunger, sexual arousal, and the like. This was supposed to be the "mammalian" layer. And over this was the cerebral cortex—the seat of thought and higher mental functions, what it is that makes us distinctly human. This picture was known as the triune brain, and it was made accessible to the public by Carl Sagan (1934–96) in his book *The Dragons of Eden* (1977).

The triune brain has a lot going for it. It is simple, appealing, and logical. Unfortunately, it is also completely wrong.

For one thing, the human brain may be different from those of other animals, but it's not *that* different. Fish have a different brain layout from ours, but pretty much all the same parts. Fish and human brains differ in the way two cars might—there's a clear difference, but they both have wheels, an engine, brakes, and so on. The fact that we have more intellectual horsepower than a fish is because we have a much larger cortex, not because the fish doesn't have one at all.

For another thing, the working of the brain is much more complex than anything that could be fitted into a simple triune model. We now know that the brain consists of many highly specialized collections of cells, and that the brain's functioning depends on these centers communicating with one another. The phrase "community of mind" is often used to convey this idea.

Take vision as a simple example of a task in which centers cooperate. The first processing of the incoming light takes place in the retina. Signals from light-gathering cells are sent to specialized neurons (*see* NERVE SIGNALS). Some of these neurons will fire if they sense a light spot with a dark surround, others will fire if they sense a dark spot with a light surround. The signal that travels to the brain is therefore a series of pulses that have broken up the visual image into a series of dark and light spots. (There are actually different kinds of processing that go on in the retina—some cells are sensitive to color, others to small differences in intensity.)

Some of the neurons from the retina connect to (the technical term is *project* to) a section at the top of the brain whose function is to form a quick and dirty picture of the visual field, and create an automatic response if there seems to be something happening. This is why everyone in a room will automatically look toward a door when it opens. Most of the signals from the neurons are relayed to the visual cortex, at the back of the brain. There, signals from different parts of the retina begin to reassemble (by a process we don't yet fully understand) into the visual image. Each neuron in the visual cortex is connected to many retinal neurons. These *cortical neurons* are highly specialized. Some of them will fire only if there is a horizontal line in the field of vision, some only if there is a vertical line, and so on. These neurons project to other parts of the brain as the process of reassembly moves to higher and higher levels. We know that there are specialized neurons in the brain that will fire only (for example) when presented with star shapes, others that will fire only when they see a circle with a bar through it, and so on. Scientists refer to the quest for understanding how the visual image is built up by the work of these specialized neurons as the *binding problem*—the problem of how the signals from the neurons are bound together to make a single image.

This kind of specialization makes sense from the point of view of EVO-LUTION. For example, the fact that some neural signals from the retina directly trigger a reflex that causes us to examine motion more closely would be an obvious advantage to an organism that lived in a dangerous environment. A quick glance might make all the difference between life and death if that motion is caused by a large predator, for example.

This specialization also explains why many scientists (the author included) are adamant that the brain is not a computer. Digital computers just don't work the way the brain does, and each is good at different tasks (*see* TURING TEST). Even a small computer, for example, can out-compute and out-memorize any human, but no computer in existence has the language ability of a five-year-old. The computer is a tool, like a hammer, to help humans in their endeavors, nothing more.

Turing Test

If a computer can act in such a way that human beings cannot tell if they are interacting with a machine or with a person, the machine is said to have passed the Turing test

The subject of conscious, human-like machines has been a staple of science fiction for decades (*see* the three laws of ROBOTICS). Ever since the beginning of modern computing, the question of whether a machine could be built that would in some way be able to replace a human being has occupied the minds of many human beings. The test devised by Alan Turing was an attempt to put this notion on a firm empirical footing.

Although his first version of it, published in 1950, was rather more convoluted, the modern version of the Turing test goes something like this: A panel of judges interacts with an unknown entity. They cannot see their correspondent, and they can communicate only through some isolating system such as a keyboard. They are allowed to ask the correspondent any questions they like, engage in any conversation they wish. If at the end they cannot tell whether they are talking to a human or to a machine, and if they are indeed talking to a machine, then the machine is said to have passed the Turing test.

Needless to say, no machine today could even come close to passing the Turing test, though some do pretty well in a very limited field. Suppose, however, that someday a machine can pass the test. Does that mean that it is intelligent or conscious?

John R. Searle (1932–), a philosopher at the University of California at Berkeley, devised an imaginary system that seems to answer this question with a "no." The system is called the Chinese room, and it works like this: You are sitting in a room with two slots in one of the walls. Through one slot someone passes you questions written in Chinese. (This choice of language serves just to indicate that, like John Searle and myself, you neither speak nor read it. If that isn't true, then pick any other language that is unknown to you.) You then consult a set of books full of instructions such as, "If you get this set of symbols, then write this other set of symbols on a piece of paper and pass it out through the other slot."

It is clear that if the instruction books were comprehensive enough, the "machine" consisting of you and the room could pass the Turing test. Yet it is also clear that you would have no understanding at all of what you were doing. According to Searle, this shows that just because a machine passes the Turing test this does not imply that it is either intelligent or conscious.

ALAN MATHISON TURING (1912–54) English mathematician. He was born in London. Turing studied at Cambridge, England, and Princeton, USA. He was a pioneer in the field of computational theory, introducing the notion of the *Turing machine*, an idealized digital computer. During World War II he was a cryptographer at Bletchley Park, the secret UK government establishment set up to crack the German Enigma military code. After the war, removed from his research position and persecuted because of his homosexuality, he committed suicide.

Tyndall Effect

*Blue light is scattered
more readily from small
particles in the air than
are other colors*

The effect that bears Tyndall's name came from studies of the interactions between light and various materials. He noted that light passing through a region containing very small particles—a smoke or mist, for example—was scattered strongly if the light was blue, less strongly if it was red. If a beam of white light, which is a mixture of all colors, is passed through such a system, the blue light will tend to scatter out while the red and green light remain in the beam. Seen from the side of the light beam, the scattered light will thus be blue, whereas if we look along the light beam, the light source will now look redder than its true color. This is why hazes (from forest fires, for example) often have a bluish cast.

The Tyndall effect arises from light scattering off particles some tens of atomic dimensions in size. Once the particles become larger than about a twentieth of the wavelength of the light (i.e., larger than about 25 nm) the scattering becomes polychromatic. That is to say, the red, green, and blue light is all scattered equally. This is why fogs, mists, and clouds, formed of water droplets from micrometers to millimeters in size, appear white.

The blue light from the sky might seem likely to be due to the Tyndall effect, but that is not the case. In the absence of clouds and smoke, the scattering of "daylight" is caused by the molecules in the air. This type of scattering is called *Rayleigh scattering,* after Lord Rayleigh (*see* RAYLEIGH CRITERION). In Rayleigh scattering the blue light is even more strongly affected than in the Tyndall effect: for example, blue light at 400 nm wavelength will be scattered nine times more strongly than red light at 700 nm. This is why the sky appears blue—light from the Sun at all wavelengths is being scattered, but the blue light far more so than the red. Ultraviolet light is scattered even more strongly than blue light, which is why sunbathers tan more or less all over, and not just on those parts exposed to direct sunlight.

JOHN TYNDALL (1820–93) Irish physicist, born in Leighlin Bridge, County Carlow. He first worked at the Ordnance Survey Office as an engineer-draftsman, but was dismissed for protesting about poor working conditions. He proceeded to educate himself, and in 1848 entered the University of Marburg, Germany, to study sciences. Back in England, he became a professor at the Royal Institution.

Tyndall was a dedicated popularizer of science. He gave many afternoon lectures, often to crowds of working men in factory yards, and Christmas lectures for young people at the Royal Institution. His renown crossed the Atlantic—the US edition of his *Fragments of Science* sold out on the day that it was published in 1871. He was accidentally poisoned by his wife, who outlived him by 47 years.

Uniformitarianism

*The Earth was shaped
by processes that are still
going on today*

One of the great developments of the late 18th and early 19th centuries was the discovery of what writer John McPhee calls *deep time*—the great antiquity of our planet. The pioneer in this development was the Scottish scientist James Hutton. He pointed out that many processes had shaped the Earth—erosion, the wearing away of rock and soil by wind and water; sedimentation, the deposition of those sediments; and uplift, one process by which mountains are raised. These processes, he argued, could, if they operated over long intervals of time, explain the present appearance of the Earth. At a time when it was almost universally believed that the past was explained by special creation and events such as Noah's flood, this idea was little short of revolutionary. It quickly gathered adherents, and Hutton and his followers found more and more evidence to support it. Hutton was thus the first to state what has become known as the principle of uniformitarianism.

Hutton's ideas were melded into a rigorous and comprehensive theory by Charles Lyell. Under the slogan "The present is key to the past," he stated the basic idea of uniformitarianism: that the Earth has been shaped gradually by forces that can still be seen operating today. For example, he measured the thickness of lava flows in Sicily to show that Mount Etna could have been built up by the slow accumulation of flows. He measured the erosion caused by Niagara Falls and argued that the present position of the falls could be understood in terms of a gradual

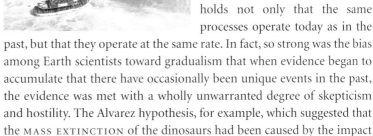

wearing down of the rocks by the Niagara river.

Lyell's doctrine quickly became an unshakable pillar of the Earth sciences, ousting a competing theory known as *catastrophism*, according to which the Earth was shaped by unique events such as Noah's flood. Uniformitarianism was sometimes even extended to a doctrine called *gradualism*, which holds not only that the same processes operate today as in the past, but that they operate at the same rate. In fact, so strong was the bias among Earth scientists toward gradualism that when evidence began to accumulate that there have occasionally been unique events in the past, the evidence was met with a wholly unwarranted degree of skepticism and hostility. The Alvarez hypothesis, for example, which suggested that the MASS EXTINCTION of the dinosaurs had been caused by the impact of an asteroid, was slow to find acceptance because of this bias. (In point

Charles Lyell used the gradual erosion caused by the Niagara Falls to argue that geological processes have operated for long periods of time.

of fact, uniformitarianism in its broadest form would have no problem incorporating a singular event like an asteroid impact, since that is just another type of natural phenomenon.)

Oddly enough, it now appears that the geologist's old nemesis—Noah's flood—may actually have a historical basis in the flooding of ocean basins after the last Ice Age. It seems to me that the anti-catastrophist bias among scientists has now largely vanished, and that we are much more willing to interpret the data and see our planet's past in terms of scattered singular events playing out against a gradualist background.

JAMES HUTTON (1726–97) Scottish geologist (left). He was born to a merchant family in Edinburgh. The city was then one of the intellectual capitals of Europe, and Hutton socialized with the likes of Adam Smith (the founder of modern political economy) and James Watt (who developed the modern steam engine). He received a degree in medicine from the University of Edinburgh, and also studied medicine at Paris and Leiden, but never practiced; later he studied law and ran a successful manufacturing enterprise. In 1754 he took over a small farm and studied agricultural methods and chemistry, which led him into mineralogy and geology. After many years of travel and study he published *The Theory of the Earth* in 1788, which established geology as a modern, scientific discipline.

CHARLES LYELL (1797–1875) Scottish geologist. He was born in Kinnordy to a prominent Scottish family; his father was a botanist. Lyell studied law at Oxford University, where lectures on geology aroused his interest in that subject. For a time he practiced as a barrister because the extensive reading bothered his eyes but he abandoned the profession to concentrate on geology. Lyell traveled widely, observing geological formations, and his book *Principles of Geology*, published in 1830, was one of the most influential scientific books ever written. Charles Darwin, for example, carried a copy with him on the *Beagle*, and drew upon it in his *On the Origin of Species*. Lyell, though, was slow to accept Darwin's theory of evolution, and refused to believe that it applied to humans.

Urea, synthesis of

The molecules in living systems can be made from inorganic materials

At the start of the 19th century, chemists were coming to realize that most materials in the world are made of molecules, and that those molecules, in turn, are composed of atoms (*see* ATOMIC THEORY). Some scientists argued that the molecules found in living systems, *organic molecules,* were different in some fundamental way from those in inanimate matter—*inorganic molecules.* This was partly a legacy of the belief that living things are somehow special (*see* VITAL FORCE). Later, chemists came to understand that organic molecules are often large and complex, unlike the inorganic molecules that chemists had been used to studying, and this added to the notion that the chemistry of life was different.

In 1828 Friedrich Wöhler settled this question once and for all by synthesizing a chemical compound known as urea, a substance routinely found in the kidneys and urine of animals, from ordinary "off the shelf" chemicals. In his words, "I found that whenever one tried to combine cyanic acid and ammonia, a white crystalline solid appeared that behaved like neither cyanic acid nor ammonia." Extensive testing showed that the "white crystalline solid" was chemically identical to urea extracted from living animals. With this experiment, Wöhler showed that organic molecules can be formed by the same processes, and from the same atoms, as inorganic ones. One more artificial dividing line between living and non-living systems was removed.

FRIEDRICH WÖHLER (1800–82) German chemist. He was born in Eschersheim, near Frankfurt, the son of a veterinary surgeon who attended the horses owned by the rulers of Hesse in Germany. He received a medical degree from the University of Heidelberg in 1823, but then switched to chemistry, spending the next year in Sweden with the chemist Jöns Berzelius (the two remained friends for life). In 1836 Wöhler was appointed professor of chemistry at the University of Göttingen, where he remained for the rest of his life.

An avid collector of minerals as a youth, Wöhler spent most of his career studying ways to synthesize and extract various mineral compounds. After a period of studying organic compounds he decided it was too complicated and returned to inorganic chemistry. A lasting achievement was to establish Göttingen as the leading European center for chemical research. There, he trained many students who went on to staff universities in Europe and North America. Germany would dominate research chemistry until the 1930s.

Van Helmont Experiment

The biomass of plants is not drawn from the soil

The color of life on our planet is green—green from the chlorophyll molecules in the plants that form the basis for all life and convert the energy of incoming sunlight into the materials of living things. It is little short of astonishing, then, that such little interest has been shown in past centuries in the basic mechanism by which this energy conversion happens—the process we now call PHOTOSYNTHESIS. The fact of the matter is that human beings understood the movements of planets and stars long before they had even an inkling of the working of the grass beneath their feet.

The first serious study of the mechanism of plant growth was done by the Flemish aristocrat Jan Baptista van Helmont. He weighed the dirt in a pot, then planted a tree in it. He watered the tree for several years, then weighed the tree and the dirt again. He found that the tree had gained 164 lb (74 kg) while the soil has lost only a few ounces. Obviously, the material that had been incorporated into the growing tree had not been drawn from the soil.

Van Helmont actually drew the wrong conclusion from his finding—he argued that the extra weight came from the water. The notion that the carbon in the tree came from the processing of carbon dioxide in the atmosphere would be two more centuries in coming, and a molecular understanding of photosynthesis would take another century after that. Nonetheless, van Helmont had established firmly that the bulk of what we call biomass did not come from the soil, but from somewhere else, a finding that formed the basis of our later understanding of plants.

JAN BAPTISTA VAN HELMONT (1579–1644) Flemish physician and chemist. He was born in Brussels into an aristocratic family. He studied medicine and chemistry at the Catholic University of Louvain, but did not accept a degree and embarked on private research. He was the first person to use the word "gas" to describe that state of matter, and identified four kinds of gas which we now know as carbon monoxide, carbon dioxide, nitrous oxide, and methane.

Van Helmont lived at a time when chemistry was a fledgling science, and the influence of alchemy was still strong. Although he was no respecter of the sanctity of ancient teaching, he still believed in the philosopher's stone, for example. But he saw the value of experiment, as reflected in his study of the growing willow tree. In another episode, he ran foul of the Church when he questioned the prevalent belief that a wound could be healed by treating the weapon that had caused it.

Vital Force

There is a special force that forms molecules in living systems

In the early 19th century, chemists were making great progress in developing their science. The ATOMIC THEORY made sense of much of the complexity of materials found in the world. There remained one problem—many molecules seemed to exist only in living systems. Chemists began to talk of a "vital force"—something that living systems alone possessed, and allowed them to create molecules that couldn't be duplicated in other parts of nature.

The substance known as urea is a typical example of what are called *organic molecules.* The molecule has a chemical formula $CO(NH_2)_2$, and is used by most animals to excrete unused nitrogen that they ingest in their food. Human urine, for example, contains 2–5% urea.

In 1828, Friedrich Wöhler made an important breakthrough when he made urea in his laboratory from ordinary "off the shelf" chemicals. This proved conclusively that no vital force was necessary to create organic molecules—that the laws that governed their creation were the same as those that governed the creation of any other molecule. After Wöhler's work, the vital force quietly faded from the scene.

With humor that doesn't fit the stereotype of the Teutonic academic, Wöhler announced his discovery in a letter to a friend with the words, "I can no longer, as it were, hold back my chemical urine, and I have to let out that I make urea without needing a kidney, whether of man or dog."

Actually, ideas like that of vital force, or vitalism, don't die easily—mere facts are seldom enough to bury them forever. Scarcely disguised vitalism informs much of "new age" thinking today, for example. In the 1930s, in a somewhat more respectable form, something very much like vitalism showed up in the debate that followed the discovery that many molecules in living systems are very complex (DNA is a good illustration of this point). Perhaps, it was argued, the laws that govern the behavior of atoms in very complex molecules are different from those that govern atoms in simpler molecules. Since scientists at the time had very little experience with complex molecules, this was not an unreasonable hypothesis. As it turned out, the hypothesis was wrong—the laws that govern a hydrogen atom in DNA are exactly the same as those that govern that same atom anywhere else. In this case at least, nature turned out to be simple.

Water Cycle

Water on the Earth's surface moves through a complex cycle, but the total amount is approximately constant

Any primeval atmosphere the Earth might have had shortly after it formed over 4 billion years ago would have been blown off by streams of energetic particles from the newborn Sun. We can think of the early Earth as a hot, naked ball floating in space. Over time, volcanic eruptions and other geothermal processes injected gases, including water vapor, from the Earth's interior into what was becoming the new atmosphere. Since then, the total amount of water on the planet's surface has stayed roughly constant. At any time, however, that water can be found in many places and in many forms. The description of how the Earth's water moves around in its biosphere is called the water cycle.

We can start thinking about the cycle by imagining that we are lying on a beach by the sea on a warm summer's day. The heat of the Sun evaporates water from the surface of the ocean, and this water vapor rises into the atmosphere, where it may form clouds. Eventually the water returns to the surface in the form of precipitation (as rain, snow or hail) and begins its long journey back to the sea or to a lake. It can flow across the surface in rivers, or it can flow underneath the ground in subsurface streams. It may get taken up and stored temporarily by living things. If it falls as snow near the poles or on a high mountain, it may be incorporated into a glacier or icepack and stay there until the ice melts. In the end, though, its fate is the same. Eventually it comes back into the sea, where it waits for the Sun's heat to lift it back into the atmosphere so the cycle can start again.

The fact that the total amount of water on the Earth is roughly constant has some interesting consequences. In the last ice age a lot of the Earth's water supply was locked up in the ice caps that moved down from the poles, so the level of water in the oceans was much lower than it is now. If you had lived 18,000 years ago, you could have walked from England to Europe or from Asia to Alaska on dry land, and the western coast of England was 100 miles (150 km) farther west than it is today.

Water is cycled through the Earth's biosphere from the oceans to clouds to the land, then back to the oceans.

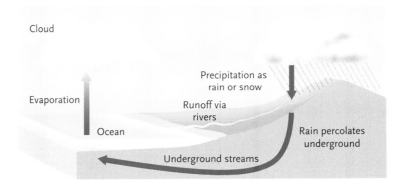

Cloud

Evaporation

Ocean

Precipitation as
rain or snow

Runoff via
rivers

Rain percolates
underground

Underground streams

Zeeman Effect

Lines in atomic spectra will split when the atom is in a magnetic field

There was a long tradition, going back to Michael Faraday, of scientists looking for the effects of a magnetic field on the light emitted by atoms. That such an effect ought to be there might seem obvious now we know that electrons and other particles spin on their axes, and hence can act like tiny magnets (*see* STERN–GERLACH EXPERIMENT). In the mid-1890s, Pieter Zeeman set up some experiments to see whether he could detect magnetic effects. He placed a small sample of sodium between the poles of an adjustable magnet, then looked at the frequencies of light emitted by the sodium atoms (*see* SPECTROSCOPY). He found that as he turned up the magnetic field, the band of frequencies associated with each spectral line began to broaden. This was the first unambiguous detection of what would soon come to be called the Zeeman effect.

The easiest way to understand this effect is to think about how light is emitted in the BOHR ATOM. An electron makes a quantum jump from an upper to a lower energy level, emitting a photon in the process. Now, if the electron is really a tiny magnet, and if the atom is placed in a magnetic field, then the energy of an electron in a particular orbit will depend on whether that tiny magnet is "north pole up" or "north pole down" in the orbit. Each of these orientations will have a slightly different energy, so that each energy level will, in effect, be split into two closely spaced levels. What was a single transition from one energy level now splits into four possible transitions between the split energy levels—a single frequency, in other words, becomes four closely related but separate frequencies.

In Zeeman's original experiment, these four close frequencies were all blurred together, so all he saw was an apparent broadening of the spectral line. Later on, though, he was able to separate the frequencies and establish that atomic energy levels were split in the presence of magnetic fields, the stronger fields producing the greatest splitting.

The Zeeman effect has turned out to be very important in astronomy, because from the splitting of lines in the spectra of astronomical objects, it is possible to determine the magnetic fields associated with those objects. It was, for example, the study of Zeeman splitting in spectral lines from atoms located in sunspots that scientists first determined that those structure resulted from intense magnetic fields in the Sun.

PIETER ZEEMAN (1865–1943) Dutch physicist. He was born in Zonnemair into a clerical family, and rarely left his native Holland. After completing his studies at Leiden, under Hendrick Lorentz (1853–1928), he joined the University of Amsterdam, becoming professor of physics in 1900 and staying there until he retired. A highly skilled experimentalist, using state-of-the-art instrumentation to make measurements no one else could. He and Lorentz shared the 1902 Nobel Prize for Physics. In 1918, he confirmed the EQUIVALENCE PRINCIPLE.

Zeno's Paradox

Motion is impossible. For example, it is impossible to cross a room, because to do so you must first cross half, then half of what's left, then half of that, then half of that …

Zeno of Elea was a member of a group of Greek philosophers who held that change in the world was all illusion, that there was only the Unchanging One. His paradox (which was posed in four basic forms, and has since spawned some forty different versions) is designed to show that motion, an example of "apparent" change, is not logically possible.

The form of Zeno's paradox most familiar to modern readers is the one stated above (it's sometimes called the *dichotomy*). In order to cross a room, you first have to go halfway. But then you have to cross half of what's left, then half of what's left after that, and so on. These divisions by half go on forever, so, the argument goes, you can never cross the room.

A more picturesque form is the *Achilles paradox*. The Greek hero Achilles is about to race against a tortoise. If the tortoise has a slight head start, then Achilles will first have to reach the tortoise's starting point. But by the time he gets there, the tortoise will have moved on, and Achilles will have to travel to the new spot. Again, the infinite number of divisions would prevent Achilles from ever catching up.

Here's another form, in Zeno's own words:

If anything is moving, then it is either moving in the place where it is or in the place where it is not. However, it cannot move in the place where it is (for the place in which it is at any moment is the same size as itself …) and it cannot move in the place in which it is not. Hence motion is impossible.

This is known, appropriately enough, as the *arrow paradox*.

Finally, there is the fourth form, in which two columns of men of equal length are passing each other in opposite directions. Zeno argued that the time it would take the two columns to pass each other is only half of the time it would take for one body to pass an entire column.

Of these four forms of the paradox, the first three are the best known and are the most paradoxical. The fourth is simply a misunderstanding of the nature of relative motion.

The most brutish and inelegant way to disprove Zeno's paradox is to get up and cross a room, pass a tortoise, or shoot an arrow, but that does nothing to address Zeno's lines of argument. It wasn't until the 17th century, however, that thinkers found the key to undoing his tortuous logic. It took the development of the calculus by Isaac Newton and Gottfried Leibniz, which introduced the concept of a *limit*; an appreciation of the subdivison of space as distinct from the subdivision of time; and a better understanding of the infinite and the infinitesimal to resolve the issue.

Take walking across a room. It's true that at each point you have to cross half the remaining distance, but *it only takes you half as long to do so.* The smaller the distance to cross, the shorter the time it takes to make the crossing. Thus, in calculating the time it takes to cross the room, we are adding up an infinite number of intervals which are becoming infinitesimally small. The sum of all those intervals is not infinite, which would make it impossible to cross the room, but some finite number—in which case we *are* able to cross the room in a finite amount of time.

This argument is similar to the one underlying the concept of taking a limit in calculus. We can understand the idea of a limit in terms of Zeno's paradox. If we divide the distance across part of the room by the time it takes to cross that distance, we get the average speed at which that interval is crossed. As both the distance and the time get smaller (and, ultimately, go to zero), the ratio between the two can be finite—in fact, it will be the velocity with which you are moving. This ratio is called the limit of the velocity as both the distance and time approach zero. The basic misunderstanding contained in this form of the paradox is the assertion that while the distance approaches zero, the time does not.

My own favorite refutation of Zeno's paradox comes not from Newtonian calculus, but from a skit by Second City, a comedy troupe in my native Chicago. A lecturer has been describing various philosophical problems. After explaining the Achilles paradox, he says this:

> But this is ridiculous. Anyone in this room could win a race with a tortoise. Even an old, dignified philosopher like Bertrand Russell—he could win a race with a tortoise. And if he couldn't beat it, he could outsmart it!

That sums it up nicely.

ZENO OF ELEA (*c.* 490–420 BC) Greek philosopher. Little is known of his life, and what we know of his work, including his famous paradoxes, is derived from the writings of later philosophers. He was a disciple of Parmenides of Elea (*c.* 515–450 BC), who taught that the true reality had to be eternal and unchanging, accessible only to reason and logic. According to legend, Zeno was tortured and killed by Nearchus, the tyrant of Elea, for taking part in a conspiracy to overthrow the government.

	Antiquity–1599	1600–1649	1650–1699
Astronomy	16th C. Copernican Principle	1609, 1619 Kepler's Laws	1687 Newton's Law of Gravitation
Chemistry			
Earth Sciences			1666 Superposition, law of
Life Sciences		1624 Van Helmont Experiment	1663, 1839 Cell Theory
Mathematics	5th C. BC Zeno's Paradox 1202 Fibonacci Numbers	1630 Fermat's Last Theorem	
Miscellany	EARLY 14th C. Occam's Razor		
Physics	Antiquity States of Matter 3rd C. BC Archimedes' Principle c. 420 BC Atomic Theory c. AD 100 Reflection, law of 1537 Compound Motion	1600 Magnetism 1604, 1609 Accelerated Motion, equations of 1621 Snell's Law	1650 Fermat's Principle; Centrifugal Force 1662 Boyle's Law 1668 Linear Momentum, conservation of 1678 Hooke's Law 1687 Angular Momentum, conservation of; Newton's Laws of Motion 1690 Huygens' Principle
Rear View Mirror	Antiquity Spontaneous Generation		1683 Phlogiston

1700–1749	1750–1799	1800–1819	1820–1839	1840–1859
1742, 1823 Olbers' Paradox	1755 Nebular Hypothesis 1766 Titius–Bode Law 1783 Black Holes			
	LATE 18th C. Flame Test	1801 Dalton's Law 1811 Avogadro's Law	1828 Urea, synthesis of 1829 Graham's Law 1834 Faraday's Laws of Electrolysis	1854 Catalysts and Enzymes
	1783 Carbon Cycle 1788 Uniformitarianism LATE 18th C. Rock Cycle			1852 Acid Rain
1729, MID-20th C. Circadian Rhythms c. 1730 Linnaean System of Classification	1779, 1905 Photosynthesis 1798 Exponential Growth			1852, 1878 Mimicry 1859 Evolution, theory of
1742 Goldbach Conjecture			1822 Fourier Analysis	
1738 Bernoulli Effect 1747 Electrical Charge, conservation of	1761 Changes of State 1785 Coulomb's Law 1787 Charles's Law 1798 Heat, atomic theory of	19th C. Electrical Properties of Matter; Heat Transfer EARLY 19th C. Surface Tension c. 1800 Thermal Expansion, law of 1801 Henry's Law 1807 Interference 1811 Brewster's Law 1813 Gauss's Law 1818 Diffraction	1820 Ampère's Law; Biot–Savart Law; Oersted Discovery 1824 Carnot's Principle 1826 Ohm's Law 1827 Brownian Motion 1831 Faraday's Laws of Induction 1834 Clausius–Clapeyron Equation; Ideal Gas Law 1835 Coriolis Effect; Lenz's Law	1842 Doppler Effect; Thermodynamics, first law of 1845 Kirchhoff's Laws 1849 Kinetic Theory 1850 Thermodynamics, second law of 1851 Terminal Velocity 1854 Rydberg Constant 1859 Kirchhoff–Bunsen Discovery; Tyndall Effect; Spectroscopy
EARLY 18th C. Balance of Nature		1809 Lamarckianism		c. 1850 Social Darwinism

	1860–1879	1880–1899	1900–1919
Astronomy		1887 Michelson–Morley Experiment	20th C. Stellar Evolution 1905–13 Hertzsprung–Russell Diagram 1912 Period–Luminosity Relation 1917 Cosmological Constant
Chemistry	1860s Periodic Table 1868, 1895 Helium, discovery of	1887 Acids and Bases, theories of 1888 Le Chatelier's Principle 1892 Argon, discovery of	1919 Octet Rule
Earth Sciences	1863 Greenhouse Effect	1886 Nitrogen Cycle LATE 19th C. Water Cycle 1890, 1940s Radiometric Dating	1910s Milankovič Cycles
Life Sciences	1865 Mendel's Laws 1873 Mutuality, principle of 1877 Allen's Rule; Germ Theory of Disease; Symbiosis	c.1895 Cohesion–Tension Theory 19th–20th C. Molecules of Life; Nerve Signals, propagation of	c.1900 Animal Territoriality; Ecological Succession 1908 Hardy–Weinberg Law
Mathematics			
Miscellany			1913 Bohr Explanation
Physics	1864 Electromagnetic Spectrum; Maxwell's Equations 1867 Maxwell's Demon 1877 Boltzmann Constant 1879 Hall Effect; Stefan–Boltzmann Law	1883–4 Reynolds Number 1887 Shock Waves 1891 Equivalence Principle 1895 Curie Point; Curie's Law 1896 Rayleigh Criterion; Zeeman Effect 1897 Electron, discovery of; Elementary Particles 1899 Photoelectric Effect	1900 Black-Body Radiation; Free Electron Theory of Conduction; Planck Constant; Radioactive Decay 1905, 1916 Relativity 1905 Thermodynamics, third law of 1911 Rutherford Experiment 1912 Bragg's Law 1913 Bohr Atom; Millikan Oil-Drop Experiment 1917, 1934 Nuclear Fusion and Fission
Rear View Mirror		1896 Cope's Law 1899 "Ontogeny recapitulates phylogeny" 19th–EARLY 20th C. Vital Force	

1920–1939	1940–1959	1960–1969	1970–1999	2000–
1929 Hubble's Law 1931 Chandrasekhar Limit 1933 Dark Matter	1940s Giant Impact Hypothesis 1948 Big Bang 1950 Fermi Paradox	1961 Anthropic Principle; Drake Equation	1980s Early Universe 1981 Inflationary Universe 1990s Cosmic Triangle	
c.1920 Aufbau Principle LATE 1920s Molecular Orbital Theory 1930s Chemical Bonds				
c. 1930, 1980 Mass Extinctions	1953 Miller–Urey Experiment	1960s Plate Tectonics	1979 Gaia Hypothesis 1985 Ozone Hole	
1920s Genetic Drift 1926 Predator–Prey Relationships 1928 Penicillin, discovery of 1934 Competitive Exclusion, principle of 1937 Glycolysis and Respiration	1947 Antibiotics, resistance to 1950s Green Revolution EARLY 1950s Proteins 1952 Hershey–Chase Experiment 1953 DNA 1954 Maximum Sustainable Yield 1958 Molecular Biology, central dogma of	1960s Stem Cells EARLY 1960s Kinship Selection 1961 Genetic Code 1964 Coevolution MID-1960s Immune System 1966 Optimal Foraging Theory 1967 MacArthur–Wilson Equilibrium Theory	1970s Differential Resource Utilization; Molecular Clock 1976 Marginal Value Theorem 1995 Cloning	2000 Human Genome Project
1931 Gödel's Incompleteness Theorems		1965 Moore's Law	1980s Deterministic Chaos	
	1942 Robotics, the three laws of MID-1940s Murphy's Law 1950 Turing Test			
1921 Stern–Gerlach Experiment 1923 Compton Effect; Correspondence Principle 1924 Dispersion, atomic theory of; De Broglie Relation; Pauli Exclusion Principle; Quantum Tunneling 1925 Quantum Mechanics 1926 Band Theory of Solids; Schrödinger's Equation 1927 Complementarity Principle; Davisson–Germer Experiment; Heisenberg's Uncertainty Principle c.1930 Antiparticles 1931 Magnetic Monopoles 1934 Cherenkov Radiation	1957 BCS Theory of Superconductivity; Lawson Criterion	1961 Quarks and the Eightfold Way; Standard Model 1962 Josephson Effect 1964 Bell's Theorem 1968 String Theories	1972 Quantum Chromodynamics 1980s Chronology Protection Conjecture	21st C.? Theories of Everything
	1948 Steady-State Theory MID-20th C. Triune brain			

accelerator A machine designed to accelerate elementary particles, often to speeds near that of light. High-speed collisions between particles help reveal their properties.

amplitude The maximum displacement of a wave from its average position—the height of a peak or the depth of a trough.

atom The smallest bit of a chemical element that can retain the element's chemical identity; a nucleus with electrons in orbit around it.

bacterium A simple, one-celled microorganism.

binary star A system consisting of two stars in orbit around each other.

biosphere The narrow shell on the outside of the Earth that supports life; it consists of the surface, subsurface, oceans and atmosphere.

dissociation A chemical reaction in which a molecule is broken down into smaller ones.

DNA The most essential molecule of life, containing the genes that are inherited by successive generations.

ecosystem All the living things that occupy a particular place, together with the physical environment found there.

electron A light, negatively charged elementary particle. In atoms, electrons orbit the nucleus.

element (chemical) A substance that cannot be decomposed chemically.

elementary particle A fundamental subatomic particle from which others are made. There are two classes: the quarks and the leptons (which include the electron).

energy level The energy of groupings of electrons within an atom.

food chain A chain of living things, each eating ones below and being eaten by ones above.

force That which causes a change in the state of motion of any object.

fossil fuel Any fuel, such as coal, oil, or natural gas, which was formed by geological processes long ago.

frequency The number of crests (or troughs) of a wave that pass a given point in one second.

galaxy A collection of stars, dust, and dark matter. The matter in the universe is organized into galaxies.

gamma rays Electromagnetic radiation of very short wavelength and very high energy.

gene A segment of the DNA molecule which codes for a protein that acts as an enzyme to run one chemical reaction in a cell; the "unit of inheritance."

hydrocarbon An organic molecule, found for example in gasoline, that is made up of carbon and hydrogen atoms.

inertia The property of matter that keeps it in a state of motion unless it is acted on by a force.

infrared Electromagnetic radiation whose wavelength is slightly longer than that of visible light.

ion An atom or molecule that has gained or lost one or more electrons, thus acquiring, respectively, a negative or positive electrical charge.

isotope Two nuclei that have the same number of protons but different numbers of neutrons are isotopes of each other.

kinetic energy Energy associated with motion.

light year The distance that light travels in one year. It is NOT a measure of time..

mass The quantity that measures an object's resistance to acceleration, a measure of the amount of matter in the object. An object's **weight** is the force that acts on it in a gravitational field.

mineral An inorganic substance having a regular crystalline form. **Rock** is a mixture of minerals.

mole The amount of material that contains the same number of atoms as there are in 12 grams of carbon.

molecule A combination of two or more atoms held together by a chemical bond.

momentum The mass of an object multiplied by its velocity.

neutron A heavy, electrically neutral elementary particle.

nucleus (of an atom) The central part of an atom, consisting of protons and (usually) neutrons.

nucleus (of a cell) The structure in advanced cells that contains the cell's DNA.

orbital A region in an atom or molecule in which electrons may be found. The likelihood of an electron being in a given region is given by an expression known as the wave function. A **shell** is a collection of electron orbitals.

organic compound A compound containing carbon. By convention, the category excludes simple carbon compounds, such as oxides and carbonates; these, and all other compounds, are termed **inorganic compounds**.

osmosis The process by which atoms in solution pass through a membrane.

phase (of a wave) A measure of how synchronized a wave is with respect to another of similar type.

photon The particle associated with light and other forms of electromagnetic radiation.

plasma A state of matter in which electrons are separated from the nuclei of atoms.

potential energy The energy an object has position in a gravitational field.

prime number A number greater than 1 which is not divisible by any number other than 1 and itself.

proportional One quantity is proportional to another if the first increases when the second does. If the one quantity decreases when the another increases, the first is **inversely proportional** to the second.

proton A heavy, positively charged elementary particle.

proton number (atomic number) The number of protons (and therefore of positive charges) in the nucleus of an atom.

quark One of several elementary particles from which other particles, such as the proton and the neutron, are constructed.

radiation Particles or high-energy waves emitted in the process of radioactive decay or, more generally, particles or waves emitted from any source.

radio waves Electromagnetic radiation of long wavelength.

reaction (chemical) A change in one or more substances that produces one or more different substances

redshift The displacement in the wavelength of radiation (usually light) toward the long-wavelength (red) end of the spectrum due to the motion of the source away from an observer.

relative atomic mass (atomic weight) The mass of the atoms (actually, the average mass of the isotopes) of a chemical element as they occur in nature, measured in units of one-twelfth of the mass of carbon .

rock *see* **mineral**

shell *see* **orbital**

SI units Short for "Système Internationale", the system of units of measurement based on the meter, the kilogram, and the second.

spacetime The four-dimensional space in which three of the dimensions are ordinary three-dimensional space and the fourth is time.

species A group of organisms which can interbreed freely.

spectrum Most commonly, the electromagnetic spectrum: the entire range of electromagnetic radiation, from gamma rays to radio waves.

speed The distance traveled by an object divided by the time it takes to travel that distance. The object's **velocity** is its speed together with the direction in which it is moving.

supernova The catastrophic explosion of a star, a process in which chemical elements are returned to the interstellar medium.

ultraviolet Electromagnetic radiation whose wavelength is slightly shorter than that of visible light.

virus An organism consisting of DNA or RNA wrapped in a coat of proteins.

wavelength The distance between the crests of a wave.

weight *see* **mass**

X-rays Electromagnetic radiation of short wavelength.

Brownian motion 29,
    59–60
bulk properties 54
Bunsen, Robert 233, 234
bunsen burner 233
buoyancy 22
butterflies 271

Calvin, Melvin 322, 323
Calvin cycle 322–3
carbohydrates 279–80
carbon cycle 61–2
carbon dating 343–4
carbon dioxide 61–2
carbon sink 62
Carnot, Sadi 63, 64, 400
Carnot's principle 63–4,
    400
carrying capacity 352
Carter, Brandon 16
catalysts 65
catalytic converter 65
catastrophism 411
cathode 160
cathode rays 144–5
cells
    membranes 280
    stem cells 382
    theory 66
cellulose 280
central dogma of molecular
    biology 261, 273–5
centrifugal force 67–8
Cepheid variable stars
    211–12, 317
CFCs see chlorofluoro-
    carbons
Chain, Ernst 314
chain reactions 301
Chandrasekhar,
    Subrahmanyan 69, 70
Chandrasekhar limit 46,
    69–70, 312
Chandra X-ray
    Observatory 70
chaos theory xxii,
    xxv–xxvi, 118–20
Chargaff, Erwin 125
charges
    electrical 137, 174–5
    electrons 269
Charles, Jacques Alexandre
    73

Charles's law 73, 218
Charnov, Eric 251
Chase, Martha 205
chemical bonds 74–6, 242,
    278, 305, 389
chemical elements see
    elements
chemical equilibrium 240
chemical evolution 151
chemical potential energy
    397
chemiostomic coupling 183
Cherenkov, Pavel 77
Cherenkov radiation 77,
    364
Chéseaux, Philippe de 308
Chinese room 409
chlorofluorocarbons
    (CFCs) 311
chlorophyll 321, 414
chloroplasts 321
chromosomes 274
chronology projection
    conjecture 78
circadian rhythms 79
circuits, electrical 235–6
circular motion 294–5
circumstellar disks 288
citric acid cycle 181
cladistics 245
Clapeyron, Émile 80
classification, Linnaean
    244–5
Clausius, Rudolf 80, 399
Clausius–Clapeyron
    equation 80
climate 266
cloning 81–2, 382
closed systems 401
cloud chamber 21
Cocconi, Giuseppe 131
codons 176
coevolution 83, 285
cohesion–tension theory 84
Collins, Francis 216
color gluons 332
Coma Cluster of galaxies
    110
combustion 28
comets xv, xxii–xxiii
commensalism 390
competition between
    species 154
competitive exclusion 85

complementarity 86–7, 113
complex adaptive systems
    88
complexity xviii–xix, 88,
    171, 184
compound motion 89–91
Compton, Arthur 92
Compton effect 92
computer models 119
computers 282, 409
condensates 375
condensation 72
conduction 172, 194
conduction band 31
conductors, electrical 31,
    138
conservation laws 14, 137,
    243, 397–8
constructive interference
    223
continental drift 328
control of experiments xi
convection 195–6
    Earth 325–6
Cooper, Leon 32
Cooper pairs 32, 225
Cope, Edward 95
Copernican principle 93–4
Copernicus, Nicholas xiii,
    5, 93, 94, 226
Cope's law 95
core
    planet 325
    star 380–1
Coriolis, Gaspard Gustave
    de 97
Coriolis effect 96–7
Coriolis force 67, 96
coronium 203
correlation 303
correlation coefficients 37
correspondence principle
    98
cortical neurons 408
cosmic microwave back-
    ground x–xi, 41
cosmic rays 146, 196, 221
cosmic triangle 99–101
cosmogony 286
cosmological constant
    102–3
cosmology 16, 41–2, 94,
    99–103
Coulomb, Charles 104

Coulomb's law 74, 104–5,
    258
covalent bonds 74
Crick, Francis 125, 127
critical angle 366
crystals 56–7
Curie, Marie 107
Curie, Pierre 107, 108
Curie point 106–7, 249
Curie's law 108
current, electrical 43, 172,
    235–6, 241

Dalton, John 25, 50, 109
Dalton's law 109, 218
dark matter 99–101,
    110–12
Darwin, Charles 152, 157
Davisson, Clinton 113
Davisson–Germer experi-
    ment 113
De Broglie, Louis 115
De Broglie relation 113,
    114–15
decay see radioactive decay
deceleration parameter 100
decidability 185
decomposers 61
degeneracy pressure 69, 380
Delbrück, Max 205
Democritus 24
dendrites 289
dendrochronology 343–4
density 99
deoxyribonucleic acid
    see DNA
Descartes, René 164
destructive interference 223
determinism xxv–xxvi,
    116–17
deterministic chaos 118–20
deuterium 238
diamagnetism 250
Dicke, Robert 16
differential resource
    utilization 121
differentiation (of planets)
    325
diffraction 122, 217, 345–6
dimensionless numbers
    353–4
dimensions of spacetime
    385

First published in the United States
by Houghton Mifflin Company, 2003
First published in the United Kingdom
by Cassell, 2002

For information about permission to reproduce
selections from this book, write to Permissions,
Houghton Mifflin Company, 215 Park Avenue
South, New York, New York 10003.

Visit our Web site:
www.houghtonmifflinbooks.com.

LIBRARY OF CONGRESS
CATALOGING-IN-PUBLICATION DATA
Trefil, James S., date.
The nature of science: an A–Z guide to the
laws and principles governing our universe /
James Trefil.
p. cm
Includes index.
1. Science-Encyclopedia. I. Title.

Q121 .T74 2003
503–dc21

ISBN 0-618-31938-7

DESIGN BY
Ken Wilson
EDITED BY
John Woodruff
ARTWORK BY
Raymond Turvey and Colin Fearon
PICTURE RESEARCH BY
Elaine Willis and Vanessa Fletcher

COLOUR REPRODUCTION BY
CK Digital, Whitstable, Kent
PRINTED AND BOUND IN SLOVENIA
by arrangement with
Prešernova Družba d.d., Ljubljana

Picture Credits

AKG 3, 54, 73, 103, 108, 115, 175, 200, 245,
263, 291, 324, 328, 344, 349, 362, 396, 405, 413;
AMERICAN INSTITUTE OF PHYSICS 69;
BABSON COLLEGE ARCHIVE 293; BRIDGEMAN
ART LIBRARY 97; CALIFORNIA INSTITUTE OF
TECHNOLOGY 20; CAMERON COLLECTION 4,
93, 109, 132, 153 *upper*, 157, 178, 228, 260, 355;
CERN 38, 146; CORBIS/TED SPIEGEL 402;
EUROPEAN SOUTHERN OBSERVATORY 379;
GETTY IMAGES 85, 118, 271 *upper*, 296, 392;
GOODYEAR 202; HUBBLE HERITAGE TEAM
(AURA/STScI/NASA) 211; INSTITUTE OF
ADVANCED STUDY, PRINCETON, USA 185;
MARY EVANS PICTURE LIBRARY 95, 314;
NASA 190, 214, 286, 311; OMAR LOPEZ-CRUZ
& IAN SHELTON/NOAO/AURA/NSF 110;
OXFORD SCIENTIFIC FILMS 271 *lower*, 322,
371; *POPULAR HISTORY OF SCIENCE 1881*
13, 55, 94, 217, 236, 288; PRIVATE COLLECTION
5, 22, 24, 97, 164, 173 *right*, 214; SCIENCE
PHOTO LIBRARY 6, 15, 29, 33, 34, 39, 45, 53, 56,
66, 70, 78, 82, 83, 127, 129, 130, 135, 145, 153
lower, 161, 162, 165, 169, 170, 173 *left*, 208, 209,
223, 225, 237, 241, 249, 254, 255, 256, 264, 273,
279, 309, 310, 318, 323, 326, 337, 345, 351, 366,
378, 386, 388, 401, 409, 411, 412; UNIVERSITY
OF FRANKFURT 363; CHANTAL & MIRANDA
WOODRUFF 183; JOHN WOODRUFF 124.